KU-325-450

PUBLIC DUTIES AND PUBLIC LAW

Public Duties and Public Law

A. J. HARDING

CLARENDON PRESS · OXFORD
1989

Oxford University Press, Walton Street, Oxford OX2 6DP
Oxford New York Toronto
Delhi Bombay Calcutta Madras Karachi
Petaling Jaya Singapore Hong Kong Tokyo
Nairobi Dar es Salaam Cape Town
Melbourne Auckland
and associated companies in
Berlin Ibadan

Oxford is a trade mark of Oxford University Press

British Library Cataloguing in Publication Data
Harding, A. J. (Andrew J.)
Public duties and Public Law
1. Public duties
I. Title
342
ISBN 0-19-825607-8

Library of Congress Cataloging in Publication Data
Harding, A. J.
Public duties and Public Law / by A. J. Harding.
1. Judicial review of administrative acts—Cases. 2. Mandamus--Cases. 3. Public interest law—Cases. 4. Judicial review of administrative acts—Great Britain—Cases. 5. Mandamus—Great Britain—Cases. 6. Public interest law—Great Britain—Cases.
I. Title.
K3423.A58H37 1989 342—dc19 [342.2] 88-25318
ISBN 0-19-825607-8

Set by Oxford Text System
Printed in Great Britain
at the University Printing House, Oxford
by David Stanford
Printer to the University

For Kun Bek
and
Miranda Su Lan

Preface

AN analysis of public-law issues over the last twenty years or so reveals an interesting shift of emphasis. We are no longer merely concerned with setting limits to the power of government, but are concerned also with what the proper function of government is. To put it another way, we recognize that public law has a positive and not merely a negative aspect; that there are occasions when justice can be done not by preventing government from acting but by compelling it to act in a particular manner. This shift in emphasis is not wholly surprising. Our conception of the state has changed considerably, as has our conception of justice. It is no longer an adequate theory of public law which says that all will be well so long as large areas of life are left free of government intervention, and it is the job of the law to ensure that those areas remain free; while we remain rightly mistrustful of too much government power, we also realize that the very rights which we hope to preserve by limiting the power of government cannot in fact be preserved unless the law takes an interest in coercing as well as prohibiting certain kinds of government action. In addition of course our notions of what rights are worth preserving (or creating) has undergone great change, and is likely to undergo greater change with the ever increasing tempo of social and technological development.

One of the interesting pieces of evidence for this trend has been a change in the emphasis in case law. Many cases decided over the last fifteen or twenty years have concerned judicial review of the performance of public duties, situations in which a citizen or group of citizens has tried to compel a public authority to perform an act in his or its interest, or in the public interest. As a result there seems to be a greater appreciation of the fact that public law has its positive as well as its negative aspect.

The purpose of this book is to examine the case law on judicial review of public duties in order to discover on what basis the courts do in fact review such duties, and to discuss critically the doctrinal approaches which are revealed by the case law in the light of modern notions of public law.

Since no study of this kind has previously been undertaken, some words of justification are necessary.

What is quickly apparent from looking at the case law of public duties is that the reasoning processes adopted and the concepts used are not the same as those which appear in the more traditional type of case, in which the court is concerned with such questions as whether a decision has been validly taken, whether natural justice has been observed, whether an error of law has been made, or whether a discretion has been exercised reasonably. The law of public duties does of course have something to do with this kind of reasoning, and in fact the remedy of mandamus, which was invented to enforce public duties, has played a crucial role in compelling the making of decisions and the exercise of discretion in accordance with law, as the traditional formulation has it. However, most of the law relating to public duties is not concerned with the propriety of public decision-making, but with whether authority should be compelled to do some particular act. To the extent that the law is concerned with duties to act rather than duties to decide things, the traditional concepts and modes of reasoning are irrelevant.

Of course it is possible to express all judicial control of administrative action in terms of the propriety of decisions. In fact the Administrative Decisions (Judicial Review) Act 1977, which governs judicial review of federal administrative acts in Australia, seeks to do precisely that, but only by stretching the concept of a decision to cover acts which would not ordinarily be regarded as such (see Chapter 5). To do this, however, is to disregard the obligatory nature of some important aspects of administrative action, and carries with it the danger that the law will be reduced to impotence just where it needs to be decisive. It is not very helpful, for example, to decide a case in which the issue is whether industrial action excuses failure to deliver the mail, or whether an individual can get an order forcing the police to enforce a statute, on the basis of whether the authorities have made a valid decision about the matter. Moreover, as will be apparent throughout this book, the courts in very many public-duty cases, most of which are quite recent, have not in fact used the traditional modes of reasoning to decide such issues. They have instead extemporized, making new law in the process, and it is this law which the book seeks to uncover and discuss. It is my central thesis that there is now, as a result of such cases, a jurisprudence of public duties which is worth discussing.

Central to the book is of course the remedy of mandamus. In addition to specific discussion of this remedy in Chapter 3, most of the cases discussed elsewhere in the book are mandamus cases. I recognize, however, that there are other available remedies for enforcing public duties, in particular the modern statutory judicial-review procedure, the injunction and the declaration, and the action for damages in negligence, and that, unlike mandamus, these remedies can be directed towards administrative action generally. This fact entails some difficulty, which is not present in the old law of mandamus, in presenting a clear picture of the law of public duties, because to do so involves taking a cross-section, as it were, of these aspects of the law, which deal with powers as well as duties. One of the important purposes of the book, however, is to see in what way powers and duties are to be distinguished from each other, and the taking of this cross-section has, I suggest, its revelations as well as its awkwardnesses.

As a final point of justification it must be said that the literature on public duties is remarkably thin. Most of it is contained in textbook chapters on mandamus which are more concerned with the nature and incidents of that remedy than with the concept of a public duty as such, or in articles about various aspects of judicial review, such as standing or remedies, which are unilluminating with regard to public duties.

No apology is made for the emphasis on case law. This area of judicial review has not been subjected to careful case analysis since 1848, when Thomas Tapping published his treatise on mandamus; even this resembled more a catalogue than a conceptual analysis of the case law.

I have found that the most useful and interesting case law on public duties is from England and Australia; for this reason most of the material which is relevant to a particular jurisdiction, particularly in Chapters 5 and 6, concerns England and Australia. However, other common-law jurisdictions have also provided some interesting case law. Due to shortage of space and some doctrinal differences I have not discussed United States and Indian material.

I would like to thank my erstwhile employers, the National University of Singapore, who granted me sabbatical leave at Monash University in 1984–5, during which much of the manuscript was researched and written. I cannot name all the various colleagues, friends, and library staff at the Singapore and Monash law schools who have helped me in various ways. None the less I wish to thank them for their comments and words of encouragement. In particular I would like to thank my Ph.D. supervisors Ron McCallum and Peter Hanks of Monash University for their careful reading of the manuscript and for their helpful comments and advice.

English and Australian law is stated as at 1 May 1988.

A.J.H

School of Oriental and African Studies,
University of London,
May 1988

Contents

Abbreviations

A. & T.	Adolphus & Ellis' Reports
AC	Appeals Cases (1891-)
affd.	affirmed
All ER	All England Reports
ALJ	*Australian Law Journal*
ALJR	Australian Law Journal Reports
ALR	Australian Law Reports
App. Cas.	Appeals Cases (1875-90)
APR	Atlantic Provinces Reports
AR	Alberta Reports
B. & S.	Best & Smith's Queen's Bench Reports
BCR	British Columbia Reports
Can. BR	*Canadian Bar Review*
Cap.	chapter (i.e. of edition of laws)
Chit.	Chitty's Bail Court Reports
Chanc.	Chancery
Chanc. D.	Chancery Division
Civ. JQ	Civil Justice Quarterly
CLR	Commonwealth Law Reports
CLJ	*Cambridge Law Journal*
CLY	*Current Law Yearbook*
CMLR	Common Market Law Reports
Cmnd.	Command Paper
Co. Inst.	*Coke's Institutes*
Crim. LR	*Criminal Law Review*
DLR	Dominion Law Reports
Dowl.	Dowling's Bail Court (Practice) Cases
Eq.	Equity
ER	English Reports
Fam. Law	*Family Law*
FC	Federal Court
FLR	*Federal Law Review*
HCA	High Court of Australia
HCR	High Court Rules (Cwlth)
HL	House of Lords
HLE	*Halsbury's Laws of England*
HLR	*Harvard Law Review*
IA	Indian Appeals
ICR	Industrial Court Reports
ILJT	Irish Law Journal Times
ILT	Irish Law Times
Imm. App. Cas.	Immigration Appeal Cases
IR	Irish Reports

JPL	*Journal of Planning Law*
JP Jour.	*Justice of the Peace Journal*
Jur.	Jurist Reports
LGR	Local Government Reports
LGRA	Local Government Reports of Australia
LJKB	Law Journal King's Bench
LJMC	Law Journal Magistrates' Cases
LJQB	Law Journal Queen's Bench
Ll. Rep.	Lloyd's List Law Reports
LQR	*Law Quarterly Review*
LT	Law Times
LTOS	Law Times Originating Summons
Melb. ULR	*Melbourne University Law Review*
MLJ	*Malayan Law Journal*
MPR	Maritime Provinces Reports
N & PEIR	Newfoundland and Prince Edward Island Reports
New Sess. Cas.	New Sessions Cases
NI	Northern Ireland
NLR	New Law Reports
NSR	Nova Scotia Reports
NZ	New Zealand
NZLR	New Zealand Law Reports
NZULR	*New Zealand Universities Law Review*
OLR	Ontario Law Reports
Ont.	Ontario
OR	Ontario Reports
P & CR	Property and Compensation Reports
QBD	Queen's Bench Division
QJPR	Queensland Justice of the Peace Reports
QL	Queensland Lawyer
Qld	Queensland
Qld R	Queensland Reports
Qld SR	Queensland State Reports
Que.	Quebec
Que. SC	Quebec Supreme Court
revd.	reversed
RSC	Rules of the Supreme Court
RTR	Road Traffic Reports
SA	South Australia
SASR	South Australian State Reports
SCC	Supreme Court of Canada
SCI	Supreme Court of Ireland
SCR	Supreme Court Reports
SLT	Scots Law Times
SR	State Reports (only with SR (NSW))
Tas. LR	Tasmanian Law Review
TLR	Times Law Reports

UQLJ	*University of Queensland Law Journal*
UTLR	*University of Tasmania Law Review*
Va. LR	*Virginia Law Review*
Vic.	Victoria (state)
VLR	Victorian Law Reports
VR	Victorian Reports
VUWLR	Victoria University of Wellington Law Review
WA	Western Australia
WALR	Western Australian Law Reports
WAR	Western Australian Reports
WIR	West Indian Reports
WLR	Weekly Law Reports
WN	Weekly Notes
WWR	Western Weekly Reports
Yale LJ	*Yale Law Journal*

We live in an age when Parliament has placed statutory duties on government departments and public authorities—for the benefit of the public—but has provided no remedy for the breach of them.

(per Lord Denning MR in *Attorney-General, ex rel. McWhirter v. Independent Broadcasting Authority* [1973] QB 629, 646)

'Tis clear,' cried they, 'our Mayor's a noddy;
  And as for our Corporation—shocking
To think we buy gowns lined with ermine
For dolts that can't or won't determine
What's best to rid us of our vermin!
You hope, because you're old and obese,
To find in the furry civic robe ease?
Rouse up, sirs! Give your brains a racking
To find the remedy we're lacking,
Or, sure as fate, we'll send you packing!
At this the Mayor and Corporation
Quaked with a mighty consternation.

(Robert Browning, *The Pied Piper of Hamelin*, 21–34)

I

Creation of Public Duties

1.1 THE CONCEPT OF A PUBLIC DUTY

The concept of a public duty is an important but elusive one. It has been consistently ignored by jurists, avoided by public-law scholars, and rendered marginal for much of its history by the judges. It may be worthwhile considering why this has happened.

If we take the jurists first, we find that although much has been written about the nature of legal obligations, no important treatise on the subject has devoted attention to the problems of public duties in the sense of duties owed by the state to its citizens. Obligation is treated as something obtaining between individuals. Individuals have rights and obligations *inter se* and against the state, but the state has no obligations with regard to individuals except the negative obligation not, in the exercise of its powers, to trespass on his rights. It may be that command theories of law derived from Austin[1] bear some of the responsibility. If one conceives of law as something in the nature of a command given by a sovereign to his subjects the notion of a public duty is either incomprehensible or is used to describe the mundane obligations of the public servant to the sovereign in the execution of his wishes or else the obligations of the citizen to obey his laws. Hearn[2] writing in 1883 used the concept of a public duty in this sense only. Hohfeld[3] too with his right/duty correlation leaves entirely out of his calculations the sphere of public duties, where the correlation may exist or may not, depending on the kind of duty which is in question. Presumably this kind of view would dismiss public duties as duties without sanction and therefore lacking the force of law *stricto sensu*.

[1] *The Province of Jurisprudence Determined* (1954 edn., with introduction by H. L. A. Hart), pp. 14–15. Austin saw duties as correlatives of commands.

[2] *The Theory of Legal Rights and Duties* (1883), pp. 28–9, 60.

[3] See *Lloyd's Introduction to Jurisprudence*, 5th edn. (1985), pp. 441–7; one could perhaps hold that the right/duty correlation exists wherever a court enforces a public duty at the instance of an individual. However, this is merely to describe a conclusion (and to describe it very misleadingly in some cases) rather than a reasoning process; see Harris, *Legal Philosophies* (1980), pp. 81–3, where the unhelpfulness of Hohfeldian analysis in this context is convincingly demonstrated; Hohfeld, *Fundamental Legal Conceptions as Applied in Judicial Reasoning* (1919); Allen, *Legal Duties* (1931), pp. 156 et seq.

Public lawyers too have not expounded the concept of a public duty. This may be due to a particular view of the nature of constitutional law which would see the task of law as being one of circumscribing the legitimate field of activity of the legislative and executive branches. Positive public duties have no place in such a view except in so far as they may, interstitially, be necessary to support the operation of the constitutional system; this is dealt with in Chapter 2. One might none the less argue that some of the traditional notions of constitutional law do entail some propositions concerning public duties; for example the supremacy of Parliament and the hostility of the law towards the prerogative of suspension of laws, and the concept of equality before the law, entail the proposition that the laws must be enforced equally. This also is pursued in Chapter 2. If one is looking, however, for a general theory of public duties one will not find it in Dicey or the writings of any other constitutional jurist.

The source of public duties is to be found not in the law but in public policy.[4] The task of the law commences only when the duty has been imposed by legislation. This task has not, however, been adumbrated authoritatively either by judges or by academic writers.[5]

There is therefore no obvious starting-point for a study of this kind. I have taken as my premises in analysing the case law the propositions (1) that an effective law of public duties would be one which is prepared to recognize the existence of a duty where the intention of the legislature is plainly to that effect or where the social conditions of a modern democratic society would seem to demand it; (2) that the law should enable a public duty to be effectively enforced by individuals, provided that latitude is given for 'legitimate' administrative action according to the canons of the modern administrative-law system, which typically allows broad judicial scrutiny without usurpation of the functions of the executive branch.

The method adopted here of analysing public duties is as follows. Chapter 1 is concerned with the questions: What is a public duty? What kinds of public duty are there? When does a public duty arise? Chapter 2 deals with the question, what constitutes performance of a public duty, using the analysis indicated in Chapter 1. Chapters 3, 4, and 5 are concerned with enforcement: how in practice is a public duty enforced? Chapter 3 deals with mandamus, Chapter 4 with other available remedies, and Chapter 5 with the modern statutory procedures for judicial review. Chapter 6 deals with the problem of standing to enforce public duties: who has the right to invoke the

[4] See Cranston, *The Legal Foundations of the Welfare State* (1985), p. 2.

[5] Cranston's book does, however, deal with welfare duties; see below, Ch. 2.3.

remedies dealt with in Chapters 3, 4, and 5? Chapter 7 deals with the question of damages for failure to perform a public duty. A short summation sets out the main conclusions reached.

As is indicated above, the purpose of Chapter 1 is to examine the concept of a public duty, and it is therefore necessary to provide a skeletal definition which will be given greater content in Chapters 1 and 2. A public duty in the sense in which the term is used here, and indeed generally in law reports and legal literature, is simply a duty of a public body arising in public law, so that it is really the terms 'duty', 'public body' and, 'public law' which require definition.

The term 'public law' is a difficult one, because it can be used in many different senses. It can be used in a *substantive* sense to draw a distinction between legal rules applying to public bodies and legal rules applying to private persons; it can be used in a *procedural* sense to draw a distinction between the method of applying for relief against a public body and the method of applying for relief against a private person; or it can be used in a *jurisdictional* sense to indicate that a court has jurisdiction only over public bodies as opposed to private persons.[6] No doubt there are other senses, but these senses are probably the only ones which serve any practical purpose. There are of course books on 'public law' which deal with constitutional- and/or administrative-law questions, in which the term is useful merely for the purpose of exposition of the law relating to the organs of state, but has no practical legal consequences; to avoid confusion I shall use the terms 'constitutional law' and 'administrative law' when referring to public law in this sense. It appears that, in England at least, the term 'public law' is now a term of art with definite legal consequences,[7] and I shall therefore use it, in an English context, in a substantive and procedural sense. In relation to Australia the position is quite complex, and is discussed, together with the English concept of public law, in Chapter 5.[8]

Broadly speaking, however, the notion of public law does have a core meaning, at least in the context of public duties, because the law

[6] See Craig, *Administrative Law* (1983), pp. 12–14.

[7] See *O'Reilly* v. *Mackman* [1982] 3 WLR 604, and below Ch. 5. The important consequence of the designation is that all 'public-law' cases must be brought under O. 53 procedure.

[8] Not every jurisdiction uses the distinction between public and private law. In Australia federal law does in a sense use it, because the Administrative Decisions (Judicial Review) Act 1977 (Cwlth) provides a statutory code for judicial review and the courts are left with the task of deciding which decisions or acts are subject to it (see below Ch. 5.3). The jurisdictional sense of 'public law' is not so important in common-law jurisdictions as it is in civil-law systems, but of course jurisdictional problems may flow from the adoption of 'public law' in the procedural sense.

of public duties has developed around the prerogative writ of mandamus, in which the public-law/private-law distinction is, and always has been, fundamental. Whatever may be the arguments for or against adopting the distinction as a basic fact of the legal system, its adoption for the purposes of the law of public duties seems inevitable because of the way in which that law has developed. This is not of course to say that the way in which, and the purposes for which, the distinction is being drawn by judges at the present time are necessarily right, but merely to adopt the distinction as a starting-point for the discussion of duties of public bodies. Even those who reject the distinction entirely would accept that there are many duties which are simply not capable of being imposed on any private individual, but only on organs of state, and it is those duties with which we are concerned. Furthermore, the adoption of 'public law' as a term of the definition of a public duty is necessary because not all duties owed by public bodies are, in a relevant sense, public duties: most duties of public bodies which arise in tort and contract and some statutory duties are duties in private law, not merely because these are traditionally regarded as areas of private law, but because the rules of liability do not differ, or do not differ significantly, according to whether the duty is sought to be imposed on a public body or a private individual.[9] Conversely we cannot define public duties simply as duties arising in public law, because many duties in public law, such as the duty to pay taxes, are owed by private individuals (these are 'public duties' in Hearn's sense); these duties are important, but demand an entirely different kind of discussion from the duties of public bodies and are not dealt with here.

I have defined a 'public body' as a body amenable to public law, i.e. public law in the sense indicated above. Again the term 'public body' must be construed narrowly, for not all bodies which we consider 'public', i.e. directed to the promotion of the general welfare and acting on behalf of the community, are in fact amenable to public law.[10] It may be that the law of public duties is too restrictive in this respect, a point which is pursued below.

We are left then with the term 'duty'. In administrative law, duties are distinguished from powers, and the two taken together comprise all the functions of a public body. What distinguishes a duty from a power is the prescriptive nature of a duty as opposed to the

[9] None the less, because of the kinds of fact situation which occur in cases involving public bodies, there can be said to be a public-law gloss on private-law duties: see e.g. *Anns* v. *Merton London Borough Council* [1978] AC 728 (HL), and below Ch. 6.3.

[10] See below Ch. 1.2(iii–iv).

discretionary nature of a power. Dias, having in mind no doubt duties in private law, described a duty as

> a prescriptive pattern of conduct that is recognized by the courts . . . To the vital question: how is one to know whether [it is] recognized? the answer is: by knowing the law. For only in this way can one know whether a duty exists in a given case or not and what its scope is, and it is for this reason that the question of duty is always one of law for the judge.[11]

This will suffice also for public law. Some might prefer a definition which stresses the imperative nature of a duty. It is, one might say, a command by the legislature that something be done. This Austinian-type formulation, in addition to its conceptual rigidity, adverted to above, overlooks two factors: first, not all duties are in fact imposed by the legislature; and secondly, some of those that are imposed by the legislature are not in fact enforceable in the courts and should not therefore come within our definition. The phrase 'recognized by the courts' is very apt for the purpose of public duties because we are concerned here with enforcement, not with mere moral obligations or exhortations which have no relevant legal consequences. Moreover, as will be seen,[12] the conferment of powers and the imposition of duties by statute does not in itself delimit the ambit of a duty, for the simple reason that there are instances when the conferment of a power involves the imposition of duty to exercise it, or to perform some other incidental act, such as obedience to the principles of natural justice. Many public duties are implied by the courts rather than commanded by the legislature; some can even be said to be assumed voluntarily. We can also note that some statutory public duties are 'prescriptive patterns of conduct' in the sense that they are treated, like private-law duties of the Atkinian variety, as duties to act reasonably, so that the prescription in these cases is indeed provided by the courts, not merely recognized by them.

Our definition can thus be summarized as follows:

1. There is, for certain purposes (particularly for the remedy of mandamus or its equivalent), a distinct body of public law.
2. Certain bodies are regarded under that law as being amenable to it.
3. Certain functions of these bodies are regarded under that law as prescribing as opposed to merely permitting certain conduct.
4. These prescriptions are public duties.

[11] *Jurisprudence*, 4th edn. (1976), p. 305.
[12] See below Ch. 1.3.

The next three sections are devoted to answering the question, when and how does a public duty arise?

1.2 PUBLIC DUTIES AT COMMON LAW

As already indicated, there are public duties which arise from sources other than a statute. These duties may be more important than they are often thought to be, and deserve more than a passing reference before we consider statutory public duties. They will be dealt with according to the nature of the source of the duty, namely prerogative, franchise, charter, and contract. The ensuing discussion omits consideration of public duties arising under the law of tort, which are dealt with in Chapter 7 in the context of damages.

(i) *Prerogative*

Duties arising under the prerogative are still important in England, though rarely elsewhere, and the question of their justiciability has given rise to some problems.

Modern ideas of the relevance of the source of a public duty derive from the constitutionally important decision in *R.* v. *Criminal Injuries Compensation Board, ex p. Lain*,[1] which has received the enthusiastic approval of the House of Lords in *Council of Civil Service Unions* v. *Minister for the Civil Service* (known as the *GCHQ* case).[2] The applicant in *Ex p. Lain* sought *certiorari* to quash a decision of the Board, a body set up by the British government under the prerogative which, in pursuance of a scheme announced in Parliament, was to determine the amount of compensation to be paid *ex gratia* to the victims of crimes of violence. It was argued for the Board that it was beyond the scope of *certiorari* because it had no statutory authority and its determinations gave rise to no enforceable rights but only to an opportunity to receive the bounty of the Crown. Although *certiorari* was refused on the facts, it was made very clear that the argument that the Board was beyond the reach of *certiorari* was untenable. Diplock LJ said this:

> The jurisdiction of the High Court . . . has not in the past been dependent upon the source of the tribunal's authority to decide issues submitted to its

[1] [1967] 2 QB 864, followed by the Court of Appeal in *R* v. *same*, *ex p. Tong* [1976] 1 WLR 1237.

[2] [1984] 3 WLR 1174; see also *R* v. *Panel on Take-overs and Mergers*, *ex p. Datafin PLC* [1987] All ER 564; Walker, 'Review of the Prerogative: the Remaining Issues' [1987] PL 62.

determination, except where such authority is derived solely from agreement of parties to the determination. . . . The earlier history of the writ of certiorari shows that it was issued to courts whose authority was derived from the prerogative, from Royal Charter, from franchise or custom as well as from Act of Parliament . . . I see no reason for holding that the ancient jurisdiction of the Court of Queen's Bench has been narrowed merely because there has been no occasion to exercise it.[3]

While Diplock LJ denied the relevance of the source of authority, Lord Parker CJ stressed the public-law element:

At one time [*certiorari*] only went to an inferior court. Later its ambit was extended to statutory tribunals determining a *lis inter partes*. Later again it extended to cases where there was no *lis* in the strict sense of the word but where immediate or subsequent rights of a citizen were affected. The only constant limits throughout were that it was performing a public duty. . . . We have as it seems to me reached the position when the ambit of certiorari can be said to cover every case in which a body of persons of a public as opposed to a purely private or domestic character has to determine matters affecting subjects providing always it has a duty to act judicially.[4]

The court also referred to a number of considerations which established the public nature of the Board's duties: it spent large amounts of public money voted by Parliament; the public were affected by its functioning; the scheme under which it operated was promulgated and amended in Parliament.[5] The argument that its determinations were not binding was circumvented by saying that its determinations did have legal effect in that they rendered lawful and irrecoverable a payment of money which would not otherwise be such.[6]

Although *Ex p. Lain* is concerned with *certiorari*, the reasoning adopted leads to the conclusion that mandamus can be awarded to compel the Board to consider an application in accordance with law,[7] though not actually to make a payment, since there is no obligation to make a payment.[8]

The 'public-law' test applied in *Ex p. Lain* is clearly a general one, and can be regarded as an acid test for the availability of public-law

[3] [1967] 2 QB 884.
[4] Ibid. 882.
[5] Ibid. per Ashworth J. at p. 891.
[6] Ibid. 889.
[7] See *Ex p. Tong*, above n. 1, p. 1242, per Lord Denning MR; *R.* v. *Criminal Injuries Compensation Board, ex p. Clowes* [1977] 1 WLR 1353. These decisions seem to be inconsistent with *R.* v. *Secretary of State for War* [1891], 2 QB 326; *Griffin* v. *Lord Advocate* 1950 SC 448 which were doubted in the *GCHQ* case, above n. 2.
[8] [1967] 2 QB 885.

remedies under the rubric of the prerogative orders or the statutory application for judicial review.[9] Its importance for public duties is obvious: duties cannot be created under prerogative powers so as to avoid judicial review of their performance, for such duties are still public notwithstanding their extra-statutory origins. *R.* v. *Panel on Take-overs and Mergers, ex p. Datafin PLC*[10] goes even further. The Panel was essentially a self-regulatory body set up as an act of government; its duties could not be described as deriving from statute or common law, though it was 'supported and sustained by a periphery of statutory powers and penalties',[11] nor could it be said to derive its duties from prerogative or contract. None the less it was subject to judicial review because its code and rulings applied equally throughout the country to all who wished to make take-over bids or promote mergers; it performed functions which would be statutory but for the historical happenstance that the city regulated itself; its decisions affected indirectly the rights of citizens; it had a duty to act judicially in relation to a breach of its code; above all it operated wholly in the public domain and performed an important public duty, fitting surprisingly well into the format of the *Lain* case. 'I should be very disappointed', said Donaldson MR, 'if the courts could not recognize the realities of executive power and allowed their vision to be clouded by the subtlety and sometimes complexity of the way in which it can be exerted.'[12]

There are, however, clearly limitations to this analysis. First, if the duty in question arises under contract, it will be a duty in private law, even if the party owing the duty is a public body.[13] Secondly, it may be that certain decisions made under prerogative powers are regarded as intrinsically incapable of being subjected to the rigours of modern judicial review. A tribunal set up to assist the military authorities in relation to appeals concerning exemption from military service was held not to be amenable to mandamus because it had no statutory authority.[14] This ratio clearly no longer holds, but in another case prohibition against a tribunal set up *inter arma* to enforce a

[9] See below Ch. 5.2. The 'public-law' content of mandamus is not significantly different from that of *certiorari*, except of course that *certiorari* is necessarily confined to decision-making bodies.

[10] Above n. 2.

[11] Ibid. 574.

[12] Ibid. 577.

[13] This is apparent from the passages quoted above, but see also the cases cited below nn. 33–7 and Ch. 5. Donaldson MR in the *Datafin* case, above n. 2, p. 577, would exclude from judicial review a consensual submission to the jurisdiction of a public body.

[14] *Ex p. Mann* (1916) 32 TLR 479.

military proclamation was refused by the House of Lords because it was not a court or court martial in any sense, having no jurisdiction by statute or at common law.[15] The likely position is that, if any, only tribunals of this latter kind are exempt from judicial coercion, and probably only in actual wartime or emergency conditions. Any other view of the matter would make serious inroads into modern notions of the rule of law.

The *GCHQ* case affirms the irrelevance of the source of authority. It concerned the question whether the Minister was under a duty to observe natural justice before removing the union rights of civil servants employed at an institution whose functions were important for national security; the House of Lords decided that since there was evidence that the duty was breached for reasons of national security, it was immune from review. However, it was this reason and not the prerogative source of the power in question which was decisive.

We have spoken so far of tribunals and the duty to determine, and in this context the function of mandamus is merely complementary: *certiorari* is to invalidate illegality, prohibition to prevent it, and mandamus is to enforce legality; in this area the field of mandamus is probably the same as that of *certiorari* and prohibition. Of its very nature, however, mandamus extends to activities outside the ambit of *certiorari* and prohibition: according to the analysis set out below in Chapter 1.5, to the duty to provide, the duty to enforce the law, and various constitutional duties. There is no reason why the 'public-law' analysis mentioned above should not extend also to these duties, and in fact there is every reason to suppose that it does. For example the duties of the police with regard to law enforcement are enforceable by mandamus even though no statute imposes such duties;[16] duties can also arise under a charter or a franchise, as indicated in *Ex p. Lain.* Nevertheless the concept of prerogative itself can be a source of exemptions from judicial coercion as well as a

[15] *Re Clifford and O'Sullivan* [1921] 2 AC 570. In *R.* v. *Army Council, ex p. Ravenscroft* [1917] 2 KB 504, 511, Visc. Reading CJ reserved his opinion on whether mandamus would lie against a military court inquiring into the conduct of an officer. The suggestion of Avory J. (p. 514) that this would be to grant mandamus against the Crown is quite untenable. These two cases and *Ex p. Mann* all related to actual wartime conditions and would be carefully scrutinized today. The same caution is apparent in the decision of the High Court of Australia in *R.* v. *Scherger, ex p. Bridekirk* (1957) 99 CLR 486, in which the court in *obiter dicta* doubted whether mandamus lay against officers of the armed forces 'unless the real meaning of the legislative provisions which are relied upon is that the officers shall rest under a public duty the responsibility for the discharge of which is upon themselves personally and not upon the Crown' (p. 504).

[16] *R.* v. *Metropolitan Police Commissioner, ex p. Blackburn* [1968] 2 QB 108.

source of enforceable duties: the Crown's duty at common law to protect its subjects is unenforceable in the courts;[17] the granting of passports has been regarded in England (although not necessarily elsewhere) as a privilege of the Crown which was not enforceable legally; however, a recent decision at first instance indicates otherwise.[18] the prerogative of mercy similarly cannot be enforced judicially;[19] and duties arising under international treaties are not enforceable in the municipal law of common-law jurisdictions.[20] All these are in effect governmental immunities. It may none the less be that following the *GCHQ* case some of these immunities will be done away with; is there, for example, anything inherently injusticiable about a refusal of a passport, or a refusal to render assistance to a subject overseas? Why should these duties continue to hide behind the last vestiges of prerogative rather than be considered on their merits?

As will be seen later,[21] the scope of mandamus is further restricted by its inapplicability to the Crown. This principle has however been narrowly construed, and is not an insuperable obstacle to review of prerogative public duties.

(ii) *Franchise*

These days franchises are not as important as they once were, but many franchises are still operative and there is a tendency for them to devolve on to public authorities because they involve public services, for example the operation of a ferry, a port, or a market. Franchises are essentially powers or privileges granted by the Crown by charter or letters patent. Since they are in effect monopolies, they carry with them implicit, correlative duties: to operate the ferry at all reasonable times, to ensure that the port is properly maintained, or to provide adequate market facilities.[22] Failure to fulfil such duties may lead to forfeiture or indictment or the award of an injunction.[23] Where it was alleged that a public authority, the owner of a franchise, had failed to operate a ferry at all reasonable times, the court was prepared

[17] *Mutasa* v. *Attorney-General* [1980] QB 114.

[18] *Home Secretary* v. *Lakdawalla* [1972] Imm. App. Rep. 26; *Loh Wai Kong* v. *Government of Malaysia* [1979] 2 MLJ 33; *R* v. *Secretary of State for Foreign Affairs, ex p. Everett*, The Times, 10 Dec. 1987.

[19] *Horvitz* v. *Connor* (1908) 6 CLR 38 (HCA); *Hanratty* v. *Lord Butler, The Times*, 12 May 1971; *Superintendent of Pudu Prison* v. *Sim Kie Chon* [1986] 1 MLJ 494.

[20] De Smith, *Constitutional and Administrative Law*, 5th edn. (1985), pp. 152–4.

[21] Below Ch. 3.4.

[22] *HLE*, vol. 21, para. 987; vol. 8, para. 1016; vol. 29, para. 620; *Nicholl* v. *Allen* (1862) 121 ER 954; *Hammerton* v. *Dysart (Earl)* [1916] 1 AC 37 (HL).

[23] *Attorney-General* v. *Colchester Corporation* [1955] 2 QB 207, 214–15.

to assess the reasonableness of the hours of operation in all the circumstances, including public demand and private hardship.[24] The uncertainty of the scope of such duties indicates that they should in general be replaced with statutory provisions, as is the case with modern public utilities. The High Court of Australia, rejecting American authorities to the contrary, has held that such utilities, although exercising in effect exclusive monopolies, are not subjected to obligations to provide the relevant service at common law, but only by statute.[25]

(iii) *Charter*

Many of the early cases of mandamus involved duties arising from charters granted to local corporations or universities,[26] and as indicated in *Ex p. Lain* there is no reason to suppose that duties arising from charters are now unenforceable. The matter is not entirely of antiquarian interest, because many universities, particularly in England, still derive their duties from charters, though of course the concept of a freehold office has disappeared from local-government law, and local authorities are governed by statute. The position of universities is still moot because, according to some, students of a chartered university have only contractual status and therefore their remedies for breach of duty by the university are private-law, not public-law, remedies.[27] The better view, however, and one more in keeping with modern notions of university education as well as precedent, is that the duties owed by a university with regard to student discipline and examinations are public, whether the university is chartered or statutory. The Supreme Court of New South Wales in *Ex p. Forster*[28] stressed that the university in question was established

[24] *Gravesham Borough Council* v. *British Railways Board* [1978] Chanc. 379; see also *Attorney-General* v. *Colchester Corporation*, above n. 23.

[25] *Bennett and Fisher Ltd.* v. *Electricity Trust (SA)* (1962) 106 CLR 492; and see *Holmberg* v. *Sault Sainte Marie Public Utilities Commission* (1966) 58 DLR (2d) 125; however, see below Ch. 2.3 for enforcement of such duties.

[26] See Henderson, *Foundations of English Administrative Law* (1963), chs. 1–3, and pp. 131–7; *Lyme Regis (Mayor)* v. *Henley* (1834) 5 ER 1097.

[27] See notes by Wade at (1969) 85 LQR 468 and (1974) 90 LQR 157, criticizing *R.* v. *Aston University Senate, ex p. Roffey* [1969] 2 QB 538, and Garner's note at (1974) 90 LQR 6; *McWhirter* v. *University of Alberta* (1976) 63 DLR (3d) 684.

[28] *Ex p. Forster, re University of Sydney* (1963) SR (NSW) 723, 727 (mandamus refused on the facts); see also *King* v. *University of Saskatchewan* (1969) 6 DLR (3d) 120. In *R.* v. *University of Sydney, ex p. Drummond* (1943) 67 CLR 5 the High Court of Australia left open the question whether mandamus would lie to compel admission to a faculty. In *R* v. *St John's College, Cambridge* (1694) 90 ER 1140, Holt CJ hit on the correct approach: 'these fellowships are in the nature of publick offices, in which

by a public Act, was largely supported by public funds, was open to all scholastically qualified residents of the state, and was subject in some degree to public control, and did not feel obliged to decide whether the duties in question were derived from statute or common law. The position cannot be otherwise if the university is a chartered one, because its public character is retained; it would indeed be odd if the range of remedies available to an aggrieved student depended on the fortuitous circumstance of the university being statutory or chartered. In the long run this is not a serious difficulty because the private-law remedies of declaration and injunction can cover the field, although in England there remains the difficulty as to whether an application for judicial review against a university must be made under Order 53 procedure; the likelihood is that it does.[29] If of course the university is genuinely a private one, deriving its duties from a trust deed, then only private-law remedies apply.[30]

This area too, then, is one where the source of the duty should be regarded as irrelevant and the 'public-law' test applied.

(iv) *Contract*

Duties arising from contract are not of course generally regarded as duties in public law merely because the duty attaches to a public authority. This is clearly seen in the case of domestic tribunals, which derive their jurisdiction from an agreement between the parties as to the determination of disputes. In respect of these tribunals the weight of authority, as implied in the passages from *Ex p. Lain* quoted above, suggests that the prerogative writs are inapplicable.[31] In fact public-law reasoning is liberally applied to these tribunals and enforced by the private-law remedies of declaration and injunction, because justice requires that bodies exercising considerable powers affecting individual rights should be coerced by the courts where appropriate.[32] This reasoning is generally satisfactory, but has been extended to cases which, by any standard, have a strong public element, such as dismissal from,[33] or refusal of tenure in,[34] a university faculty, dismissal

the Government is concerned, and have a trust annexed to them, for the education of youth.'

[29] See *O'Reilly* v. *Mackman* [1982] 3 WLR 604, 620 per Lord Denning MR.

[30] *Herring* v. *Templeman* [1973] 3 All ER 569.

[31] e.g. *Armstrong* v. *Kane* [1964] NZLR 369 (suspension from trade-union membership).

[32] See the interesting pronouncements in *Lee* v. *Showmen's Guild of Great Britain* [1952] 2 QB 329 and *McKinnon* v. *Grogan* [1974] 1 NSWLR 295.

[33] *Vidyodaya University* v. *Silva* [1965] 1 WLR 77 (PC); *Australian National University* v. *Burns* (1982) 43 ALR 25.

[34] *Re Vanek and Governors of University of Alberta* (1976) 57 DLR (3d) 595.

from public employment,[35] termination of apprenticeship by a professional body,[36] and where an industrial tribunal exercises arbitral jurisdiction based on contract.[37]

The same deference to contract law appears in mandamus cases outside the context of adjudication. Lord Campbell CJ in refusing statutory mandamus to enforce a contract for a lease said the remedy 'does not extend to the fulfilment of duties arising merely from a personal contract. If it did so, it would extend to every case of contract.'[38]

The reasoning in these cases has been adopted also in cases where the duty has a strong statutory flavour. Australian courts have refused mandamus to compel performance by a public authority of an agreement to give effect to mutual obligations arising under a statute,[39] and have held a public authority's contractual duty based on the provisions of a by-law to be purely private.[40]

These cases all illustrate how the blurring of the distinction between the public and private sectors, due to the expansion of both, can cause some difficulty in defining the ambit of public law and contract. The apparent conceptual distinctness of these two areas of the law is not perfectly clear in practice and can be misleading. Particularly awkward are cases of public employment which have a strong contractual and statutory element; there is authority which would support the application of public-law remedies here.[41] Whitmore and Aronson[42] have argued that the British Motor Insurers' Bureau, whose

[35] *R.* v. *Post Office, ex p. Byrne* [1975] ICR 221.

[36] *R.* v. *National Joint Council for Dental Technicians, ex p. Neate* [1953] 1 QB 704.

[37] *R.* v. *Industrial Court, ex p. ASSET* [1965] 1 QB 377; see, however, *Imperial Metal Industries (Kynoch) Ltd.* v. *AUEW* [1979] ICR 23, in which the *ASSET* case was doubted; it was held there that an arbitration made at the instance of a trade union not a party to the arbitral agreement was not a private arbitration for the purposes of the Arbitration Act 1950 (England and Wales). For the position under the English regime of 'public law', see below Ch. 5, where the Australian federal law is also discussed.

[38] *Benson* v. *Paull* (1856) 119 ER 865, 866, in the light of which the same judge's famous dictum that 'a legal obligation, which is the proper substratum of a mandamus, can only arise from common law, from statute, or from contract' (*Ex p. Napier* (1852) 118 ER 261, 263) cannot be taken to mean that mandamus lies to enforce a contract *qua* contract; and see *Re O'Rourke* (1886) 7 NSWR 64; *Koon Hoi Chow* v. *Pretam Singh* [1972] 1 MLJ 180 (doctor's report, contractual not public duty). In any case the public-law element of statutory mandamus is somewhat uncertain: see below Ch. 4.5.

[39] *Belcaro Pty. Ltd.* v. *Brisbane City Council* [1964] Qld R 302 (HCA).

[40] *Fairfax (John) Ltd.* v. *Australian Telecommunications Commission* [1977] 2 NSWLR 400.

[41] *R. (Doris)* v. *Ministry of Health* [1954] NI 79.

[42] *Review of Administrative Action* (1978), pp. 376–7.

duties derive from a contract between insurers and the Minister of Transport, should be subjected to public-law reasoning: clearly in this instance the fulfilment of these duties affects many people who are not parties to the contract, and is surely a legitimate concern of public law. It is suggested that the *instrumental* basis of the duty should not decide such cases unless it is really relevant to the requirements of justice.[43] However, similar problems have arisen under the statutory review procedures dealt with in Chapter 5, and it may be in this area where the proper mode of distinguishing contractual and public duties will be worked out.

A further point concerning contracts is that a public duty cannot be fettered by contract as to its fulfilment or the manner of its fulfilment, a principle affirmed by the High Court of Australia in *Ansett Transport Industries Ltd.* v. *Commonwealth.*[44] The practical effect of this obviously necessary rule may be reduced by the fact that the law generally does not prescribe the manner in which public authorities should contract; thus mandamus was refused where an authority agreed to employ only union members, because it had a discretion to contract with its employees on such terms as it thought fit.[45]

1.3 STATUTORY PUBLIC DUTIES

(i) *Implication of statutory public duties*

Public duties of a statutory kind can normally be traced to specific statutory commands such as 'A shall do Y', 'it shall be the duty of A to do Y', and the like; these phrases are often described as 'mandatory', 'imperative', or 'peremptory'. Sometimes, however, the courts will imply a duty even where there is no such obligatory wording; an obvious example is where the duty is expressed in terms of B's entitlement to have Y done by A; another is the construction of an enabling provision in such a way as to find a duty to act in the particular circumstances of the case (this is discussed below). In some cases, however, a duty may be implied from the broadest consideration

[43] See e.g. *R.* v. *Glamorgan County Council* [1899] 2 QB 536, where the absence of a statutory basis to compel the reimbursement of tradesmen who supplied troops called in to suppress a riot could easily have been circumvented by stressing the public nature of the obligation.

[44] (1978) 139 CLR 54, esp. per Mason J. at pp. 74–5; *Birkdale District Electric Supply Co. Ltd.* v. *Southport Corporation* [1926] AC 355, 364; *Triggs* v. *Staines Urban District Council* [1969] 1 Chanc 10, 18; *Windsor Royal Borough Council* v. *Brandrose Investments Ltd.* [1983] 1 All ER 818.

[45] *R.* v. *Greater London Council, ex p. Burgess* [1978] ICR 991.

of the statutory provisions, without basing that implication on particular obligatory or even enabling words. Cases of this kind are interesting because the courts sometimes impose burdens which have not been specifically addressed by the legislature, and are therefore open to charges of judicial legislation. Australian cases have been especially useful in this area.

The *locus classicus* is the decision of the High Court in *Bradley* v. *Commonwealth*.[1] The plaintiff was the acting director of the Rhodesia Information Centre, whose purpose was to disseminate in Australia information concerning the illegal government of Rhodesia. The Postmaster-General issued a direction to withdraw all postal and telecommunication services to the Centre. The plaintiff sued *inter alia* for a declaration that the Commonwealth was thereby in breach of its duty under the Post and Telegraph Act 1901–71 (Cwlth) to provide him with those services. The Commonwealth argued that its only duty under the Act was owed to the Crown and not to individuals. It is interesting to notice that the majority[2] (who found for the plaintiff) and the minority[3] reasoned from opposite premises. The majority reasoned that clear words were necessary to confer an arbitrary power to remove a person's right of 'real freedom of dissemination of information' and construed the Act with no difficulty so as to impose a duty to deliver mail to the Centre (in fact the Act abounded with indications that properly addressed and stamped mail should be delivered to the address indicated); regulations made under the Act concerning telecommunication services were ambiguous but were construed in favour of the existence of a duty to provide the service. The minority on the other hand reasoned that no declaration could be granted unless there was a right to the performance of the duty clearly created by the statute, and this could not be found. The rights-based decision of the majority is clearly preferable, and indeed any other decision would render a large number of public duties practically unenforceable. The case is clear authority for careful judicial scrutiny of statutes which deal with public goods and services, a scrutiny which has not always been apparent, as will be seen at many points in the ensuing chapters.

By way of contrast the High Court refused to imply a statutory duty on the part of a local authority to construct a stormwater drain on the subdivision of a piece of land where the statute, whose purpose was merely to give limited relief to a subdivider, provided for

[1] (1973) 128 CLR 557; see also *R.* v. *Arndel* (1906) 3 CLR 557; *Fairfax (John) Ltd.* v. *Australian Postal Commission* [1977] 2 NSWLR 124.

[2] Barwick CJ, Gibbs and Stephen JJ.

[3] McTiernan and Menzies JJ.

apportionment of the cost of drainage works without reference to the question whose burden the work should be.[4]

In a recent English case[5] the Court of Appeal was faced with a statutory duty imposed on the Secretary of State for Social Services 'to make arrangements with a view to securing that benefit . . . and other officers . . . exercise their functions in such a manner as shall best promote the welfare of persons affected by the exercise of those functions'. This duty was held not to include a duty to correct administrative errors by searching through 15 million files at a cost of £4.8m. to find 16,000 claimants who were entitled to a refund averaging £25. No doubt the court was influenced by the fact that the cost of the exercise would be more than ten times the amount of benefit which would be paid out. The decision is questionable because the duty to make a refund seems to be clearly implied from the statutory duty, which does not take account of administrative cost or convenience; one might well say that the unidentified claimants had a right, in the *Bradley* sense, to a refund. The court treated the matter, however, as one of review of the exercise of a power on grounds of reasonableness. On this basis the decision would be correct, but it is surely incorrect to treat a duty as if it were a power.

The decision in *Smith* v. *Commissioner of Corrective Services*[6] seems to be quite inconsistent with the approach in *Bradley*. The plaintiff, an accused person, sought a declaration that the facilities at the prison where he was held were inadequate to enable proper consultation with his lawyers because the room used for the purpose had a listening device which was capable of being plugged in without the plaintiff knowing. The New South Wales Court of Appeal held that the plaintiff's privilege of private communication with his lawyers did not mean that the Commissioner had to take steps to ensure that there

[4] *Belcaro Pty. Ltd.* v. *Brisbane City Council* [1964] Qld R 302. Educational duties have also been the subject of litigation: see *Ex p. Cornford, re Minister for Education* [1962] SR (NSW) 220, and cf. *Gateshead Union Guardians* v. *Durham County Council* [1918] 1 Chanc. 146; *Bachmann* v. *Government of Manitoba* [1984] 6 WWR 25. In this last case the court was able to find a duty to provide school transport on a non-discriminatory basis, even though the statute conferred only a power to provide; in effect this means that a decision to provide may involve a duty to provide equally.

[5] *R.* v. *Secretary of State for Social Services, ex p. Greater London Council and Child Poverty Action Group, The Times*, 8 Aug. 1985. See also *R.* v. *Secretary of State for Social Services, ex p. Child Poverty Action Group, The Times*, 15 Feb. 1988, which displays a similar reluctance to imply a duty.

[6] [1978] 1 NSWLR 317; and see *Bromley* v. *Dawes* (1984) 34 SASR 73; *R.* v. *Secretary of State for the Home Department, ex p. McAvoy* [1984] 3 All ER 417; *R.* v. *Durham Prison (Governor), ex p. Hardial Singh* [1984] 1 WLR 704; contrast, however, *Smith* v. *Commissioner of Corrective Services* (1980) 147 CLR 134 (HCA).

was no suspicion of eavesdropping; this was an 'impossible burden', and to grant the declaration would be equivalent to an order of mandamus to make alterations to the prison; it was said, further, that the Commissioner's only duty was to hold the prisoner. If a *Bradley*-style approach had been taken, the court would have implied a duty to take reasonable steps to remove suspicion of eavesdropping in order to observe the plaintiff's 'real right' of freedom of consultation with his lawyers; the relevant regulations under the Prisons Act 1952 (NSW) clearly indicated that visits by legal advisers were not to be within the hearing of a prison officer; the burden of making the necessary alterations hardly seems impossible, and the restriction of the Commissioner's duty to that of holding the prisoner, if taken literally, means that a prisoner has no rights at all with regard to the manner of his incarceration. This kind of approach to public-duty questions is extremely restrictive. It may, however, be that this case can be explained by the unwillingness of courts to encourage applications which tend to delay criminal proceedings,[7] and does not therefore represent evidence of a general restrictive approach.

(ii) *Statutory powers: the* Julius *and* Padfield *principles*

Obligatory wording of the kind indicated above is often contrasted with permissive wording such as 'A may do Y', 'it shall be lawful for A to do Y', or 'if A certifies or directs that . . .', which are described as 'enabling', 'empowering', or 'permissive'.

This semantic approach to the analysis of statutory duties is a good enough guide to deal with most cases, but not all. A simple enabling provision may in practice prescribe a pattern of behaviour. Take the standard formula 'if X, then A may do Y'. In the first place there may be a duty, whether express or implied (as in 'if X, then A may after inquiry do Y'[8]), to investigate the relevant matters, i.e. whether X exists; or if X exists, whether A ought to do Y. Secondly, a proper construction of the provision may entail a duty to exercise the power, if the circumstances warrant it; this is not to stretch the wording of the provision, because clearly all statutory powers are granted for a purpose, and, except where the entire scheme is intended to be optional, as is often the case with public-health legislation,[9] there

[7] Ibid. 322, per Moffitt P.

[8] See *R.* v. *Milk Board, ex p. Sanders* [1961] VR 196, and contrast *R.* v. *Milk Board, ex p. Tomkins* [1944] VLR 187.

[9] The optional nature of the legislation does not preclude a duty at common law once it is adopted: *Anns* v. *Merton London Borough Council* [1978] AC 728 (HL), and see below Ch. 7.3.

must be situations in which the legislature intended that the power in question should be exercised. To put it another way, the fact that a discretion is given cannot mean that the donee is entitled to refuse to exercise it in such a way as to flout the very purpose of its being conferred. For this reason a power can become obligatory, especially where the conditions for exercising the power take the form 'if A is satisfied that X, then A may do Y'. The ensuing paragraphs attempt to define some criteria for deciding when a power becomes obligatory in this sense. It must be stressed at the outset that no consistent approach has been taken. Although the courts are more prepared than they perhaps once were to coerce authority, the reasoning is highly potent and controversial in its effects.

Two key ideas run through the cases: (1) that the provision must be interpreted in the light of the purpose of the statute, and (2) that the interpretation must be consistent with individual rights. These two ideas will often produce the same result, but may, equally, pull in opposite directions, especially with regulatory statutes, where individual liberty is often determined by the width of administrative discretion, but preserved by the imposition of a duty.

Clearly the court must consider the purpose of the statute taken as a whole in the light of its history, its preamble, and the tenor of all its provisions,[10] an exercise which demands the broadest possible approach. Modern learning takes its cue from the theory formulated in *Julius* v. *Oxford* (*Bishop*).[11] In that case the House of Lords considered a provision which said: 'in every case of any clerk who may be charged with any offence . . . it shall be lawful for the bishop . . . on the application of any party complaining thereof, to issue a commission for the purpose of making inquiry as to the grounds of such charge', the bishop having refused to act on a complaint. It was held that the statute imposed a duty only to hear and determine the application, which on the facts had been fulfilled. Earl Cairns LC dealt with the case on the following basis:

The words 'it shall be lawful' are not equivocal. They are plain and unambiguous. They are words merely making that legal and possible which there would otherwise be no right or authority to do. They confer a faculty

[10] For an example of extreme conscientiousness in ascertaining these matters, which can only be commended, see *R.* v. *Martin* (*Judge*), *ex p. Attorney-General* [1973] VR 339.

[11] (1880) 5 App. Cas. 214. The thorough review of early authorities in this case, e.g. *R.* v. *Tithe Commissioners* (1849) 117 ER 179, shows that the principle was by no means a new one. In fact it can be said that *Julius* represents a conservative application of precedent; cf. *Allcroft* v. *London* (*Bishop*) [1891] AC 666 (HL) where a provision which was similar but clearly mandatory was restrictively interpreted.

or power, and they do not of themselves do more than confer a faculty or power. But there may be something in the nature of the thing empowered to be done, something in the object for which it is to be done, something in conditions under which it is to be done, something in the title of the person or persons for whose benefit the power is to be exercised, which may couple the power with a duty, and make it the duty of the person in whom the power is reposed, to exercise that power when called upon to do so.[12]

In similar vein Lord Penzance said:

[T]he true question is . . . whether, regard being had to the person so enabled—to the subject-matter, to the general objects of the statute, and to the persons or class of persons for whose benefit the power may be intended to have been conferred—they do, or do not create a duty in the person on whom it is conferred, to exercise it.[13]

Although the bishop's power was held to be entirely discretionary, it is important to realize that the court might have found differently had the statute provided suitable safeguards in the form of particular applicants named in the statute (such as the churchwardens), and security for costs.[14] Since the only safeguard was the bishop's discretion he could not be compelled to act in every case. None the less this does not mean that he could never be compelled, for it was conceded that he was under a duty to hear and determine the application.[15] Given the modern development of the notion of abuse of discretion and the recognition that this can constitute a constructive refusal of jurisdiction, this concession greatly increases the ambit of judicial coercion. This is evidenced by *Padfield* v. *Minister of Agriculture*.[16] In that case the statute established a milk marketing scheme and provided for a committee of investigation who were to be under the duty, 'if the Minister in any case so directs', of considering and reporting to the Minister on certain complaints concerning the operation of the scheme. Producers in one region complained that the differential factor in the price fixed for their region was outdated because of increased transport costs, so that they were discriminated against by the scheme. Being, naturally, unable to command a majority on the

[12] (1880) 5 App. Cas. 214, 222–3. His Lordship rejected the test of Coleridge J. in the *Tithe Commissioners'* case, that permissive words have a compulsory force where the thing to be done is for the public benefit, on the ground that all statutory powers are for the public benefit. The passages quoted clearly place the burden of proof on him who seeks to establish that there is a duty: *Southwark & Vauxhall Water Co.* v. *Wandsworth District Board of Works* [1898] 2 Chanc. 603; *Re Gleeson* (1907) VR 368.

[13] (1880) 5 App. Cas. 227.

[14] Ibid. 226.

[15] Ibid. 240.

[16] [1968] AC 997.

board administering the scheme, they complained to the Minister, who refused to act, mainly because he might be obliged to act on a finding in favour of the complainants by overriding the board's decision, although the complaint was conceded to be relevant and substantial. The House of Lords decided, in the words of Lord Reid, as follows:

> [T]he Act imposes on the Minister a responsibility whenever there is a relevant and substantial complaint that the board are acting in a manner inconsistent with the public interest, and that has been relevantly alleged in this case. I can find nothing in the Act to limit this responsibility or to justify the statement that the Minister owes no duty to producers in a particular region. The Minister is, I think, correct in saying that the board is an instrument for the self-government of the industry. So long as it does not act contrary to the public interest the Minister cannot interfere. But if it does act contrary to what both the committee of investigation and the Minister hold to be the public interest the Minister has a duty to act. And if a complaint relevantly alleges that the board has so acted, as this complaint does, then it appears to me that the Act does impose a duty on the Minister to have it investigated. If he does not do that he is rendering nugatory a safeguard provided by the Act and depriving complainers of a remedy which I am satisfied that Parliament intended them to have.[17]

In the result mandamus was granted to compel the Minister to consider the application according to law. In view of the above quotation one wonders whether the applicants might not have asked for the actual exercise of the power to direct the committee to investigate the complaint.

(iii) *Exhaustion of discretion*

The principles laid down in *Julius* and *Padfield* are not confined to the duty to exercise discretion on proper principles. The court can go further and find a duty actually to exercise the power in question, as occurred in *Car Owners' Mutual Insurance Co. Ltd.* v. *Treasurer of the Commonwealth of Australia*.[18] Statutory provisions designed to protect policy-holders required insurance companies to make and maintain a deposit with the Treasurer. If the latter was satisfied that the parent company of a subsidiary company wholly owned the subsidiary, a deposit made and maintained by the parent equal to that which would be required if it carried on the subsidiary's business in addition to its own was, if the Treasurer so certified, a sufficient compliance with the statute by the subsidiary. A parent company and its subsidiary

[17] Ibid. 1032.
[18] [1970] AC 527.

had each made and maintained a deposit and the parent maintained a deposit which satisfied the above provisions; the subsidiary applied for mandamus to compel the Treasurer to issue the necessary certificate. The Privy Council, upholding a minority in the High Court of Australia, found that the purpose of the statute was to protect policy-holders by deposits quantified in the statute and not by administrative discretion, so that the Treasurer was bound to issue the certificate, having no residuary discretion to refuse it once he found the statutory conditions satisfied.

When the conditions for the exercise of a power are spelt out in the statute conferring it, there is a tendency to confine the exercise of discretion to the conditions referred to, so that if they are fulfilled the power must be exercised in the applicant's favour. The discretionary element is thus exhausted, leaving a bare duty to be exercised.[19] In cases of this kind the duty, although initially discretionary in that the authority must exercise discretion in deciding whether particular conditions are met, becomes ministerial when it is in fact so satisfied. This situation can also be called one in which the discretion has been exhausted. It may indeed not always be clear whether the statute intends the power to be exercised on the fulfilment of the express statutory conditions without regard being had to any other conceivably relevant factors, but generally such intention will be implicit, particularly where important rights (discussed below) are in question.[20] It is perhaps surprising that the notion of exhausted discretion has not been more widely used in cases of this kind. One could even go further and argue that it applies in cases of the *Padfield* variety where it may be certain that a failure to exercise the power would be an abuse of discretion; if this position is reached the courts would take a bold step, but a justifiable one; there is little point in stopping short of ordering the exercise of the power, only to find the refusal to exercise it the subject of further litigation when the question could have been already settled. Another way of achieving the same result is to hold, where appropriate, that where an opinion has been expressed by the donee of the power, or by some body on whose

[19] See below, esp. nn. 31–2.

[20] The position is as stated by Madden CJ in *R.* v. *Watt, ex p. Slade* [1912] VLR 225, 236: 'There would be no longer anything doubtful, and there would remain only the duty to do what the Statute commanded.' See also *Murray Co. Ltd.* v. *District of Burnaby* [1946] 2 DLR 541; *R.* v. *Richmond City, ex p. May (E. B.) Pty. Ltd.* [1955] VLR 379; *R.* v. *Oakleigh City, ex p. New Gamble Brickworks Pty. Ltd.* [1963] VR 679. In none of these cases, however, was the term 'exhausted discretion' actually used, though it was, in effect, applied. A very similar notion is the *droit administratif* concept of *compétence liée*: see Brown and Garner, *French Administrative Law*, 3rd edn. (1983), p. 147.

recommendation he is to act, an issue estoppel arises so that the discretionary element of the power remains only if it is legally possible to decide on matters other than those giving rise to the estoppel.[21]

(iv) *Rights and duties*

Generally the courts only find a bare duty in the sense indicated above if they can discover a *right* to the exercise of the power, as is indicated in the passages from *Julius* and *Padfield* quoted above. The idea of a right is unsatisfactory as a general criterion for the finding of a duty; although the use of rights-based reasoning, as is mentioned above, is to be applauded, there are many duties which exist without any such Hohfeldian correlative right. None the less it is legitimate to argue from an individual right in deciding whether there is a duty to exercise a power, and the right may, as it were, be a legitimate card with which to trump a discretion. It may operate at different levels of cogency.

At the highest level, the finding of a duty rather than a power may be necessary in order to save the provision itself from constitutional invalidity.[22] This kind of situation is, however, rare.

Alternatively a right may be apparent on the face of the statute, as in 'B shall be entitled to Y',[23] or 'B may apply to A for Y'.[24] At the lowest level a right may be implied from the scheme of the statute, as in cases already mentioned, or may be implied simply because it is regarded as fundamental, regardless of the statutory scheme. For example in *Union of India* v. *Narang*[25] a court was empowered to order a fugitive to be discharged from custody if it would be unjust or oppressive to return him; the House of Lords denied that there was any discretion to exercise once it was shown that return would be unjust or oppressive. Similarly, in *Attorney-General* v. *Antigua Times*

[21] This point was canvassed by Bray CJ in *Mitchell (G. H.) & Sons (Australia) Pty. Ltd.* v. *Minister of Works* (1974) 8 SASR 7, 17–18, but the result is inconclusive.

[22] See *P. & C. Cantarella Pty. Ltd.* v. *Egg Marketing Board (NSW)* [1973] 2 NSWLR 366 (Australian Constitution, s. 92). Constitutionality was also a factor in *R.* v. *Anderson, ex p. Ipec-Air Pty. Ltd.* (1965) 113 CLR 177; and see also *Attorney-General* v. *Antigua Times Ltd.* [1976] AC 16, 30 (PC).

[23] See e.g. *R.* v. *Corby District Council, ex p. McLean* [1975] 1 WLR 735. In *R.* v. *Mahoney, ex p. Johnson* (1931) 46 CLR 131 (HCA) the words 'may renew' in relation to an applicant for renewal of a licence were held to impose a duty to grant renewal.

[24] See *Cantarella*, above n. 22 (duty to hear application); *Metropolitan Meat Industry Board* v. *Finlayson* (1916) 22 CLR 340 (HCA).

[25] [1978] AC 247, 283; and see *In re Shuter* [1960] 1 QB 142; *R.* v. *Metropolitan Police Commissioner, ex p. Hammond* [1964] 1 All ER 821.

Ltd.[26] a statute required a deposit to be lodged before the issue of a newspaper licence, but this requirement might be waived if the Minister was satisfied with specified alternative security; the Privy Council stressed in *obiter dicta* that the Minister was obliged to waive the deposit once he was satisfied with the security. Liberty of the person and liberty of expression clearly count highly whatever the purpose of the statute and the same approach is at least implicit in cases involving property rights.[27]

It is sometimes said that the courts will be more ready to find a right and therefore a duty where the power is granted to a judicial body.[28] This is because

> [J]urisdiction and powers are conferred on judicial bodies, usually for the enforcement of rights and the protection of interests, and permissive language will often in such a case be used not because it is intended to give the tribunal a discretion to grant or refuse the remedy, but because, although it is intended or contemplated that persons interested will be entitled to the remedy the tribunal is empowered to give, it is also intended, or at all events taken for granted, that the existence of the interests and the validity of the claim to the remedy of a person seeking it will be for the tribunal to determine.[29]

A good example of the application of this principle is *R.* v. *Derby Justices, ex p. Kooner*,[30] in which the justices, although merely empowered to order legal aid covering representation by counsel, were held to be under a duty so to order in the case of a murder charge, because it was clear that the statute envisaged such an order in the case of a grave charge. The principle is by no means confined to judicial bodies, however, and has been applied in numerous

[26] Above n. 22; see also *Finance Facilities Pty. Ltd.* v. *Commissioner of Taxation* (1971) 45 ALJR 615 (HCA).

[27] See e.g. *Hartnell* v. *Ministry of Housing and Local Government* [1965] AC 1134 (HL); *Car Owners' Insurance* case, above n. 18 and text; *Re Fettell* (1952) 52 SR (NSW) 221; *R.* v. *Hounslow London Borough Council, ex p. Pizzey* [1977] 1 WLR 58.

[28] See Aronson and Franklin, *Review of Administrative Action* (1987), p. 505.

[29] *Ward* v. *Williams* (1955) 92 CLR 496, 507 (HCA). See also *In re Baker* (1890) 44 Chanc. D. 262; *In re Sooka Nand Verma* (1905) 7 WALR 225; *Newmarch* v. *Atkinson* (1918) 25 CLR 381 (HCA); *Sheffield Corporation* v. *Luxford* [1929] 2 KB 180; *De Keyser* v. *British Railway Traffic and Electric Co.* [1936] 1 KB 224; *Lugg* v. *Wright* [1941] SASR 106; *McCaskill* v. *Marzo* (1944) 46 WALR 64; *Shelley* v. *London County Council* [1949] AC 56 (HL); *Re Davis* (1947) 75 CLR 409 (HCA); *In re Shuter*, above n. 25; *Annison* v. *District Auditor for St Pancras Borough Council* [1962] 1 QB 489; *Lord Advocate* v. *Glasgow Corporation* 1973 SLT 33 (HL); *Union of India* v. *Narang*, above n. 25; *Derisi* v. *Vaughan* [1983] 3 NSWLR 17; *Lamb* v. *Moss* (1983) 49 ALR 533; *Re Hassell, ex p. Pride* (1984) 55 ALR 219.

[30] [1971] 1 QB 147.

licensing cases,[31] and also to administrative decisions bearing no resemblance to the exercise of judicial powers.[32] In *R.* v. *Mahoney, ex p. Johnson*,[33] a classic application of the *Julius* principles to licensing functions, it was held that there was no discretion not to grant renewal of a waterside worker's licence because the purpose of the statute was to control misbehaviour, or to limit the number of workers, in the interests of smoothness of trade. The court also stressed that any other interpretation would tend to deprive the applicant of his right to work.

(v) *Identity of the donee of the power*

Another factor referred to in *Julius*' case is the identity of the person on whom the power is conferred. In *Julius* itself the court was influenced by the fact that the bishop was responsible for the oversight of his diocese, and in *Padfield* by the fact that the Minister had power to override the outcome of the ordinary democratic working of the board. Generally, the higher the official, the broader the width of his discretion and the considerations, including government policy, he can take into account, and therefore the less likely it is that he will be compelled to act, as opposed to merely hear and determine according to law.[34]

This factor of identity is illustrated by two contrasted cases. In *R.*

[31] See *Ex p. Nyberg, In re Nicholson* (1882) 8 VLR (L) 292; *R.* v. *Tynemouth Rural District Council* [1896] 2 QB 219; *R.* v. *Metropolitan Police Commissioner, ex p. Holloway* [1911] 2 KB 1131; *Attorney-General for Canada* v. *Attorney-General for British Columbia* [1930] AC 111 (PC); *Hartnell* v. *Ministry of Housing and Local Government*, above n. 27.

[32] See *R.* v. *Tithe Commissioners*, above n. 11; *R.* v. *Income Tax Special Purposes Commissioners, ex p. Cape Copper Mining Co. Ltd.* (1888) 57 LJQB 513; *R.* v. *Roberts* [1901] 2 KB 117; *Metropolitan Coal Co. of Sydney Ltd.* v. *Australian Coal and State Employers' Federation* (1917) 24 CLR (HCA); *R.* v. *Saskatchewan Labour Relations Board* (1964) 48 DLR (2d) 770 (affd. (1965) 51 DLR (2d) 737). In an isolated case which could be presented as a *pessimum exemplum* of narrow legal formalism the High Court of Australia refused to find a duty to make a payment under a statute designed to compensate victims of mining accidents, where there was a power to grant allowances in accordance with a schedule to the Act: *Smith* v. *Watson* (1906) 4 CLR 802. Fortunately Isaacs J. dissented. See also *Ng Bee* v. *Chairman, Town Council, Kuala Pilah* [1975] 1 MLJ 273; *Hallett* v. *Nicholson* (1979) SC 1.

[33] (1931) 46 CLR 131 (HCA).

[34] See e.g. the sharp divergence of view in the High Court of Australia in *R.* v. *Anderson, ex p. Ipec-Air Pty. Ltd.* (1965) 113 CLR 177 concerning the significance of a grant of power to a departmental head rather than a minister, esp. the views of Menzies J. (p. 202) and Windeyer J. (pp. 204–5); on the other hand, the identity of the person asking for exercise of a power may turn the power into a duty: see *Lord Advocate* v. *Glasgow Corporation*, above n. 29.

v. *Martin (Judge), ex p. Attorney-General*[35] the statute empowered a court to order the detention in custody at the Governor's pleasure of an accused person who was found unfit to plead. A majority in the Full Court of Victoria found, after analysing exhaustively the history of the provision and practice elsewhere, that the purpose of the provision was to enable the Crown, rather than the court, to take custody of subjects who were *non compos mentis*, so that the judge was obliged to make the order. In *R.* v. *Commissioners of Lunacy*[36] very similar considerations induced the court to find that the Commissioners had no duty to discharge a lunatic upon production of medical certificates.

(vi) *Volunteers*

Another principle, which was not mentioned specifically in *Julius*' case, has some support in case law. Where an authority seeks and obtains special powers, or adopts legislation, it may be obliged to exercise the powers so voluntarily assumed, at least in the sense that it cannot leave them partly exercised.[37] Thus the Supreme Court of Canada found a duty on the part of a local authority to collect garbage when the relevant by-laws had been adopted for its area,[38] and a parish vestry was compelled to complete a compulsory purchase of land at the suit of the landowner who was saddled with unmarketable property.[39] There is also support for the proposition that an undertaking to provide a statutory service may entail a duty to continue it, independently of contract.[40] It is submitted that these propositions are consistent with current trends in the law of torts as applied to statutory powers,[41] particularly where the public relies on the exercise of the powers. Subject to preserving the proper exercise of discretion not to put statutory powers into effect, this aspect of the law of public duties could be developed further.

1.4 DUTIES OF IMPERFECT OBLIGATION

Any statutory duty may in the result be unenforceable by an individual, for example because there is an equally convenient remedy

[35] [1973] VR 339.
[36] [1897] 1 QB 630.
[37] *York & North Midland Railway Co.* v. *R.* (1853) 118 ER 657, 662; cf. *Dartford Rural District Council* v. *Bexley Heath Railway Co.* [1898] AC 210 (HL); *R.* v. *Great Western Railway Co.* (1893) 62 LJQB 572.
[38] *Joseph Investment Corporation* v. *Cité d' Outremont* [1973] SCR 708 (SCC).
[39] *Birch* v. *St Marylebone Vestry* (1869) 20 LT 697.
[40] See *Fairfax (John) Ltd.* v. *Australian Postal Commission* [1977] 2 NSWLR 124.
[41] See above n. 9.

available. It is said, however, that some duties are not enforceable legally, except by the appropriate minister, but only politically, because they are too vague.[1] Since such duties are becoming rather common in modern statutes, especially those which set up various kinds of statutory board with duties to provide public goods or services, their status and effect is worth examining.

A very typical scheme of duties is this one, which relates to the Australian Postal Commission, established by the Postal Services Act 1975 (Cwlth):

6 The functions of the Commission are—
(a) to operate postal services for the transmission of postal articles within Australia and between Australia and places outside Australia; . . .

7(1) The Commission shall perform its functions in such a manner as will best meet the social, industrial and commercial needs of the Australian people for postal services and shall, so far as it is, in its opinion, reasonably practicable to do so, make its postal services available throughout Australia for all people who reasonably require those services.
(2) In performing its functions in accordance with sub-section (1), the Commission—
(a) shall comply with any directions given to it under section 8; and
(b) shall have regard to—
(i) the desirability of improving and extending its postal services in the light of developments in the field of communications;
(ii) the need to operate its services as efficiently and economically as practicable; and
(iii) the special needs for postal services of Australian people who reside or carry on business outside the cities.
(3) Nothing in this section shall be taken—
(a) to prevent the Commission from interrupting, suspending or restricting, in the case of emergency, a service provided by it; or
(b) to impose on the Commission a duty that is enforceable by proceedings in a court . . .

8(1) The Minister may, after consultation with the Commission, give to the Commission, in writing, such directions, with respect to the performance of its functions and the exercise of its powers, as appear to the Minister to be necessary in the public interest.

These duties are typically 'imperfect' because (1) they are general directives, (2) they are unconditional and go to the root of the authority's activities, (3) they relate to the provision of services, (4) they contain large elements of discretion and qualification, (5) they

[1] e.g. Wade, *Administrative Law*, 5th edn. (1983), pp. 625–6; *Mutasa* v. *Attorney-General* [1980] QB 114 (duty of Crown to protect subjects held to be one of imperfect obligation).

are expressed to be unenforceable, and (6) they are subject to ministerial control. This is not, however, to say that duties of this kind are always unenforceable by an individual. These very provisions were held in *Fairfax (John) Ltd.* v. *Australian Postal Commission*[2] to give rise to an enforceable duty to deliver mail which had been temporarily and deliberately withheld from the plaintiff as a consequence of industrial action. The apparent exclusion of this possibility under section 7(3)(b) only meant that the duty to provide a service could not be enforced by someone in a remote area; where a service had been undertaken in a large city, it had to be performed. Similarly in *R.* v. *Metropolitan Police Commissioner, ex p. Blackburn*[3] the court enforced the vaguest of duties, that imposed on the police to enforce the criminal law, when the Commissioner adopted a deliberate policy not to enforce particular provisions, even though the duty was nowhere set out in a statute.

Even where the duty is not directly enforceable it may be indirectly enforceable. Generally a minister is given powers to intervene by the giving of directions, as in section 8(1) above. Often he must first of all be satisfied that the authority has failed to discharge its duty, or is acting unreasonably, before he can intervene by giving directions or exercising other default powers.[4] There is no guarantee of course that he will act on an individual complaint, but just as his decision to intervene can be obstructed, if unreasonable, by the court refusing mandamus to enforce his directions,[5] so he can be compelled to act if he unreasonably refuses to do so.[6] The existence of such powers often means that the duty in question cannot itself be directly enforced, as will be seen.[7]

Where the duty is merely to have regard to certain matters, as in section 7(2)(b) above, it may still be enforceable indirectly in some instances, for example when a decision is invalidated for abuse of discretion because the authority has failed to take into account a statutorily relevant consideration, as in *Bromley London Borough Council*

[2] [1977] 2 NSWLR 124. See also *Bradley* v. *Commonwealth* (1973) 128 CLR 557 (HCA).

[3] [1968] 2 QB 118.

[4] e.g. Education Act 1944 (England and Wales), ss. 99 and 68 respectively, which have featured in numerous cases; see esp. *Watt* v. *Kesteven County Council* [1955] 1 QB 408 (s. 99), and *Secretary of State for Education and Science* v. *Tameside Metropolitan Borough Council* [1977] AC 1014 (HL) (s. 68). For default powers as a means of enforcing public duties, see below Ch. 4.2.

[5] e.g. in the *Tameside* case.

[6] *R.* v. *Secretary of State for the Environment, ex p. Ward* [1984] 2 All ER 556.

[7] See below Ch. 3.5.

v. *Greater London Council*,[8] where a supplementary rate to subsidize passenger transport was quashed because of failure to have regard to the relevant authority's duty to operate the system economically.

Now that the courts are more prepared than they once were to investigate questions of administrative practice and to examine statutory provisions purposively, it can be anticipated that these 'imperfect' duties will be enforced in cases of express refusal or rank incompetence, and may not therefore always be duties of imperfect obligation. If so, this would be a desirable development, because it would give the public legal access to duties commonly thought to be enforceable only administratively.

1.5 AN ANALYSIS OF PUBLIC DUTIES

We have seen that to analyse public duties according to their source is revealing, but has no specific legal consequences, because the courts have not confined themselves to reviewing statutory public duties. We have also seen that public duties may be owed by different kinds of public body and involve different degrees of coercion. At this point it is appropriate to introduce another kind of analysis which, it will be seen later, does have important consequences with regard to the legal rules applicable to a particular public duty, and perhaps more importantly, influences the attitude of the courts to public duties, particularly with regard to their enforcement. This analysis is analysis according to content. Its most important feature is that it distinguishes duties to determine questions affecting individual rights and interests from other kinds of duty.

(i) *The duty to determine*

The duty to determine questions affecting individual rights and interests is treated as the most important duty in administrative law, and around it most of the important principles of judicial review have

[8] [1982] 2 WLR 62 (HL). See also *Lee* v. *Enfield London Borough Council* (1967) 66 LGR 195, where it was held that the vague duty of education authorities under s. 76, Education Act 1944, to 'have regard to the general principle that, so far as is compatible with the provision of efficient instruction and training and the avoidance of unreasonable expenditure, pupils are to be educated in accordance with the wishes of their parents', was capable of invalidating an admission scheme where the principle had not been observed; *Wheeler* v. *Leicester City Council* [1985] 3 WLR 335 (HL) (duty to have regard to need to eliminate racial discrimination); *R.* v. *Ealing London Borough Council, ex p. Times Newspapers Ltd.* (1987) 85 LGR 316 (duty to provide a comprehensive and efficient library service).

developed. This duty encompasses a wide variety of situations, and can be subdivided according to the kind of question being determined, which may be:

1. a question whether a person should be granted a licence to do something, for example to build a house, drive a taxi, sell liquor, or enter the country; or whether such licence should be renewed, revoked, transferred, or altered in its terms;
2. a question whether a person should be provided with a service, for example a social security payment, public housing, a student grant, a mining concession;
3. a question whether a person should be granted, or deprived of, a particular status as, for example, a university graduate, a furnished tenant, a qualified optician, or a recognized trade union;
4. a question whether an order should be made which directly or indirectly, adversely or otherwise, affects persons, for example a compulsory purchase order, an order committing a person to a prison or mental hospital, or a decree absolute of divorce.

It is one of the achievements of administrative law that the duties involved in determining such questions do not depend on the kind of question being determined, the identity of the decision-maker as a court or tribunal or administrative authority, or the identity or status of the persons applying or otherwise affected, or whether there is a *lis inter partes*. The ways in which the jurisdiction to determine a question can carry with it certain duties are well known, but some neglected aspects of those duties are considered in Chapter 2.

(ii) *The duty to provide*

If the duty to determine is regarded as the most important duty in administrative law, the duty to provide is certainly the most important duty in administrative practice, because it covers the whole sphere of public provision, at least so far as such provision is compulsory. It covers the duty to provide education, health services, social welfare, housing, public-health facilities, public-protection services, highways and transportation, and a host of ancillary public goods provided for the public at large indivisibly, or services provided for particular persons or groups entitled. These duties interlock with the duties mentioned above to determine questions, but are more general in their nature and application, and may in certain cases be owed by different bodies, for example, a government department provides a prison, but a judge determines whether a person shall be committed, and a parole board determines whether the sentence shall be remitted.

(iii) *The duty to enforce the law*

This duty falls mainly on the police, but also on other enforcement authorities which have a duty to prosecute offenders, for example, consumer protection authorities. Authorities other than the police may also be under a duty to regulate an activity, for example, construction of buildings, of which part may be a duty to determine (for example whether a building permit should be granted), and part a duty to enforce the law (for example to prosecute for failure to comply with building regulations). The duty to enforce the law may also involve a duty to exercise ancillary powers other than prosecution (for example forfeiture and default powers or other administrative sanctions). Included in this category of public duty is the duty to collect taxes, which has many features in common with police or regulatory duties, notably the possibly discriminatory effect of the exercise of discretion not to collect.

(iv) *Constitutional duties*

These are duties connected with the system of government which do not fall into any other category. Examples are the duty to hold an election; the duty of a local authority to comply with its standing orders; the duty to give access to documents or other information connected with the affairs of the authority or an individual under its jurisdiction; the duty of a minister to appoint an official; the duty of a government which has lost the confidence of the legislature to resign; the duty of the head of state to act on advice.

These four kinds of duty are dealt with in more detail in Chapter 2. The courts have not expressly used this analysis of public duties, but in practice, as will be seen, they are treated somewhat differently and raise different kinds of problem.

Finally, it will be useful to advert to some common terminology in relation to public duties. Public duties are often given an extra epithet such as 'mandatory', 'absolute', 'unequivocal', 'peremptory', or even 'imperious'. It is not clear what, if any, legal content those epithets have, and generally they can be dismissed as rhetoric. More meaningful are the terms 'discretionary' and 'ministerial'. A duty described as discretionary is usually one in the manner of performance of which the authority has a wide discretion; in this sense most of the duties discussed in this work are discretionary. A ministerial duty is a duty which is not discretionary, one which is simply to be performed mechanically without any exercise of discretion. A duty may be said to be ministerial (or 'executory') where although there is a discretionary element, the discretion has been exhausted.

Occasionally one finds mention of the term *ultra vires* in relation to a public duty. At first sight the idea of an *ultra vires* breach of duty seems odd because *ultra vires* generally refers to excess of powers; a duty can also, however, be exceeded: if a statute imposes a duty on A to do X, and in purported fulfilment of that duty A does Y, or abuses its discretion in the doing of X, then A has committed an *ultra vires* breach of duty. This terminology does not occur often, and may indeed be quite unhelpful as a mode of analysis. This is because it introduces a quite unnecessary distinction between nonfeasance and misfeasance (nonfeasance cannot be *ultra vires*, but misfeasance will always be) and because given that the action is unlawful, it hardly helps to know that it is also *ultra vires*; the notion of an *intra vires* breach of duty appears, understandably, not to exist.

2
Performance of Public Duties

2.1 INTRODUCTION

In Chapter 1 we have considered the question, when and how does a public duty arise? We now turn to the question, when and how is a public duty breached, or to put it another way, what constitutes due performance of a public duty? The answers provided to this question are of course of crucial importance to public authorities because they need to know what, precisely, is the extent of the obligations imposed on them; they need to know this not merely to avoid unlawful inaction, but to be able to draw up their budgets, strike a rate, and so on.[1] Ascertaining the precise content of public duties is essential for mapping the policy space which a public body has. At a fundamental level a duty must be performed, and lack of funds is not generally an excuse for non-performance,[2] although it is also true that many duties, especially those involving considerable expenditure of funds, involve a large discretionary element. It is also true that public bodies often get away with complete or partial non-performance of a duty because of the uncertainty of the scope of the duty or the unlikelihood of legal or administrative challenge. Referring back to the analysis of public duties in Chapter 1, it is perhaps surprising that, while administrative lawyers have lavished great attention on the duty to determine, and an enormous amount has been written on what constitutes due performance of this kind of duty, for example about observance of the principles of natural justice, taking into account relevant considerations, and so on,[3] the other duties have been largely neglected, being relegated to a footnote or two. Yet the economic, administrative, and even political consequences of the performance of these duties are probably immeasurably greater than those flowing from observance of duties relating to determinations. The performance of these duties has of course traditionally been regarded as a matter for the administration, whereas the duty to determine is one more peculiarly suited to legal analysis. While this

[1] See *R.* v. *Secretary of State for the Environment, ex p. Hackney London Borough Council* [1984] 3 All ER 358.

[2] See below Chs. 2.3, 3.4.

[3] See below Ch. 2.2.

factor explains the emphasis of lawyers, it does not necessarily cater adequately for the needs of society or particular individuals or groups of individuals similarly affected. To take an example, while the law may well provide an effective mechanism for correcting an unlawful *determination* that a person is not entitled to public housing, or admission to a particular school, or a student grant, it may well not provide an effective mechanism for ensuring that a duty to provide a house, a school place, or a student grant is actually observed. These matters are becoming matters of concern for lawyers and the public, especially with the increasing welfare consciousness, democratic participation, and complex provision of numerous public services which is taking place in advanced societies, and there is evidence[4] that the law will have to deal more purposively with the problems which they raise. Since the various duties under consideration raise different problems, the analysis of public duties in Chapter 1 will be used to organize the ensuing discussion of the performance of public duties.

2.2 THE DUTY TO DETERMINE

The duty to determine questions affecting the rights and interests of individuals covers the whole field of administrative decision-making.[1] The entire machinery of judicial review of administrative decisions is designed to secure the enforcement of this duty, directly or indirectly. Since the seventeenth century the prerogative writs have been the principal means by which tribunals and administrative authorities have been compelled to exercise their jurisdiction according to law, fulfilling it exactly and not exceeding it.[2] A distinction emerged between judicial and administrative functions so that *certiorari* (but not mandamus) governed the former, but not the latter. This restriction was, happily, buried in *Ridge* v. *Baldwin*[3] in 1963, so that the duty to determine can be considered without reference to the nature of the decision-maker's functions, or indeed the identity of the decision-maker. It is one of the great achievements of administrative law that tribunals exercising jurisdiction and administrative authorities exercising statutory discretion have been subjected to global principles which are now well developed and understood. Around these principles

[4] See e.g. Cranston, *The Legal Foundations of the Welfare State* (1985).

[1] For full discussion of review of administrative decisions, see De Smith, *Judicial Review of Administrative Action*, 4th edn. by Evans (1980); Wade, *Administrative Law*, 5th edn. (1983).

[2] See De Smith, above n. 1, pp. 584–95.

[3] [1964] AC 40 (HL).

a vast legal literature has grown,[4] and it is not necessary here to do more than summarize them for the purposes of the ensuing discussion.

1. The principles of natural justice, which govern both constitution and procedure, and do not depend on whether there is a *lis inter partes*, must be observed.
2. The decision-maker must exercise his jurisdiction when properly called upon to do so; he must decide the question remitted to him and not some other question not remitted to him.
3. He must take into account all relevant considerations and ignore all irrelevant considerations.
4. The decision must be made in good faith.
5. Where the decision involves the exercise of discretion, that discretion must be exercised in accordance with the above principles by the person entrusted with it, who must not sub-delegate it, allow himself to be dictated to, or refuse to exercise it.
6. Such discretion must also be exercised reasonably and for a proper purpose.

These propositions constitute the basis of the duty to determine. It is not necessary here to discuss their content, but simply to attempt to set them in their proper context in relation to the performance (and later, in Chapter 3, the enforcement) of the duty to determine.

The first question to ask is, when does the duty to determine arise? For it is only when it arises that the above principles can even come into play.

As has been indicated, the duty to determine is one of broad application and can be said to arise whenever a public tribunal or administrative authority determines a question which it is empowered or obliged to determine and which affects individual rights or interests. This general statement will be amplified later in relation to the effect (in Australia) of the concept of a decision of an administrative character,[5] and (in England) the concept of public law.[6]

The discussion will concentrate on the narrower but important and neglected question, at what point in the process of determination the duty arises, and on the notion of refusal of jurisdiction.

[4] See above n. 1 and literature referred to in those discussions.

[5] The benchmark adopted by the Administrative Decisions (Judicial Review) Act 1977 (Cwlth); see below Ch. 5.2.

[6] See below Ch. 5.1.

(i) *Applications and determinations*

Before the duty to determine arises, the decision-maker must of course first of all be seised of, or become cognizant of, the question which he is required to determine. The way in which this occurs depends on the circumstances, in particular the statutory conditions under which the exercise of jurisdiction or discretion arises.

Administrative authorities exercising administrative functions may act on any information which they have as a result of their own researches or the advice of their officers: they have a choice here of what have been called 'detectors'.[7] Generally they are not under any duty to investigate or search for questions to determine unless specifically required to do so. For example if a statute allows an authority to make a street closing order it is under no implied duty to inquire whether any streets require to be closed; if for any reason a question arises whether a particular street requires to be closed, their duty is only to exercise their discretion in accordance with law. On the other hand if a statute enjoins an authority to provide assistance for disabled persons, they may well be under an ancillary statutory duty to discover whether there are any persons who fall within the statute.[8]

On the other hand the duty may only become operative on the application of a member of the public. In this case there is usually no statutory requirement as to the form of the application, and the authority will be free to design its own form of application or else act on purely informal applications.

In certain cases, however, the requirements for a valid application are laid down in a statute. For example, applications for building permits, which require at least outline plans, must usually conform with the statute. A person who wishes to invoke the jurisdiction of a court or tribunal also usually has to comply with some formal requirements, although sometimes the duty to exercise the jurisdiction will arise only on the reference of some intermediate arbitral authority, as is often the case with labour disputes.

In theory the initial step of invoking the jurisdiction or discretion ought to be simple. In practice, however, it is often the applicant's hardest step, because, in these days of increased administrative regulation, to get a matter even considered by the decision-maker

[7] Hood, *The Tools of Government* (1983), ch. 6.

[8] See e.g. *Wyatt* v. *Hillingdon London Borough Council* (1978) 76 LGR 272. With regard to a general choice whether to exercise newly granted powers at all, it would seem that there is a duty to consider the question in accordance with law: *Anns* v. *Merton London Borough Council* [1978] AC 728 (HL); and see below Ch. 6.3.

may entail weeks or months of waiting in a queue. At this stage the decision-maker has not considered the matter of jurisdiction and refused it: he has not considered the matter at all, though he is seised of the fact that there is a matter to be determined. In order to establish his place in the queue an applicant must succeed in lodging an application, and it is therefore necessary to know what are the legal conditions for an application, and what are the duties of an authority with regard to it. An application is after all a kind of administrative 'chose in action' upon which all the applicant's rights depend.

There do not appear to be any clear legal principles for this kind of situation, but there ought to be, because the performance of the duty at this stage can crucially affect the outcome of the matter or its effect on the applicant. The following statement of this aspect of the duty to determine is therefore necessarily tentative.

In so far as the requirements of an application are statutory, the courts tend to take a rigid view and will not find that there is a duty to determine a question unless the applicant/plaintiff has complied with all the statutory requirements. This is not merely due to fastidiousness about 'clean hands', but a sensible position based possibly on two reasons: first, if procedural requirements are important enough to merit inclusion in a statute, it is only right that those who have taken the trouble to comply with those requirements should not be placed behind those who have not and have thereby possibly jumped the queue; secondly, 'it is presumed all officers will properly discharge their official duties as they arise and at the time when such duties become incumbent on them',[9] and this is surely only when a proper application is received.[10] This strict approach has been taken in a number of Canadian cases[11] involving applications for building permits; it was, however, perhaps taken too strictly in a case where the statute required an application by the landowner and the

[9] *Karavos* v. *Toronto and Gillies* [1948] 3 DLR 294, 300 per Laidlaw JA.

[10] See *Frankel* v. *Winnipeg* (1912) 8 DLR 219; *R.* v. *New Westminster, ex p. Canadian Wirevision Ltd.* (1965) 48 DLR (2d) 219. Two Queensland cases either side of the line are to similar effect: *R.* v. *Charleville Town Council, ex p. Corones* [1928] Qld SR 155 and *R.* v. *Clerk, ex p. Roger* [1957] Qld SR 67.

[11] *Karavos*, above n. 9; *Frankel*, above n. 10; *Canadian Wirevision*, above n. 10; *R. ex rel. Ellerby* v. *Winnipeg* [1930] 3 DLR 205. Similar Australian cases are *R.* v. *Local Government Court, ex p. Pine Rivers Shire Council* [1972] Qld R 127 and *Grzybowicz* v. *Smiljanic* [1980] 1 NSWLR 627. See also *R.* v. *Northumberland Quarter Sessions, ex p. Williamson* [1965] 1 WLR 700, where the application was out of time and was therefore invalid. In the *Pine Rivers* case the court held that though there was no proper application, jurisdiction did not depend on there being an 'impeccable' application.

landowner was refused relief, although he swore the application for the permit was made on his behalf.[12] Presumably one can add to this that minor defects in the application will be ignored, and the mandatory/directory distinction[13] can be brought into play to allow this. Presumably it also follows that on receipt of a valid application the decision-maker must record the application, and if it is not in his view valid he should inform the applicant within a reasonable time to allow for appropriate correction.[14] If of course there is no question to determine, the duty being purely ministerial, the validity of the application is the only matter the court need consider, and if the application is valid, it must be granted.[15]

Once the applicant has established a place in the queue, the next stage of the duty is for the application to be heard. Is unconscionable delay in hearing an application refusal to exercise jurisdiction or discretion? On this question there is some case law which may well be developed in future.

In *R.* v. *Secretary of State for the Home Department, ex p. Phansopkar*[16] P was a patrial with a right of abode in the United Kingdom, where he lived. His wife W lived in India with their four children. W had the same right of abode in the United Kingdom as P, subject to the statutory requirement of obtaining a certificate of patriality. P went to India to obtain this certificate but was told there would be considerable delay in considering applications. He then travelled to London with W and their children, intending to apply for the certificate directly to the Home Office. The Home Secretary refused

[12] See *Ellerby*, above n. 11.

[13] See *De Smith*, above n. 1, pp. 142–6. In *R.* v. *Merthyr Tydfil Licensing Justices, ex p. Duggan* [1970] 2 All ER 540 this distinction was used to allow the validity of an application which did not specify the type of licence applied for, even though this requirement was statutory; see also *Ex p. Stocks and Parkes Investments Pty. Ltd., re the Minister* (1969) 72 SR (NSW) 104; *Schiller* v. *Southern Memorial Hospital* [1976] VR 484.

[14] See *R.* v. *Gaming Board, ex p. Benaim* [1970] 2 All ER 528, 535 per Lord Denning MR. Sometimes the validity of an application depends on the authority's response; e.g. in *R.* v. *Richmond City, ex p. May (E. B.) Pty. Ltd.* [1955] VLR 379 the applicant was prepared to pay the appropriate fee for a permit, which had to be assessed by the authority; the authority sought to impugn the application on grounds of non-payment of the fee, but mandamus was granted to assess the fee and, on payment, to issue the permit.

[15] See e.g. *R.* v. *Newcastle-on-Tyne Corporation, ex p. Veitch* (1889) 60 LT 963; *R.* v. *Preston Rural District Council, ex p. Longworth* (1911) 106 LT 37; *Attorney-General for Canada* v. *Attorney-General for British Columbia* [1930] AC 111 (PC); *R.* v. *Oakleigh City, ex p. New Gamble Brickworks Pty. Ltd.* [1963] VR 679; *Ex p. Catlett (SR) Constructions Pty. Ltd., re Baulkham Hills Shire Council* [1964–5] NSWR 1667.

[16] [1976] QB 606.

W entry on the ground that to grant it would be to sanction queue-jumping, which would be unfair to those waiting overseas; he further refused to consider W's application for a certificate of patriality because it could more conveniently and satisfactorily be dealt with in Bombay. W applied (*inter alia*) for mandamus to hear and determine the application, and the Court of Appeal considered the appeals by W and another similarly placed applicant from the refusal of the Divisional Court to grant the remedy.

The court reasoned as follows. W could enter as of right, not by leave. She was therefore entitled to have her application for a certificate of patriality dealt with fairly and within a reasonable time. The authorities were therefore under a duty to set up appropriate machinery for dealing with applications for such certificates, which by statute could be issued by the Home Office or a British Government representative overseas. It followed that W's application could properly be made to either authority and had then to be entertained by it unless there was some sufficient reason for requiring it to be made to the other. The Home Secretary was entitled to require the application to be made in Bombay only if it could be dealt with promptly. The facts were such that in Bombay, but not in London, some fourteen months would pass before an interview would be granted, and that a very important part of the inquiry had to be made in the United Kingdom; in either London or Bombay, telexes might be required to clarify queries. In view of these facts the Home Secretary was not justified in refusing to consider the application and mandamus was accordingly granted.

The judgments reveal a further point of great interest. The reason for the long delay in Bombay was that the queue comprised applicants for leave to enter as well as those claiming a right to enter, but the latter kind of application was hardly ever refused and the numbers involved were far smaller.[17] The court observed[18] that there should be two queues, not one, which would result in little delay in dealing with those with a right to enter. Needless to say, the judgments are replete with references to Magna Carta's solemn promise 'to none will we sell: to no one will we delay or deny right and justice', and to the right to respect for family life.

This case, unusual as it is in its foray into the field of administrative technology, has a considerable bearing on our inquiry. A duty to determine can be said to entail a duty to establish machinery adequate to deal with the cases or applications to which the duty gives rise.

[17] 859, as opposed to 12,864 applications for leave, in 1974.

[18] [1976] QB 622, 628.

What content can we give to the term 'adequate'? In reply to the Home Secretary's view that the application in *Ex p. Phansopkar* would be more conveniently and satisfactorily dealt with in Bombay, Scarman LJ said that '[i]f the true balance really be between the convenience of the authorities (for it cannot be convenient for the [applicants]) and respect for human rights, the Secretary of State would be misreading the scales of justice'.[19] If one follows this approach, then 'adequate' must mean not merely administratively convenient; it must mean also consistent with reasonably prompt determination of the question, fairness as between applicants in the same position,[20] and observance of the rights of applicants as generally understood or apparent from the statute.

'Reasonably prompt' also needs some explanation. Delay may be caused not simply by the weight of cases to be considered, but by extraneous factors, such as the fact that the manner of dealing with the application may be affected by the outcome of other proceedings. Cases of this kind are difficult because the effect of such proceedings may be uncertain, or the period of waiting of uncertain duration, and the status quo may be unsatisfactory, albeit convenient to the administration. The case law seems to be unduly favourable to the administration.

In *P. & C. Cantarella Pty. Ltd.* v. *Egg Marketing Board (NSW)*[21] a company sought approval of its premises as suitable for the purpose of egg-candling. The Board deferred determination of their application for a period of seven months pending the outcome of a dispute between the parties over conditions and the outcome of prosecutions of the company by the Board for breaches of regulations concerning the sale of eggs. This deferral was not considered by the court to be a refusal of the application, and it is significant that the matter of the prosecutions was considered to be relevant, even though it related to the applicants and not their premises, merely because it was a matter which could properly be considered by the Board when imposing conditions.

A cautious step towards the adumbration of some principles was taken in *Engineers' and Managers' Association* v. *Advisory, Conciliation, and*

[19] Ibid. 627.

[20] Esp. where, as in liquor licensing cases, two or more applications are conflicting ones; *Ex p. Pines, re Webb* [1970] 1 NSWR 778. However, in *Brooks* v. *Jeffery* (1897) 15 NZLR 727 (affd. (1897) 16 NZLR 276) the first of two applicants for a water right (a limited resource) obtained mandamus to compel priority registration of his right even though both applicants arrived at the office simultaneously; the law apparently really does help those who help themselves!

[21] [1973] 2 NSWLR 366.

Arbitration Service,[22] in which ACAS deferred determination of a trade-union recognition issue because of the commencement of legal proceedings which affected, indirectly, the consideration of the issue. Even though the statute required ACAS on referral of a recognition issue to examine it, consult all parties affected, and make such inquiries as it thought fit, it was held that this duty was subject to an overriding requirement of reasonableness, so that ACAS had a discretion to defer taking particular steps in fulfilment of its duty if the circumstances rendered that desirable for the purpose of enabling it properly and usefully to fulfil its function of reporting and, if appropriate, making a recommendation. Since on the facts ACAS bona fide believed itself subject to pressures both to proceed and to defer, and an early disposal of the proceedings concerned was probable, the decision was one to which it could properly come.

The theme of reasonableness was elaborated upon by the Federal Court of Australia in *Thornton* v. *Repatriation Commission*.[23] The Commission deferred T's application for a widow's pension pending a decision of the High Court of Australia which would directly affect the outcome of the application. T applied under the Administrative Decisions (Judicial Review) Act 1977 (Cwlth), section 7(1), for an order of review on the ground that there had been unreasonable delay in making the decision. Fisher J. held that the reasonableness of the delay was a matter for objective determination, the question being whether a reasonable man acting in good faith could consider the delay as appropriate or justified in the circumstances, or whether it was capricious or irrational; the delay was for a considered reason and not in consequence of neglect, oversight, or perversity, and was for a finite and not indefinite period; it was more satisfactory to determine the question once and for all rather than on a conditional basis; a decision in favour of T would entail the irrecoverable payment of a lump sum and periodic payments even if the High Court's decision made those payments unlawful; and it was undesirable to raise T's expectations when there might be a prospective reversal of the decision. For all these reasons, but not taking account of T's personal circumstances, which were in any case not impecunious, the delay could not be said to be unreasonable.

[22] [1978] ICR 875; see also *Re Civil Service Association of Alberta and Alberta Human Rights Commission* (1976) 62 DLR (3d) 531. Use of an Interpretation Act to argue that a duty must be performed from time to time as occasion requires does not appear to be a helpful development; see *R.* v. *Ealing London Borough Council, ex p. McBain* [1986] 1 All ER 13.

[23] (1981) 35 ALR 485; and see *R.* v. *Coolangatta Shire Council* (1930) 24 QJPR 156; *Re O'Reilly, ex p. Australena Investments Ltd.* (1983) 58 ALJR 36 (HCA).

These decisions all show a reluctance to place on a rational footing an authority's discretion as to the ripeness of an application for consideration. The test applied by Fisher J., whose application to the facts is perhaps questionable, amounts to asking whether a reasonable man would consider the delay reasonable: in fact the test indicated in the statute, which is presumably identical to the position at common law, is merely whether the delay is reasonable, and the court could have formulated clear principles upon which to determine the matter. Naturally there must be some flexibility in the matter of delay, but the level of scrutiny seems to have been fixed a notch too low, unless the test applied can be taken to mean that the court itself is to decide what is reasonable. Cases in which the courts have allowed a reasonable delay pending proper investigation of facts display a similar tendency,[24] but of course refusal to determine merely on the ground that the decision might be challenged in the courts has not been allowed.[25]

A case which sheds some further light on the matter of applications is *R.* v. *Tower Hamlets London Borough Council, ex p. Kayne-Levenson.*[26] A husband and wife (H and W) both applied for a street trader's licence for a certain pitch in Petticoat Lane in London, which was vacant on the death of H's mother (L). H already held a licence for another pitch, so that his application was treated as invalid. W's application was put on a waiting list, which consisted of applicants who had applied for pitches allocated to someone else, and who would be offered in turn, in the order in which they joined the list, any vacant pitch which subsequently arose; this meant waiting probably for several years. The statute allowed a licence-holder to nominate a relative as successor, but L had not nominated H as her successor because the authority's application form (wrongly, it turned out) prohibited nomination of another licence-holder. Since the terms of the statute prevented W from being nominated, L had tried to circumvent the difficulty by nominating the couple's baby daughter (S). S's application was, however, also rejected because she was too young. S appealed from this decision, but her appeal had been adjourned pending determination of the applications of H and W.

[24] See *R.* v. *Comptroller-General of Customs, ex p. Woolworths Pty. Ltd.* (1935) 53 CLR 308 (HCA); *R.* v. *Central Professional Committee for Opticians, ex p. Brown* (1949) 65 TLR 599.

[25] *Woolworths*' case, above n. 24, followed in *R.* v. *Collector of Customs (Vic.), ex p. Berliner* (1935) 53 CLR 322 (HCA).

[26] [1975] QB 431; cf. *Re Bukit Sembawang Rubber Co. Ltd.* [1961] 27 MLJ 269; *Bhatnager* v. *Minister of Employment and Immigration* [1985] FC 315; *Re Leo Boh Boey* [1986] 1 MLJ 469.

The authority claimed that they had not refused these two applications. H and W then applied for mandamus to compel the issue of notices of refusal so that they could appeal. Mandamus was granted and the authority's appeal was dismissed by the Court of Appeal, which held that both applications had on the facts been refused rather than simply not determined. In the event, as a result of the litigation, the authority granted the licence to H.

This case shows that a rejection of an application can also be an administrative 'chose in action', at any rate if there is a right of appeal from it. The queue of applicants was not really a queue of persons whose applications awaited attention, as in *Ex p. Phansopkar*, but a queue of persons whose applications had in effect been rejected and transferred to another queue of applications for no particular pitch. It is also implicit in the decision that the giving of a reason why the application should not succeed is in effect a refusal of the application, not a statement of its invalidity as an application, even if it had not been 'heard' in the normal way. A distinction must therefore be drawn between an application which is bad in form and one which is bad in substance. An application which is bad in form can be returned on the basis that no valid application has been received and therefore there is no duty to consider its substance. An application which is good in form but bad in substance is none the less a valid application, even if it must be rejected at the determination stage, and any refusal of it must be treated as determination which is subject to the usual legal principles. If of course there is some doubt as to whether the application does in fact conform with procedural requirements, the applicant can still pursue his case by applying for mandamus to hear and determine, thus testing the validity of the application.[27] Where the procedural requirements are extra-statutory, however, it seems doubtful whether an authority would be right in law in refusing to consider an application which did not comply with them.[28] In *Shah* v. *Barnet London Borough Council*[29] the House of Lords treated the rejection of an application for a mandatory student's grant as importing a duty to consider the application as one for a discretionary grant under the same statutory provisions.

So far we have dealt with the duty to assume jurisdiction or exercise

[27] Of course it is preferable to correct the defect, but that may involve the applicant losing his place in the queue.

[28] There is no authority for this assertion, but it seems wrong to deprive an informal applicant of his place in the queue when the defect in his application is not a statutory one. The cases referred to above, nn. 9–11, make mention only of statutory requirements.

[29] [1983] 1 All ER 226, 239.

discretion up to the point of establishing machinery, which complies with all the relevant principles, for dealing with cases or applications arising. There may, however, be some doubt about the precise point at which a determination takes place which is subject to the incidents of the duty to determine. A good illustration of the problem (and of an imperfect solution) is the case of *R.* v. *Barnstaple Justices, ex p. Carder.*[30] The justices were empowered to grant cinematograph licences in respect of premises, but there was no power in the statute to grant provisional licences, so that an applicant who constructed a cinema ran the risk of rejection of his application on the completion of the building. The justices adopted a sensible expedient: they considered the plans of a proposed cinema informally, hearing objections, the formal granting or refusal of the application on completion being understood to follow their informal decision as a matter of course. When a disappointed applicant sought judicial review of an 'informal' refusal of his application the court held that the refusal was an extra-judicial or administrative proceeding in a matter not known to the law, which proceeding the judges disapproved on the grounds that it sought to commit another body (the justices acting judicially) to a particular decision, but which was not held to be unlawful. As a result the justices had not determined any matter and there was nothing for the court to review. These days it seems unlikely that the courts would draw such a rigid distinction between a judicial and an administrative proceeding but would be alert to review any proceeding which materially affects individual rights and interests, even if the duty to determine the question cannot strictly be said to be ripe for fulfilment.[31]

(ii) *Refusal of jurisdiction*

We come now to the determination itself, which is subject to the principles stated earlier.

All duties to determine questions pursuant to a statute are expressed in terms which can be reduced to one of the two paradigms 'if X exists, then A may do Y' or 'if X exists, then A shall do Y'. As was seen in Chapter 1, the former may carry with it a duty to exercise discretion wherever X exists, or even to do Y where it would be an abuse of discretion not to;[32] the latter carries with it a duty to do Y whenever X exists, and conferment of jurisdiction falls into this

[30] [1938] 1 KB 385.

[31] See the decisions of the Federal Court of Australia under the Administrative Decisions (Judicial Review) Act 1977 (Cwlth) below Ch. 5.3.

[32] See above Ch. 1.3.

paradigm, for exercise of jurisdiction is not discretionary: it cannot be refused if the case falls within the jurisdiction. Thus we cannot ascertain whether there is a duty merely by examining the main verb of the statutory provision and deciding whether it is permissive or obligatory. Questions which arise at the 'X level', as Craig[33] has called it, are often characterized as preliminary, collateral, or jurisdictional questions; questions arising at the 'Y level' on the other hand go to the substance of the determination. The determination of questions arising at the X level may, however, in fact form the substance of the determination, especially of course where there is a duty to do Y *because* X exists. Craig has also pointed out[34] the complexity involved in questions arising at the X level, such that X may be made up of a number of different elements, factual, legal, or discretionary; it may involve determining pure questions of fact, questions of law, mixed questions of fact and law, and questions which involve the decision-maker exercising his discretion.

These considerations have some important consequences for non-performance of the duty to determine, and it is necessary to advert here to the concept of jurisdiction as developed by the courts.

Over the years the scope of collateral or jurisdictional questions has been gradually expanded to cover virtually all, if not all, questions which a decision-maker has to determine. Lord Reid summarized the duties in relation to these questions in the famous *Anisminic* case in the House of Lords:

> there are many cases where, although the tribunal had jurisdiction to enter on the inquiry, it has done or failed to do something in the course of the inquiry which is of such a nature that its decision is a nullity. It may have given its decision in bad faith. It may have made a decision which it had no power to make. It may have failed in the course of the inquiry to comply with the requirements of natural justice. It may in perfect good faith have misconstrued the provisions giving it power to act so that it failed to deal with the question remitted to it and decided some question which was not remitted to it. It may have refused to take into account something which it was required to take into account. Or it may have based its decision on some matter which, under the provisions setting it up, it had no right to take into account. I do not intend this list to be exhaustive. But if it decides a question remitted to it for decision without committing any of these errors it is as much entitled to decide that question wrongly as it is to decide it rightly.[35]

[33] *Administrative Law* (1983), pp. 23–7, 299–345, an excellent discussion whose method, though not conclusions, I gratefully adopt.

[34] Ibid. 301–4.

[35] *Anisminic Ltd.* v. *Foreign Compensation Commission* [1969] 2 AC 147, 171.

It is necessary to say 'virtually all, if not all, questions' because it has been seen that, although Lord Reid's list of errors is supposed by him not to extend to deciding the question wrongly, it seems hard to find any error at all which does not fall within the list.[36] Thus jurisdiction seems to have ceased to be a preliminary matter and the distinction between legality and merits has been virtually obliterated, at least as concerns questions of law. In fact the House of Lords has now stated in the clearest terms that that distinction has, in England at least, been obliterated.[37] Thus not only 'X questions' but also 'Y questions' can be regarded as involving an error of law which takes the decision-maker beyond his jurisdiction. In Australia the distinction between jurisdictional and non-jurisdictional errors still apparently holds as far as review at common law is concerned, but not in the case of statutory review,[38] where error of law simple is regarded as a ground of review.

The position thus established is significant for our purposes because, as one might expect, the principles with regard to failure to exercise jurisdiction, that is to say non-performance of the duty to determine, are an exact mirror image of the principles applicable to excess of jurisdiction. To put it another way, if a tribunal exceeds its jurisdiction by, say, determining a question not remitted to it, it has also failed to exercise its jurisdiction by failing to answer the question remitted to it, and can be compelled to do so. If we compare with Lord Reid's dictum the following passage from the decision of the High Court of Australia in *R.* v. *War Pensions Entitlement Appeal Tribunal*, *ex p. Bott*, we find that the language of excess of jurisdiction is the same as the language of refusal of jurisdiction. They are in fact two sides of the same coin.

In the case of a tribunal, whether of a judicial or an administrative nature, charged by law with the duty of ascertaining or determining facts upon which rights depend, if it has undertaken the inquiry and announced a conclusion, the prosecutor who seeks a writ of mandamus must show that the ostensible determination is not a real performance of the duty imposed by law upon the tribunal. It may be shown that the members of the tribunal have not applied themselves to the question which the law prescribes, or that in purporting to decide it they have in truth been actuated by extraneous

[36] See e.g. Gould, 'Anisminic and Jurisdictional Review' [1970], *PL* 358; Rawlings, 'Jurisdictional Review after Pearlman' [1979], *PL* 404; Craig, above n. 33.

[37] *Re Racal Communications Ltd.* [1981] AC 374, esp. per Lord Diplock at pp. 382 et seq. The Privy Council apparently thinks otherwise: *South East Asia Firebricks Sdn. Bhd.* v. *Non-Metallic Mineral Products Manufacturing Employers Union* [1981] AC 363, 370.

[38] See below, Ch. 5.3.

considerations, or that in some other respect they have so proceeded that the determination is nugatory and void.[39]

Cases involving the duty to determine are therefore dealt with in the following way. There are two kinds of duty involved in determining a question: first, the duty to assume jurisdiction by determining the question remitted in a case which falls within the decision-maker's jurisdiction; and secondly, the duty to determine that question in accordance with law, i.e. in the manner indicated above. A tribunal which expressly refuses jurisdiction will be compelled to exercise it by mandamus to hear and determine if the facts warrant it.[40] If, however, a tribunal actually does determine the question remitted, thereby apparently assuming jurisdiction, but determines it other than in accordance with law, it will be regarded as having constructively refused jurisdiction,[41] and will be compelled by mandamus to hear and determine according to law, i.e. to determine the matter again, this time observing the correct principles or procedures. By this reasoning mandamus has regularly been used to correct errors of law as well as to compel the assumption of jurisdiction. Thus it can be concluded that all breaches of the duty to determine involve an error of law of some kind, and such an error can be committed at any stage of the inquiry, not just at the preliminary stage;[42] and that most, if not all, errors of law, whenever committed, will constitute a breach of duty by the decision-maker. The time may not be far away when the duty to determine can be simply stated as a duty to observe at every juncture the law as laid down in the statute and interpreted by the courts.

We have been considering jurisdiction in this discussion. It is also clear, however, that discretion is dealt with similarly. If for example an authority is mistakenly under the impression it has no discretion, or acts under dictation, or fetters itself by a self-created rule of policy,[43] or refuses to exercise its discretion in a case in which it would

[39] (1933) 50 CLR 222, 242.

[40] See *R.* v. *Blakely, ex p. Association of Architects, etc.* (1950) 82 CLR 54 (HCA); *R.* v. *Pugh (Judge), ex p. Graham* [1951] 2 KB 623; *R.* v. *Briant (Judge), ex p. Abbey National Building Society* [1957] 2 QB 497; *R.* v. *Clerkenwell Metropolitan Stipendiary Magistrate, ex p. DPP* [1984] 2 All ER 193.

[41] See the principles laid down in *Ex p. Bott*, above n. 39; *R.* v. *Commonwealth Court of Conciliation and Arbitration, ex p. Ozone Theatres (Australia) Ltd.* (1949) 78 CLR 389; *Manning* v. *Thompson* [1979] 1 NSWLR 384 (PC).

[42] Thus even an imposition of an unlawful penalty at the *end* of the inquiry would be an excess of jurisdiction.

[43] See *R.* v. *Stepney Corporation* [1902] 1 KB 317; *R.* v. *London County Council, ex p. Corrie* [1918] 1 KB 68; *R.* v. *Port of London Authority, ex p. Kynoch Ltd.* [1919] 1 KB 176; *Ex p. Donald, re Murray* (1969) 89 WN (Pt. 1) (NSW) 462; *R.* v. *Willson, ex p. Jones* [1969] SASR 405. See further De Smith above n. 1, pp. 309–20.

be unlawful not to, it will be compelled by mandamus to do so. If it exercises its discretion on incorrect legal principles it will be held to have abused its discretion or exceeded its powers, and, by the same token, to have constructively refused to exercise its discretion,[44] a failure which again is corrigible by mandamus. *Padfield*'s case, discussed in Chapter 1,[45] is a good example of the application of these principles in the context of irrelevant considerations.

We can now state summarily the content of the duty to determine. The decision-maker must create machinery adequate to deal with all proper applications or references within a reasonable time. When the matter is ripe for a determination he must ascertain whether he has jurisdiction to hear the case. If he has not, he must refuse jurisdiction, and prohibition will issue to prevent him from proceeding without jurisdiction.[46] If he has, he must determine the case according to the usual principles. Having determined the case he must make an appropriate award[47] and inform the parties affected of the decision.[48] Once he has determined the matter he is *functus officio*[49] (that is unless the determination is invalid[50]) and cannot rehear the matter unless it is remitted on appeal; it follows that he cannot be compelled to hear a second application on the same facts as the first.[51] If the determination is held on review to be invalid he must, if directed to do so, rehear the matter in accordance with law.

[44] See *Randall* v. *Northcote Corporation* (1910) 11 CLR 100 (HCA); *Wade* v. *Burns* (1966) 115 CLR 537 (HCA); *R.* v. *Birmingham Licensing Planning Committee, ex p. Kennedy* [1972] 2 QB 140.

[45] *Padfield* v. *Minister of Agriculture* [1969] AC 997 (HL). See above Ch. 1.3.

[46] In *R.* v. *Warden at Cloncurry, ex p. Sheil* [1971] Qld R 406 mandamus was issued to compel rejection of an application; the proper remedy, however, is prohibition.

[47] The duty is to hear *and determine*, not simply to hear. The fear of Darling J. in *R.* v. *Nicholson* [1899] 2 QB 455, 461 that unless there was a distinction between not hearing according to law and not determining according to law, mandamus being applicable only to the former, every decision wrong in law would be challengeable, represents not only an outdated theory of jurisdiction but defective logic; an unlawful award cannot be within jurisdiction.

[48] *R.* v. *Tower Hamlets London Borough Council, ex p. Kayne-Levenson*, above n. 26; *R.* v. *Smith* (1873) LR 8 QB 146; *R.* v. *Bolton Metropolitan Borough Council, ex p. B* (1986) 84 LGR 78 (duty to inform parent of refusal of access to child in care without reasonable delay, right of application for access order).

[49] A matter can be reconsidered at any time before the determination: *R.* v. *Victorian Licensing Court, ex p. Rubinstein* [1958] VR 384; once the determination is made, i.e. lawfully made, it cannot be changed: *R.* v. *Essex Justices, ex p. Final* [1963] 2 QB 816.

[50] *R.* v. *South Greenhoe Justices, ex p. DPP* [1950] 2 All ER 42.

[51] *Ex p. Miller, re NSW Ambulance Transport Service Board* (1950) 67 WN (NSW) 179; *Ex p. Blackburn* [1956] 1 WLR 1193; *Delahaye* v. *Oswestry Borough Council, The Times*, 29 July 1980.

In some instances, the nature of the decision itself will entail the performance of a further duty; for example a decision as to entitlement may involve a duty to provide. If the decision is invalid, there may still be a duty if a valid decision would necessarily be such as to entail such a duty.[52]

(iv) *Actual refusal of jurisdiction*

For the notion of 'constructive refusal' we can refer to the usual principles, discussed above. What, however, is an *actual* refusal of jurisdiction? The most obvious case is where a point is taken as to jurisdiction and a decision made upon it,[53] but there are some that are not so obvious. A refusal of jurisdiction can be apparent from the conduct of the decision-maker, as in the case where a party's agent and witnesses were simply asked to withdraw.[54] Adjournments are more problematical, because whether an adjournment constitutes a refusal of jurisdiction depends on the width of the powers of the decision-maker to order such adjournment. The same principles seem to be applied as are applied to delay in assuming jurisdiction, so that the issue is essentially whether there is a good reason for the adjournment: thus an adjournment arising from an impending breach of the *nemo iudex* rule is valid,[55] but one based on a mistake of law is not.[56] Refusal to admit evidence is not necessarily a refusal of jurisdiction, but if it arises from a misconception of the nature of the question to be determined, it will be. As Lord Esher MR once put it:

> The distinction between the two is sometimes rather nice; but it is plain that a judge may wrongly refuse to hear evidence upon either of two grounds: one, that even if received the evidence would not prove the subject-matter which the judge was bound to inquire into; the other, that whether the evidence would prove the subject-matter or not, the subject-matter itself was one into which he had no jurisdiction to inquire. In the former case the

[52] See above Ch. 1.3(iii); *R.* v. *Hillingdon London Borough Council, ex p. Streeting* [1980] 3 All ER 413 is a good example; see also *In re Islam* [1981] 3 WLR 942 (HL).

[53] In *In re Ross-Jones and Beaumont* [1979] 53 ALJR 259 the High Court of Australia held that there was no refusal to exercise jurisdiction where the judge made a consent order that he had no jurisdiction.

[54] *R.* v. *Thames and Isis Navigation Commissioners* (1839) 112 ER 1080; and see *R.* v. *Assessment Committee of St Mary Abbots, Kensington* [1891] 1 QB 378.

[55] Where the power to grant the adjournment is discretionary and statutory, the usual principles apply, so that regard must be had to relevant considerations: *Bilbao* v. *Farquhar* [1974] 1 NSWLR 377.

[56] *R.* v. *Kelly, ex p. Victorian Chamber of Manufactures* (1953) 88 CLR 285 (HCA).

judge would be wrongly refusing to receive evidence, but would not be refusing jurisdiction, as he would in the latter.[57]

At all stages statutory procedural requirements must be observed, if they are mandatory ones.[58]

(v) *Compelling a particular result*

As a final question we might ask whether the courts can compel a decision-maker to make a particular decision, so that failure to do so will be a breach of his duty. Is there ever a duty to determine *in a particular fashion*?

The courts have always adhered to the principle that they will not compel inferior jurisdictions to decide in a particular fashion, but only correct excess or want of jurisdiction, or abuse of or failure of discretion. This theory of limited review is still adhered to: the courts are concerned not with merits but with legality. Litigants on the other hand are generally concerned with merits but not so much with legality, except where it affects the merits. Since the distinction between the two seems to be fast disappearing, as concerns questions of law, the expansion of the notion of jurisdiction has entailed a corresponding contraction of the notion of merits. There must indeed be cases where the decision-maker is left in effect with only one decision which he can validly make: for example if the decision is based on an interpretation of a particular term such as 'in the course of employment' or some other statutory expression at the X level or the Y level, which has been held by the courts to be incorrect, the decision will probably have to be reversed on a rehearing. If, however, the decision involves a breach of natural justice or some other procedural error it may properly be adhered to on a rehearing if the defect does not in the event affect the substance of the decision. Whether a particular decision can be compelled depends therefore on the nature of the illegality established and the part played by it in the original decision. It also depends on the statutory context: for example in *Padfield*, although the minister was in effect compelled to refer the complaint to the committee of inquiry, he was not in the

[57] *R.* v. *Marsham* [1892] 1 QB 371, 378. To similar effect is the dictum of Lord Goddard CJ in *R.* v. *Licensing Authority for Goods Vehicles, ex p. B. E. Barrett Ltd.* [1949] 2 KB 17, 22. See also *WA Amalgamated Society of Railway Employees* v. *Commissioner of Railways for WA* (1906) 3 CLR 67 (HCA); *R.* v. *Fowler, ex p. McArthur* [1958] Qld SR 41. *A fortiori* refusal to find a fact when faced with conflicting evidence is within jurisdiction: *R.* v. *Board of Commissioners of Public Utilities, ex p. Halifax Transit Corporation* (1970) 15 DLR (3d) 720.

[58] See above nn. 9–11.

result compelled to act on their recommendation.[59] Formally of course the courts cannot expressly compel a particular decision, for this would be seen as a breach of the doctrine of the separation of powers; this has not, however, prevented them from doing precisely that in some cases involving the making of an order or the granting of a licence in the applicant's favour, where there is no possibility of a valid decision to the contrary, for example where the duty is purely ministerial or the decision-maker's discretion has been exhausted.[60]

2.3 THE DUTY TO PROVIDE

The duty to provide services, public works, and the like is potentially a considerable problem for administrative lawyers, and one which has only been the subject of litigation in recent years. The ambitions of public authorities and the demands made on them are great, and there is also a big question mark against the community's ability to support extensive schemes to provide services for its members. Although we have hopefully moved far from the time when the courts frowned on 'eccentric principles of socialistic philanthropy',[1] there are still situations in which the courts have to look carefully at the validity of the provision of the service,[2] or listen to cogent reasons why a duty should not be enforced against an authority because they say 'we have been doing our best', or 'we have no money for this', or 'we are being prevented'. The statute itself is usually of little help because either the duty is couched in absolute terms which appear to admit no excuses for non-performance, or else the draftsman has covered the situation with a phrase such as 'so far as is reasonably practicable' or 'shall make suitable provision' which still leaves the choice to be made by the courts as to the precise content of the duty.

Although these duties have figured in the law reports in recent years, development of a jurisprudence to deal with them has been seriously limited by three factors: the possibility of lack of standing

[59] See the discussion in Bailey, Cross, and Garner, *Cases and Materials in Administrative Law* (1977), pp. 284–5; and above Ch. 1.3.

[60] See above Ch. 1.3.

[1] *Roberts* v. *Hopwood* [1925] AC 578, 594, per Lord Atkinson (HL).

[2] See *Bromley London Borough Council* v. *Greater London Council* [1982] 2 WLR 62 (HL); *Prescott* v. *Birmingham Corporation* [1955] Chanc. 210; *Taylor* v. *Munrow* [1960] 1 WLR 151. These decisions proceed on the basis that local authorities owe a general duty to their ratepayers not to squander public funds.

on the part of the applicant;[3] the possibility that the courts will not intervene, even in the case of a clear breach of the statute, because a minister is given a power to enforce compliance;[4] and the possibility that the courts will in any event be inhibited from interfering in what are seen as purely administrative matters. In fact none of these factors, upon examination, ought really to preclude the courts from making crucial choices in suitable cases. Recent cases indicate that the courts are both able and willing to consider on their merits arguments based on administrative prudence,[5] and the suggestion that the courts are intrinsically incapable of making the right choices or considering the wider implications of their decisions is quite palpably false. Even the lesser argument that the minister is more qualified to choose because he stands at the head of a great department of government with all the facts and arguments at his fingertips is becoming rather thin; in many cases the existence of default powers is almost an embarrassment to the minister, and he has refused to act, even where he obviously should, precisely *because* of his position as the political head of a great department.[6]

Duties to provide are characterized by their discretionary element. Although performance is not optional, the authority has considerable leeway with regard to the manner of performance, so that substantial and bona fide compliance with the statute is generally regarded as sufficient: it is no argument for an aggrieved citizen to say that the authority ought to have done more, provided what has been done is reasonable in the circumstances.[7] The case law is riddled with express or implied references to a flexible standard of reasonableness. Occasionally one finds a duty referred to as 'absolute'; but on closer inspection this is merely to say that the court takes a strict view of the excuses which can legitimately be made for failure to perform, although it is true that some duties which are purely ministerial, such

[3] See below Ch. 6; e.g. in *Kensington and Chelsea London Borough Council* v. *Wells* (1973) 72 LGR 289 it was held that even if there was a breach of duty the applicant might not benefit from its performance and relief was accordingly refused.

[4] See below Ch. 3.4.

[5] See above Ch. 2.2, and *Inland Revenue Commissioners* v. *National Federation of Self-Employed and Small Businesses Ltd.* [1982] AC 617, 652 per Lord Scarman. However, the statement in the text is not always true. The courts still steer clear of complex questions of public finance, for which see *R.* v. *Secretary of State for the Environment, ex p. Hackney London Borough Council* [1984] 3 All ER 358; *Roberts* v. *Dorset County Council* (1976) 75 LGR 462.

[6] See e.g. *Padfield* v. *Minister of Agriculture* [1968] AC 997 (HL); *Meade* v. *Haringey London Borough Council* [1979] 1 WLR 637; *R.* v. *Secretary of State for the Environment, ex p. Ward* [1984] 2 All ER 556.

[7] See the cases discussed below.

as the appointment of certain local-government officers,[8] could genuinely be described as 'absolute'; generally, however, this kind of terminology is merely unhelpful rhetoric which adds nothing to our understanding of the nature of the duty. The discretionary element involved in a duty depends on the kind of duty involved, and the extent to which the courts, as a matter of policy, are prepared to restrict that element. The ways in which it is likely to be restricted are discussed in the following paragraphs. It must be borne in mind that, whatever latitude is given to the authority, the mere assertion of a discretion as to performance of the duty will not be allowed to outweigh the fact that a duty, not a power, is in question, and can never therefore excuse complete failure to perform it.[9]

(i) *Equality*

One very important aspect of decisions on performance of duties to provide is the concept of equality, that is, treating like cases alike, a principle which is fundamental both to legal thought and to administrative practice. The notion of equality can be assumed as part of the judicial apparatus in dealing with cases involving a duty to provide public goods on a collective basis.[10] On the whole, however, it seems that there is no very clear concept, such as *droit administratif*'s

[8] *R.* v. *Leicester Union Guardians* [1899] 2 QB 623 (vaccination officers); *Peart* v. *Westgarth* (1765) 97 ER 1007 (parish overseers).

[9] *R.* v. *Marshland, Smeeth, and Fen District Commissioners* [1920] 1 KB 155; and see *R.* v. *Bristol Dock Co.* (1827) 108 ER 420. The principle that an authority cannot fetter its performance of statutory duties by undue expenditure has recently been discussed in *R.* v. *Secretary of State for the Environment, ex p. Hackney London Borough Council* above n. 5. Although it was held that the minister, in issuing guidance to local authorities as to the level of their expenditure, need not have regard to whether they could attain the level indicated without being in breach of their statutory duties, there seems to be no reason why the principle should not receive judicial endorsement in other contexts, for it is clearly fundamental. In the cases cited above n. 2, the courts made use of the idea of a fiduciary duty owed to the ratepayer; a duty can properly be regarded as a kind of trust personal to the authority and the courts will insist if necessary on actual performance rather than merely ensuring that it is effected by someone else: *Docherty (T.) Ltd.* v. *Burgh of Monifieth* 1971 SLT 13; *R.* v. *Wyre Borough Council, ex p. Parr* (1982) 2 HLR 71; *City of Birmingham District Council* v. *O* [1983] 1 All ER 497 (HL) (Council responsible in spite of delegation of function to committee); *R.* v. *Independent Broadcasting Authority, ex p. Whitehouse, The Times*, 14 Apr., 1984. Similarly they will not enforce a contract made in derogation of a statutory duty: see above Ch. 1.2.

[10] See the remarks of Lord Scarman in relation to tax cases in the *National Federation* case, above n. 5, pp. 951–2. The notion of equality of access to public services is of course a familiar aspect of *droit administratif*: see Brown and Garner, *French Administrative Law*, 3rd edn. (1983), p. 140.

equality of access to public services, which would serve to focus attention on this important aspect of administrative law.

Many statutes, especially those concerned with public utilities, specifically provide for equal treatment of consumers, and such provisions can be every bit as much a sword for the consumer as a shield for the administrator. Two Scottish cases show how far the courts can go to enforce them, in spite of their vague character. In *South of Scotland Electricity Board* v. *British Oxygen Co.*[11] the House of Lords upheld a complaint of discrimination against a large industrial consumer based on a statutory provision which prevented the authority from showing undue preference or exercising any undue discrimination against any person or class of persons in fixing their tariffs, the substance of the complaint being that a tariff which discriminated fairly between high-voltage (low-tariff) and low-voltage (high-tariff) consumers had become unfair due to increased fuel costs which were distributed equally regardless of the differential. By way of contrast, in *Attorney-General of Victoria* v. *Melbourne Corporation*[12] a similar claim failed because the statute gave the consumer a choice between fixed-rate and differential tariffs, and the 'no preference' provision on which the claim was based was interpreted as preventing preference as between consumers dealing in similar circumstances. In the second Scottish case, *Wilson* v. *Independent Broadcasting Authority*,[13] a statutory provision imposed a duty on the IBA to ensure that its programmes maintained a 'proper balance' in their subject-matter. The IBA gave each of four Scottish political parties media time during a referendum campaign based on the votes received by each in a general election. Some voters moved the court to prevent the programmes being broadcast because three of them were directed to securing a 'yes' vote and only one to securing a 'no' vote. The IBA argued that their duty was to secure a proper balance of subject-matter rather than point of view, but this argument was rejected, the court holding that there was a duty to maintain a balance between the proponents of 'yes' and the proponents of 'no', not simply to maintain a balance between different subject-matters, and the parties' votes at a general election were irrelevant when a referendum was in question.

Of course equality will also be read into the statute where necessary, even if there is no express prohibition of discrimination. Thus the owner of a utility must continue to provide the service for all who

[11] [1956] 1 WLR 1069.
[12] [1907] AC 469 (PC).
[13] 1979 SLT 279; see also *Mitchell* v. *New Zealand Broadcasting Corporation* [1970] NZLR 314; *R.* v. *Broadcasting Complaints Commission, ex p. Owen* [1985] 2 WLR 1025.

continue to pay the rates,[14] and the court will strike down a by-law which allows for a discount on the rates in favour of consumers who pay promptly.[15] Similarly school transport must be provided for French-speaking pupils on an equal footing with English-speaking pupils, even though there is no obligation to provide it at all.[16]

When there is competition for a scarce resource or where the provision of a service is restricted due to industrial action or some other emergency, equality becomes even more pressing, and is regarded as a legitimate argument *in favour* of an authority which refuses to provide. It is essential here to have regard to the rights of consumers who are not before the court and not to grant a remedy which has the effect of giving the applicant before the court an unfair advantage.[17] In *R.* v. *Bristol Corporation, ex p. Hendy*[18] the authority made a closing order in relation to a flat, as a result of which the tenant lost possession of it by a court order in favour of the landlord. The authority then came under a duty to provide the tenant with suitable alternative residential accommodation on reasonable terms, and offered him temporary accommodation with the possibility of an offer of accommodation on terms offered by them to their own prospective tenants. In the circumstances the Court of Appeal refused to interpret the duty so as to give the applicant preference to those already on the authority's housing list, and held that the duty was being fulfilled.

The tension between individual aspiration and collective provision can be seen in some cases concerning the Education Act 1944 (England and Wales). Under section 8(1) of the Act local education authorities are under a duty to secure that there shall be available for their area sufficient schools for providing primary and secondary education; and under section 76 they must, in performing their powers and duties, have regard to the principle that, so far as is compatible with the provision of efficient instruction and training and the avoidance of unreasonable public expenditure, pupils are to be educated in accordance with the wishes of their parents. In *Watt* v. *Kesteven County Council*[19] the plaintiff, a Roman Catholic, declined to send his sons to

[14] *St Lawrence Rendering Co. Ltd.* v. *City of Cornwall* [1951] 4 DLR 790.

[15] *Attorney-General for Canada* v. *Toronto* (1893) 23 SCR 514 (SCC); and see *Hamilton* v. *Hamilton Distilling Co.* (1907) 38 SCR 239 (SCC).

[16] *Bachmann* v. *Government of Manitoba* [1984] 6 WWR 25; see also *Ridings* v. *Elmhurst School, Trustees,* [1972] 3 DLR 173.

[17] *R.* v. *Bristol Corporation, ex p. Hendy* [1974] 1 WLR 498, 503 per Scarman LJ. The same notion is probably present in *R.* v. *Secretary of State for Social Services ex p. Hincks* (1981) CLY 274 (waiting list for orthopaedic surgery).

[18] Above n. 17; see also *Roberts* v. *Dorset County Council*, above n. 5; *Din* v. *Wandsworth London Borough Council* [1981] 3 WLR 918, 941 per Lord Bridge.

[19] [1955] 1 QB 408.

an independent grammar school in the area, for which the authority were prepared to pay the fees, there being no state grammar school in the area, but sent them instead to Roman Catholic boarding-schools outside the area. The authority agreed to make a contribution, based on his means, towards the fees, which were less than for the grammar school, but refused to pay the full amount. The plaintiff, alleging a breach of section 76, brought an action to enforce the duty and to secure repayment of the fees paid by him. The Court of Appeal held that the duty of the authority was only to make free places available in their own schools or independent schools, and to contribute to the fees payable for a place at a school, chosen by the parent, with which they had no arrangements. On the facts they had had regard to the parent's wishes, but it was open to them to have regard also to other matters. If they granted the plaintiff's request they would have to pay all the fees for every child in their area sent to a boarding-school, no matter how rich the parents; they could not do for him what they did not do for others.[20] In *Cumings* v. *Birkenhead Corporation*[21] the authority decided that all the children leaving Roman Catholic primary schools could comfortably be accommodated in Roman Catholic secondary schools, but children leaving non-Roman Catholic primary schools could not be accommodated in non-Roman Catholic secondary schools. They sent a circular to all parents of children affected saying that Roman Catholic primary pupils would be considered only for Roman Catholic secondary schools, save in exceptional cases, and then only when accommodation was available in the school preferred. In the instant case C applied for his son, who had attended a Roman Catholic primary school, to attend a non-Roman Catholic secondary school, and the application was refused on the ground that the son was below some other applicants in his attainments. C and others similarly placed sued for declarations, alleging that the authority's action was in breach of section 76. The Court of Appeal held that the authority had to have regard to the wishes of the parents of other children and groups of children for whom they had to cater, and their action was described by Lord Denning MR as a 'sound administrative policy decision to which no

[20] In *Winward* v. *Cheshire County Council* (1978) 77 LGR, 172 the authority's insistence on provision of special education for a handicapped child was allowed to outweigh the parents' preference for ordinary primary education, and s. 76 was held not to apply. See also *Wood* v. *Ealing London Borough Council*, [1966] 3 All ER 514, in which s. 76 was held not even to give parents as a body a right to be consulted on a development plan.

[21] [1972] Chanc. 12.

objection could be taken'.[22] It could, however, be doubted whether the policy would have survived a constitutional guarantee of equal protection, because it cannot be denied that the plaintiffs were restricted in their choice on the basis of an irrelevant criterion of religious affiliation. Although it was asserted that discrimination, for example on the basis of colour, would be struck down, this decision does not give one confidence that the courts will in fact intervene where necessary.

(ii) *Inconvenience and unreasonableness*

An obvious corollary of the principle of equal treatment is that the general rate fund should not be overburdened by provision for a particular individual, for the whole community will suffer as a result of the consequent cutting of other expenditure.[23] Generally, of course, a duty will not be enacted by the legislature unless it is clear that money will be available for its performance;[24] none the less in some cases provision can involve unreasonable expenditure either in itself or by setting an awkward precedent. This can be seen in *Watt*'s case, in which equal treatment would have resulted in unreasonable expenditure, if the plaintiff's demands had been met, and the authority would have fallen foul not only of the qualifications to section 76, but also no doubt of the district auditor. Some useful principles were laid down in *Gravesham Borough Council* v. *British Railways Board*, in which the plaintiffs failed to establish that reduced hours of operation of a ferry were unreasonable. Slade J. said this:

> . . . it is necessary to strike a balance between (a) the extent of the demand for the ferry service at particular hours and the inconvenience and hardship, if any, to the public which would be caused if such service were not provided and (b) the extent of the burden in terms of money and otherwise that would fall on the ferry owner in providing such service at the hours in question.[25]

In similar vein the Court of Appeal has held that a duty to maintain footpaths involves a duty to remove snow only when sufficient time

[22] Ibid. 38.

[23] *Cholak* v. *Westok* (1944) 1 WWR 139 (affd. [1944] 3 WWR 256); *Hey and Croft Ltd.* v. *Lexden and Winstree Rural District Council* (1972) 70 LGR 531; cf., however, *William Leach (Midlands) Ltd.* v. *Severn-Trent Water Authority*, *The Times*, 5 June 1981.

[24] See above n. 9.

[25] [1978] Chanc. 379, 393. See also *Attorney-General* v. *Colchester Corporation* [1955] 2 QB 207; and *Darlaston Local Board* v. *London & North Western Railway Co.* [1894] 2 QB 694, where it was held to be unreasonable to keep open an uneconomic branch line when the duty was 'to afford all reasonable facilities for the receiving and forwarding and delivering of traffic'; cf., however, *Morton* v. *Eltham Borough* [1961] NZLR 1.

has elapsed to render it unreasonable, in the light of its other highway duties, for the authority not to take remedial measures;[26] and that the content of a duty to provide adequate accommodation for gypsies residing in or resorting to their area is to be determined by reference to the numbers of gypsies known to reside in or resort to the area and the length of time necessary to provide accommodation.[27] It is suggested, however, that the House of Lords in *R.* v. *Governor of Brixton Prison, ex p. Walsh*[28] failed to strike a proper balance between individual rights and public burdens when it held that the prison governor was not in breach of any duty when he failed to produce an accused person before a magistrates' court due to staff shortages. This decision is very surprising when one considers the importance of the liberty of the subject as against the mere inconvenience of not being able to use a ferry at certain hours.

Mere inconvenience should not be, and is not generally regarded as being, a sufficient excuse for non-performance, unless the inconvenience is so great as to involve totally unreasonable expenditure of money or effort.[29] The limits are fairly well set by a case in which the court refused to enforce a judge's order that compensation for criminal damage to property be raised from the rates on the ground that a rate had already been struck and the authority had provided for the compensation in the estimates for the following year.[30]

(iii) *Impossibility and duress*

Financial or administrative difficulty must be distinguished from actual physical impossibility. Just as it is physically impossible to keep a footpath completely clear of snow and ice[31] or the streets completely clear of garbage,[32] and the duty will be confined to reasonable

[26] *Haydon* v. *Kent County Council* [1978] QB 343; to similar effect are *R.* v. *Secretary of State for Social Services, ex p. Hincks*, above n. 7 and *R.* v. *same, ex p. Greater London Council and Child Poverty Action Group, The Times*, 8 Aug. 1985.

[27] *West Glamorgan County Council* v. *Rafferty* [1987] 1 All ER 1005, 1017, 1023–4 (*obiter*).

[28] [1985] 1 AC 154.

[29] See below Ch. 3.4; and esp. *R.* v. *Paddington Valuation Officer, ex p. Peachey Property Co. Ltd.* [1966] 1 QB 380.

[30] *R.* (*Bennett*) v. *Kings County Council* [1908] 2 IR 176. *A fortiori* the court will not compel prison authorities to redesign a prison merely to ensure that a prisoner will not suspect that his freedom of communication with his lawyers might be impeded; *Smith* v. *Commissioner of Corrective Services* [1978] 1 NSWLR 317 (see, however, above Ch. 1.3).

[31] *Haydon* v. *Kent County Council*, above n. 26.

[32] *R.* v. *Kensington and Chelsea London Borough Council, ex p. Birdwood* (1976) 74 LGR 424. See also the remarks of Lord Brightman concerning the difficulties of local

attempts to perform, so the courts will not insist on performance of a duty to provide some medical care and assistance for a handicapped woman when her husband causes the nurses to refuse to visit because of his abusive conduct and refuses to give assurances as to his future conduct.[33]

Again physical impossibility must be distinguished from duress. There are occasions when authorities and the courts are faced with very difficult decisions as to whether performance of a duty is to be expected in the face of industrial strife or the like, which may be caused or exacerbated if the authority attempts to perform the duty in the usual way. Some of these cases involve actual physical impossibility, in that the authority may be unable to act other than through their employees, who refuse to obey orders;[34] thus in *Fairfax (John) Ltd.* v. *Australian Telecommunications Commission*[35] it was held that a statutory duty to repair the plaintiff's teleprinters was not breached when the defendants' employees refused to repair them because of an industrial dispute involving the plaintiffs.[36] In other cases performance is possible but the authority decides not to act. These cases are more difficult because the courts may, by ordering performance, cause unnecessary disruption or violence, or may, by refusing to intervene, flout the will of the legislature and weaken the hand of the authority in dealing with the situation.[37] While caution

authorities in fulfilling their housing duties and the consequences for judicial review, in *Pulhofer* v. *Hillingdon London Borough Council* [1986] 1 All ER 467, 474 (HL).

[33] *R.* v. *Hillingdon Area Health Authority, ex p. Wyatt, The Times*, 20 Dec. 1977; *Wyatt* v. *Hillingdon London Borough Council* (1978) 76 LGR 727 (the latter case arose out of the former, and was an action for damages for breach of statutory duty); see also *R.* v. *Kent County Council, ex p. Bruce, The Times*, 8 Feb. 1986.

[34] See *Hackney Borough Council* v. *Doré* [1922] 1 KB 431, in which the refusal of workmen to carry out the statutory duty to supply electricity did not excuse the authority employing them from paying a statutory penalty because they were only excused for *force majeure*, which meant actual physical restraint. Contrast *Re Richmond Gas Co.* v. *Richmond Borough Council* [1893] 1 QB 56 where a statutory penalty for failure to supply gas was not imposed because the failure was due to exceptionally severe frost which was beyond the authority's power to deal with; a similar position was taken in an action for negligence: *Blyth* v. *Birmingham Waterworks Co.* [1856] 156 ER 1047.

[35] [1977] 2 NSWLR 400; cf. *Parker* v. *Camden London Borough Council* (1985) 84 LGR 16.

[36] Nevertheless one wonders whether the 'impossibility' is real, bearing in mind the possibility of employing other persons to do the job, difficult as that might be: see *Meade* v. *Haringey London Borough Council* [1979] 1 WLR 637, 642 per Lord Denning MR.

[37] The problem is well illustrated by an exchange between judge and counsel in the Supreme Court of NSW in *Fairfax (John) Ltd.* v. *Australian Postal Commission* [1973] 2 NSWLR 124, 134: 'HIS HONOUR: . . . You say that the inevitable consequence

is necessary, the courts cannot abdicate their responsibility any more than the authority itself can. Some guidance can be gleaned from two important cases.

The first case, *Fairfax (John) Ltd.* v. *Australian Postal Commission*, arose out of the same dispute as the case mentioned above, in consequence of which the authority refused for an indefinite period to deliver any mail to the plaintiffs, on the basis that the whole mail service of New South Wales would be brought to a standstill if they delivered, even though it was conceded that delivery could be effected merely by allowing the plaintiffs to collect their mail. The Supreme Court of New South Wales held on appeal by a majority that the authority had breached its duty. Moffitt P., with whom Reynolds JA agreed, said this:

> If the commission, which has a statutory monopoly, refuses to continue to supply a postal service to some person so that it does not perform its duty to provide services without discrimination in accordance with the statute of the Parliament, but instead provides services to such persons only as is dictated by others, there is provided a basis for the exercise by the Court of declaratory and injunctive relief.[38]

Mahoney JA, in a carefully reasoned judgment, reached the opposite conclusion:

> What, short of physical impossibility, will excuse non-performance of a statutory duty cannot be stated in general terms; it will be affected, inter alia, by the nature of the duty to be performed and the circumstances and extent of the non-performance. In relation to the delivery of mail, there are, I do not doubt, grounds upon which delivery may at least be suspended, even in a case such as the present. A threat by a lunatic or a terrorist could be sufficient, though it would not necessarily be so. The credibility of the

of your seeking to obey this order by delivering to taking steps to see that the delivery is made—and either you stand people down who don't deliver or use some other labour to do the delivery—in either of those events you say there would be a stoppage throughout the whole of the service? MR. SHELLER: Yes. HIS HONOUR: Supposing I am otherwise persuaded that he is legally entitled, is the court to shrink from the task of carrying out its duty because of wrong doing, which would be wrong doing, even if it is industrial wrong doing? MR. SHELLER: In the ordinary case, one might say of course not. Indeed, one would say that. But one has to bear in mind the consequences to other people—other people who are recipients of our services—and that is why we are here today, because in exercising our function we have to take into account under our Act all those consequences. If one assumes that the effect of this is to cut off mail deliveries for a period of time to the community at large, the people being hurt are the community. HIS HONOUR: That is true. Are they going to be more hurt by that than by saying that the law has got the stage where it can't be enforced?'

[38] [1977] 2 NSWLR 124, 136; cf. *Inland Revenue Commissioners* v. *National Federation of Self-Employed and Small Businesses Ltd.* [1982] AC 617, 644 (HL).

threat, its immediacy, and the effect of it, would be relevant, but I do not think that it would be beyond the defendant's power to suspend delivery of mail to the plaintiffs in the event that, for example, a lunatic, with the desire of injuring the plaintiffs, threatened to blow up the General Post Office, unless there were such suspension. Nor, if this be accepted, do I doubt that suspension or refusal of performance could, in some circumstances, be justified by actual or threatened industrial action. Once it be accepted that non-performance of a duty may be excused by reference to, e.g., the effect of an act or threatened act, then I do not think that industrial action, or the threat of industrial action, can be put aside, unless it can be shown that industrial matters are to be treated as, in this regard, special. I do not think that they are to be so treated. For present purposes, actual or threatened disruption of public services is not to be relevantly distinguished according as to whether it arises from a threat by employees or suppliers or any other social or economic group.[39]

In the second case, *Meade* v. *Haringey London Borough Council*,[40] the school caretakers in the authority's area went on strike and informed the authority that all schools would be closed from the commencement of the strike. The authority instructed all headmasters to advise parents not to send their children to school, and said that no one should attempt to open a school. They also expressed support for the caretakers' claim and said they were doing everything possible to lessen the impact of the strike by negotiations, which resulted in some minor concessions. After four weeks, in which all education in the borough was suspended, some parents, after complaining unsuccessfully to the minister, started an action to force the authority to reopen the schools, and this resulted in the caretakers agreeing to the opening of four-fifths of the schools. The judge refused an interlocutory injunction, but the parents appealed, and as the hearing of the appeal commenced in the Court of Appeal all the schools opened fully for the first time in six weeks. Lord Denning MR started from the premise that, the authority being under a duty under section 8(1) of the Education Act 1944, mentioned above, to provide schools, the court would intervene in the event of any school closure made without just cause or excuse. He concluded:

It seems to me that if the borough council closed the schools—at the behest of the trade unions—or in agreement with them, they were acting unlawfully. The trade unions had no right whatever to ask the borough council to close the schools. The borough council had no business whatever to agree to it. Instead they should have kept the schools open—and risked the consequences of the dispute escalating. Or they should have moved the court for an

[39] Ibid. 150.
[40] [1979] 1 WLR 637.

injunction to restrain the leaders of the trade unions from interfering with the due opening of the schools. I am confident that the people at large would have supported such a move and expected the trade union leaders to obey it; and they would have obeyed it.[41]

The other two judges were unwilling to commit themselves to this view of the case when interlocutory relief was in question, and went little further than saying that there was a triable issue. Eveleigh LJ considered that if the authority made a legitimate choice of the various options open to them they would not be in breach of their duty, but some of his other statements seem to indicate that if they were influenced by considerations not relevant to education, or if they allowed themselves to be dictated to, their choice would not be a legitimate one. Sir Stanley Rees' judgment indicates support for Lord Denning's premise, but did not deal with the crucial question of how it might be applied to the facts, and concludes with a warning of the dangers involved in the courts interfering with 'the delicate mechanism of industrial disputes'.[42]

(iv) *Some conclusions*

These two last cases go to the root of the problem of non-performance and it is unfortunate that they do not give any clearer guidance as to how the courts should deal with such situations. It is tempting to apply Lord Denning's remedy of grasping the nettle, but while it was perhaps appropriate for the reprehensible abdication of responsibility by the authority in *Meade*'s case, it will not necessarily be suitable in all other cases. It is not really possible to lay down precisely how such cases should be decided, because they depend very much on the facts, as Mahoney JA indicated in the *Fairfax* case, but it is possible to examine the legal premises from which the decisions should start. It has been assumed that a 'reasonable excuse' proviso can be read into duties of the kind in question. This, it is suggested, is a correct assumption, but it would be well to be a little more certain what is being assumed. The following observations are tentatively proffered:

1. The 'reasonable excuse' proviso applies to duties to provide, which are necessarily subjected to it because of the nature of the duty and the impossibility of setting rigid standards to be met in all

[41] Ibid. 648.

[42] His Lordship quoted from Geoffrey Lane LJ in *Harold Stephen & Co. Ltd.* v. *Post Office* [1977] 1 WLR 1172, 1180. In that case an order for delivery of mails to a party not involved in the dispute giving rise to the failure to deliver was refused because, *inter alia*, it would have enabled the Post Office workers to flout disciplinary action by the Post Office.

circumstances.

2. The burden of proving such excuse is on the authority because the facts are within their knowledge. Consequentially, the applicant must show some prima-facie refusal or neglect of the duty, such as failure to open schools, deliver mail, or collect garbage.

3. The question of reasonable excuse is generally to be considered as an aspect of non-performance rather than an aspect of the court's discretion to refuse relief. Although the courts have sometimes in effect upheld the excuse by exercising their discretion so as to refuse relief,[43] this approach is, for a number of reasons, not to be encouraged, except perhaps where the argument is not so much that there is a reasonable excuse for not performing the duty, as that it is impossible to comply with the court's order.[44] First, it is for the applicant to show that the court's discretion should be exercised in his favour, whereas, as we have seen, the burden should be on the authority to show that there is a reasonable excuse. Secondly, the court's discretion is exercised on clearly established principles, and to open up a new ground might lead to unjust refusal to put the statute into operation. Thirdly, for the question of discretion to arise it must be already established that the duty has been or is being breached; if the court decides that the duty has been breached but in its discretion refuses relief, the finding that there is a breach of duty can be used in subsequent cases where the remedy does not involve any discretion; thus in a case in which a breach of duty could result in an action for damages, the authority might be exposed to open-ended liability for a purely 'technical' breach of duty;[45] or in a case where mandamus

[43] See below Ch. 3.4. The position suggested in the text has however recently been taken by Sir John Megaw in *West Glamorgan County Council* v. *Rafferty*, above n. 27, p. 1024.

[44] In the case of an excuse which continues to be valid, both arguments can be based on the same facts, but the discretion of the court will not arise if there is no breach of duty. There are many cases where the two arguments are severable: e.g. in *R.* v. *London & North Western Railway Co.* (1851) 117 ER 1113 compliance with mandamus was impossible because the company had no longer any power to act, though it had at the time performance was demanded.

[45] This is a serious problem for authorities with a duty to provide. In each of the *Wyatt* cases (above, n. 33) the result was the same, but the plaintiff would have succeeded in the action for breach of statutory duty if the court in the first case had found a breach of duty but used its discretion to refuse relief. That case involved no other potential plaintiffs, but in *Meade*'s case 37,000 children were affected and it was alleged that they had suffered disadvantages through lack of facilities to prepare for examinations. The problem of open-ended liability does not of course disappear as a result of the suggestion in the text; in *Meade*'s case there is little doubt that an injunction would have issued if the case had been tried, but that would have involved a finding that the authority had breached its statutory duty. The cases involving failure to deliver mail do not involve this problem because postal authorities are

is applied for under statutory provisions, as for example where a minister applies to the court for enforcement of his directions in a case of failure by an authority to perform its duty, the court would not be able to refuse relief even though the authority was not really guilty of failing to perform its duty.[46]

4. What is a reasonable excuse naturally depends on all the circumstances. This matter should be determined on the basis of the usual principles of judicial review of the exercise of discretion, so that the authority must act for a proper purpose and have regard to relevant considerations and so on, but with the important difference that a duty is in question, not a power, and this consideration must, quite obviously, be paramount.[47] In cases of emergency obviously much depends on the nature of the emergency and the dangers involved, as indicated by Mahoney JA in the *Fairfax* case. In particular it will be important to ensure that in emergencies, as indeed in normal times, exceptional burdens are distributed fairly.[48]

In conclusion it is suggested that the concept of reasonableness, for all its vagueness, is a necessary concept in relation to the construction and performance of duties to provide; as with the law of negligence, it enables the courts to maintain oversight of the performance of duties consistently with legitimate expectations of the community. Only an accretion of experience through case law will give sufficient guidance as to the proper content of these duties.

2.4 THE DUTY TO ENFORCE THE LAW

Enforcement of the duty to enforce the law has always been regarded as one of the principal objects of the writ of mandamus. Lord

generally exempt from liability. For further discussion of this problem, see below Chs. 5.6, 6.2.

[46] See below Ch. 3.4.

[47] *West Glamorgan County Council* v. *Rafferty*, above n. 27, p. 1022 is to this affect: 'legal duty must be given proper weight'.

[48] In *Harold Stephen & Co. Ltd.* v. *Post Office*, above n. 42, an order for delivery of mail during an industrial dispute was refused because, *inter alia*, it would have led to discriminatory action by Post Office workers, a criminal offence. The duty to deliver without discrimination was also mentioned in the *Fairfax* case, above n. 37, in which it was suggested by Mahoney JA, at p. 144, that it would be open to an individual to prevent the authority from discriminating against his mail. Presumably to allow oneself to be dictated to with regard to who should receive mail would be discrimination, but restricting the service to e.g. the city rather than the country, or essential rather than non-essential services during an emergency, would not be discrimination.

Mansfield described 'defect of police' as one of the wrongs rectified by mandamus.[1] In the eighteenth century, magistrates were compelled to put into effect the Statute of Forcible Entries,[2] and indicted for failing to quell a riot.[3] Such cases were very rare in the nineteenth century, but of late the principles involved have been tested in a number of cases. The basic principles are quite clear. Where an authority has a duty to take legal action to protect public rights or public security, it would defeat the purpose of the statute if it were allowed to take no action when action is demanded; at the same time there are many good reasons, both of justice and expediency, which will excuse a refusal to put the law into operation. The problem is to discover whether these reasons or the general duty to enforce the law will prevail in a given case.

The essential point is that, like all other duties, the duty to enforce the law is not absolute, but contains a large discretionary element. It is fundamental to the decisions in the *Julius* and *Padfield* cases, discussed earlier,[4] that an authority charged with a function of oversight has a discretion not to act on a complaint, as well as a duty to act on it if it is substantial and relevant. The position cannot essentially be otherwise when a duty rather than a power is involved and the function is one of enforcement rather than mere oversight.

(i) *Police duties*

The case which broke new ground in this area was *R.* v. *Metropolitan Police Commissioner, ex p. Blackburn*,[5] decided in 1968. The Commissioner had issued a policy direction which was in effect an instruction not to enforce the gaming laws in London, and the applicant sought mandamus to compel him to enforce the law. The Commissioner's reason for issuing the direction was that the expense and manpower involved in keeping observation in gaming clubs could not be justified in view of the uncertainty in the law existing at the time, but the case presented in the Court of Appeal rested mainly on the proposition that the Commissioner was under no duty to the public to enforce

[1] *R.* v. *Barker* (1762) 97 ER 823, 824–5.

[2] *R.* v. *Montagu* (1729) 93 ER 107; cf. the vaccination cases: *R.* v. *Keighley Union Guardians* [1874] 40 JP 70; *R.* v. *Lewisham Union Guardians* [1897] 1 QB 498; *R.* v. *Leicester Union Guardians* [1899] 2 QB 632.

[3] *R.* v. *Kenneth* (1781) 172 ER 976.

[4] See above Ch. 1.3.

[5] [1968] 2 QB 118. The questions raised here have received considerable academic attention however. See Davis, *Police Discretion* (1975); Lustgarten, *The Governance of Police* (1986); Williams, 'Letting off the Guilty and Prosecuting the Innocent' (1985), *Crim.* LR 115.

the law. Faced with such an extreme proposition it is hardly surprising that the court rejected it very decisively:

> Is our much-vaunted legal system in truth so anaemic that, in the last resort, it would be powerless against those who, having been appointed to enforce it, merely cocked a snook at it? The very idea is as repugnant as it is startling, and I consider it regrettable that it was ever advanced.[6]

It therefore became important to decide exactly what the duty of the Commissioner was and whether he had breached it. Lord Denning MR described the duty in the following terms:

> it is for the Commissioner . . . to decide on the disposition of his force and the concentration of his resources on any particular crime or area. No court can or should give him direction on such a matter. He can also make policy decisions and give effect to them, as, for instance, was often done when prosecutions were not brought for attempted suicide. But there are some policy decisions with which, I think, the courts in a case can, if necessary interfere. Suppose a chief constable were to issue a directive to his men that no person should be prosecuted for stealing any goods less than £100 in value. I should have thought that the court could countermand it. He would be failing in his duty to enforce the law.[7]

The court was not in the event constrained to find a breach of duty because the Commissioner had reversed the direction and taken some steps to enforce the law, but the judgments indicate that, but for this, a breach of the duty would have been found.[8] Although the Commissioner was influenced (and rightly) by the uncertainty of obtaining convictions in view of the state of the law, the direction had not been withdrawn for some considerable time after the doubts in the law had been removed by a decision of the House of Lords.

In *R.* v. *Metropolitan Police Commissioner, ex p. Blackburn* (*No. 3*)[9] in 1973, the same parties were before the court in relation to a similar application regarding the enforcement of obscenity laws. The manpower resources allocated to the enforcement of these laws had recently been increased from fourteen officers to eighteen, and it was concluded that the Commissioner was doing all that could reasonably be expected. While deploring the state of affairs which led to the

[6] Ibid. 148 per Lord Edmund-Davies.

[7] Ibid. 136; cf. *R.* v. *Rathmines Urban District Council* [1928] IR 260 (SCI); *Belliveau* v. *Legislative Assembly, Nova Scotia* (1978) 31 NSR (2d) 346; *Adams* v. *Metropolitan Police Commissioner* [1980] RTR 289.

[8] See esp. Salmon LJ at [1968] 2 QB 144.

[9] [1973] 1 QB 241; and see *R.* v. *Metropolitan Police Commissioner, ex p. Blackburn*, *The Times*, 7 Mar. 1980, in which the same issue was raised but again the Commissioner was found to have been doing his best.

application, the court blamed the law itself as being ineffective to prevent the widespread sale of hard pornography; not only were there procedural uncertainties, but the statute contained widely drawn defences and inadequate penalties, and was not directed at the persons really responsible.

A more recent case, *R.* v. *Devon and Cornwall Chief Constable, ex p. Central Electricity Generating Board*,[10] illustrates the difficulties involved in taking the extreme step of compelling the police to take specific action. The Board were conducting a survey of possible sites for a nuclear power station. Objectors obstructed the Board's survey of a particular site; they did so peacefully, but in a manner which amounted to criminal contravention of planning laws. The Board sought assistance from the Chief Constable in removing the obstructors from the site, but he refused to act on the ground that the police had no definitive legal mandate to act, and to comply with the Board's request would harm the relationship of his men with the public locally. Accordingly the Board sought mandamus to compel him to remove the obstructors from the site. The Court of Appeal refused mandamus, but set out carefully the legal principles applicable to the position of the police, the Board, and the obstructors, which in effect gave the Chief Constable the definitive legal mandate which he thought was lacking to intervene in the event of a threatened breach of the peace. In spite of the Chief Constable's possibly mistaken view of the law, the Court did not suggest that he had breached his duty, but considered that the decision not to intervene was a policy decision which he was entitled to make. All the court could do in the circumstances was to clarify the law as it did and leave the handling of the matter to the parties, with heavy hints that the police should not refuse to act when necessary:

> Police constables are no one's lackeys; but they do have a duty to preserve the peace no matter how unpopular that may make them with some sections of the community . . .[11]

> the court cannot judge the explosiveness of the situation or deal with the individual problems which will arise as a result of the activities of the obstructors . . . [but] can and does indicate that the time has come for the board and the police to exercise their respective powers so that the survey may be completed.[12]

In *R.* v. *Oxford, ex p. Levey*[13] the judge refused to grant mandamus

[10] [1981] 3 WLR 967.
[11] Ibid. 980, per Lawton LJ.
[12] Ibid. 985, per Templeman LJ.
[13] *The Times*, 18 Dec. 1985.

to compel the Liverpool police to send police cars into the Toxteth area when, in accordance with the Scarman Report, they had adopted a policy of not doing so; this application was brought by one who had had his briefcase stolen from his car by masked robbers in the area concerned. It was shown, however, that the policy adopted had reduced the incidence of crime.

In the result these cases have done much to clarify the duties of the police. It seems clear that the courts will generally be reluctant to compel action except in the case of the adoption of a general and clearly erroneous policy as in the first *Blackburn* case. A factor not mentioned in the cases is the difficulty of deciding what exactly to order the police to do. The best that can be done is to order the proper exercise of discretion on certain implicit principles rather than the taking of specific action. Whether the courts grant or refuse an order is often not crucial, since in either event the crucial decisions will still rest with the police as to what resources to allocate to which task, and whether to act in particular instances.

The cases discussed so far do not tell us whether a law-enforcing authority can be compelled to take *particular* legal proceedings. There is no recorded instance of an application of this kind being granted, with respect to a prosecution, probably mainly because it is open to an aggrieved person to launch a private prosecution.[14] None the less a number of factors suggest that such action is possible: the requirements for standing have been relaxed of late;[15] the proper manner of exercise of the discretion to prosecute is fairly well settled[16] and therefore potentially, at least, justiciable; private prosecution is not in favour and is noticed judicially to be very hazardous;[17] and actions of this kind can clearly succeed against law-enforcing authorities other than the police, as is shown below. There seems indeed to be

[14] This of course is hardly a beneficial remedy; see below n. 17. In *Clyne* v. *Attorney-General* (*Cwlth*) (1984) 55 ALR 92 a decision of the Attorney-General not to prosecute was held to be unreviewable; *Gouriet* v. *Union of Post Office Workers* [1978] AC 435 (HL); *Barton* v. *R.* (1980) 147 CLR 75 (HCA).

[15] See below Ch. 5.

[16] See Williams, 'Discretion in Prosecuting' [1956], *Crim. LR* 222; Wilcox, *The Decision to Prosecute* (1972); Moody and Tombs, *Prosecution in the Public Interest* (1982). The principles involved have also been noticed judicially, e.g. by Visc. Dilhorne in *Smedleys Ltd.* v. *Breed* [1974] AC 839, 855–6 (HL); see also *Buckoke* v. *Greater London Council* [1971] Chanc. 655. Visc. Dilhorne was especially concerned to reject the idea that the decision not to prosecute is a dispensing power.

[17] See *R.* v. *Metropolitan Police Commissioner, ex p. Blackburn*, above n. 5, p. 145 per Salmon LJ, who regarded the remedy as 'fantastically unrealistic' and per Edmund-Davies LJ at p. 149; see also below Ch. 4.8; and the *Report of the Royal Commission on Criminal Procedure*, Cmnd. 8092 (UK), pp. 160–2.

no reason why, in a blatant case of refusal to prosecute in a particular instance for an improper or irrelevant reason, the police could not be compelled to act.

(ii) *Other law-enforcing authorities*

Local authorities in England have recently been under attack in the courts with regard to their duties of law enforcement. In *R.* v. *Braintree District Council, ex p. Willingham*[18] a group of shopkeepers sought mandamus to compel the authority to prosecute certain persons for alleged Sunday trading. At first instance mandamus was granted because the authority had decided not to act merely because they thought it would be both unpopular and expensive to proceed, neither of which reasons were regarded as legitimate. This decision was reversed on appeal,[19] but in a later case in the House of Lords Lord Roskill, commenting on this case *per curiam* (the other Law Lords expressly concurred), indicated the duty of the authority in such circumstances to be as follows: (1) to consider whether there is a prima-facie contravention of the Act; (2) if so, to consider whether it is necessary to take proceedings to secure observance of the Act; but (3) in deciding this latter question, to have regard to the financial consequences of proceeding, for example where a serious or doubtful question of law is involved which may entail a series of expensive appeals.[20] His Lordship conceded, however, that an authority cannot say it will *never* carry out its duty because of the expense involved. To these statements can be added the firm views of Viscount Dilhorne in another *per curiam* statement in the House of Lords. In relation to a prosecution under food and drugs laws of the producers of a can of peas containing a perfectly harmless caterpillar his Lordship said:

> . . . although this Act imposes on the food and drugs authorities the duty of prosecuting for offences . . . it does not say and I would find it surprising if

[18] (1982) 81 LGR 70; see *Hayes* v. *Montreal Nord* [1944] Que. SC 415; *Re Civil Service Association of Alberta and Alberta Human Rights Commission* (1976) 62 DLR (3d) 531; *Turmel* v. *Ottawa Crown Attorney* [1981] 2 FC 593.

[19] It is unfortunate that the Court of Appeal's impromptu judgment in this case has not been reported. See the comments of Lords Roskill and Diplock in *Stoke-on-Trent City Council* v. *B. & Q. (Retail) Ltd.* [1984] 2 All ER 332, 334, 335.

[20] Ibid. 336; cf. *Visy Board Pty. Ltd.* v. *Attorney-General (Cwlth)* (1983) 51 ALR 705, 711, where Woodward J. identified the following factors as relevant: the desirability of proceeding, the precedent set by the prosecution, the likelihood of success, the cost and the availability of staff; and see *R.* v. *Lancashire County Council, ex p. Guyer* (1977) 34 P & CR 264 (no duty to take proceedings to enforce a public right of way, where the landowner might successfully challenge the status of the footpath); *Re Guyer's Application* [1980] 2 All ER 520; *Lynch* v. *Mudgee Shire Council* [1981] 46 LGRA 203.

it had that they must prosecute in every case without regard to whether the public interest will be served by a prosecution.

What this litigation has cost I dread to think. A great deal of the time of the courts has been occupied. I cannot see that any advantage to the general body of consumers has or will result, apart, perhaps, from the exposition of the law.[21]

These observations are salutary as far as they go. They do ignore, however, what seems to be the crucial question. The duty imposed on a regulatory authority to enforce the law is not the same as the duty imposed on the police. The duty of a regulatory authority is to ensure that the object of the statute is met, using, judiciously, the powers conferred on it; the duty of the police is ordinarily to prosecute in every case, subject to well-known exceptions.[22] A regulatory authority generally has a wide variety of remedies open to it, from persuasion and negotiation, through administrative sanctions such as suspension of a licence or service of a notice demanding compliance, to civil proceedings, for example for an order or injunction, and only in the last resort to criminal prosecution.[23] Thus while it is assumed in the case of the police that the public interest is best served by securing convictions, in the case of regulatory authorities the public interest may only be served by taking into account all the circumstances, including other possible methods of enforcement and expense.

An interesting Canadian case shows the difficulties which can be involved in compelling enforcement agencies to act. In *Re Canadians for the Abolition of the Seal Hunt and Minister of Fisheries and the Environment*[24] some citizens (not in fact a body with legal identity) applied for mandamus to compel the Minister to enforce regulations designed to eliminate, so far as possible, cruelty in the hunting of

[21] *Smedleys Ltd.* v. *Breed* [1974] AC 839, 856–7. Similar principles are applied to civil proceedings, for which see *R.* v. *Southampton Port Commissioners* (1870) LR 4 HL 449; *Elwood* v. *Belfast Corporation* (1923) 57 ILJT 138; *Hughes* v. *Henderson* (1963) 42 DLR (2d) 743.

[22] See above n. 17. There may perhaps be some instances where refusal to prosecute could be justified by reasons other than those regarded as accepted exceptions.

[23] See Dickens, 'Discretion in Local Authority Prosecutions' [1970], *Crim. LR* 618. In a survey the author of this article found that English local authorities, being highly selective in their prosecutions, nearly always secure a conviction, which finding, it is submitted, supports the point made in the text. The fact that the method of enforcement is one for the discretion of the authority was seen as a reason for holding, in *Re Cornenki and Township of Tecumseth* (1971) 17 DLR (3d) 655, that enforcement of building regulations could not be compelled by mandamus. In view of the more recent English authorities this decision appears to be mistaken.

[24] [1980] 111 DLR (3d) 333; cf. *Re North Vancouver District and National Harbours Board* [1978] 89 DLR (3d) 704.

seals. The case went off on the question of standing, but on the following facts the judge held that the duty to enforce had not been breached. The authority had eighty-three officers covering an area of 420,000 square miles; in 1979 they took proceedings in forty-four cases, twenty of which were against observers interfering with the hunt; in nineteen cases the hunter's licence was suspended. The judge was able to say that this was a small number of cases, considering the likely number of infractions. There was also evidence that they had refused to allow observers on terms which would enable them to observe the hunt effectively. It was necessary, however, that there should be a deliberate turning of a blind eye or adoption of a policy of non-enforcement before mandamus would issue.

A court has some difficulty in assessing the relative merits of different methods of enforcement. Even so, it seems too restrictive to require a deliberate policy of non-enforcement before the courts will interfere. Where a regulatory statute is enacted it serves a purpose, and while obviously a broad latitude must be given to enforcement agencies as to the resources applied and the method of enforcement, it should not be at all impossible for a court equipped with appropriate evidence (and it is all a matter of evidence) to decide whether the authority is doing its best to fulfil the purpose of the statute or is merely going through the motions of enforcement without any enthusiasm; whether the inaction is deliberate or not is beside the point. In the seal-hunt case the judge was unwilling to balance economics against cruelty; however, it was the statute which did the balancing act, and to allow the authority to get away with inadequate enforcement would be in effect to disturb that balance. On the facts it is difficult to say whether the authority was doing its best, but more evidence should have been required of them with regard to difficulties in enforcing the statute.

(iii) *Enforcement of fiscal laws*

The duty to collect taxes or rates is in many ways analogous to the duty of police and other enforcing authorities to enforce the law. Just as, by failing to prosecute an offender or a particular group of offenders, the enforcing authority is denying all other persons similarly placed their right of equal treatment before the law, so, by failing to collect from a person or group of persons taxes or rates properly due, the collecting authority is denying the same right to all persons similarly placed. In both instances it is understood that the law will not be enforced in every case, and a practice has built up whereby the authority, by invoking 'prosecutorial discretion' or 'extra-statutory concession', assumes the right not to enforce in certain cases which are defined with various degrees of precision, ranging from a general understanding (for example that persons who attempt to commit suicide will not be prosecuted) to detailed rules (for example published rules governing extra-statutory tax concessions). There is of course nothing inherently objectionable in such practices; indeed the law would be lacking in the quality of mercy if they did not exist. The problem is that given this kind of 'dispensing power' there is, in the absence of judicial review, no guarantee that the power will be used for proper and reasonable purposes rather than to discriminate arbitrarily between citizens. The assumption is that laws enacted by the legislature should be enforced to the letter unless there are very pressing reasons of expediency or justice which require the rigours of the law to be relaxed; given this assumption it is imperative that there should exist some objective and effectively enforceable criteria for deciding when the law has been properly mitigated and when its purpose has been flouted; it is suggested that ultimately only judicial review of the performance of the duty to collect can fulfil the purpose of providing such criteria. Strange to say, however, the duties in question have rarely been litigated, and enforcing authorities have tended to take the view that they are not justiciable.

In the case of fiscal authorities the problem of justiciability presents itself in the form of the problem of standing; while it is not unreasonable to say that every citizen has an interest in the law being enforced, it is not immediately obvious that any citizen has a legitimate interest in any other person's tax or rate liability; furthermore the House of Lords in the *National Federation* case[25] has linked the duty to collect taxes with standing to enforce that duty in such a way as to make discussion of the former impossible without a reference to the latter; this discussion will proceed on the basis that the existence and nature of the duty are independent of the question who has standing to enforce the duty, but the reader is referred to Chapter 6 for a discussion of standing in this context.

The facts of the *National Federation* case were as follows. Some 6,000 casual workers in Fleet Street avoided paying income tax by giving false names and addresses on their call slips, with a resulting loss to the Inland Revenue of about £1m. a year. The Inland Revenue Commissioners, after discussions with the casuals' employers and unions, reached an arrangement with them whereby future tax would be assessed but the Revenue would not investigate previous years'

[25] *Inland Revenue Commissioners* v. *National Federation of Self-Employed and Small Businesses Ltd.* [1982] AC 617.

taxes owing. The applicants, who represented the self-employed and small businesses, claiming that their members, by way of contrast, were treated with severity by the Revenue and that the Revenue were acting under pressure from the unions, sought a declaration that this 'amnesty' was unlawful, and mandamus to compel the Revenue to collect taxes due from the casual workers. The application ultimately failed on the question of standing, but it is important for the present purposes to examine the approach taken by the House of Lords to the substantive issue of judicial review of performance of the duty to collect taxes, because the case affords the only judicial guidance on this issue.

The taxing statute merely imposed a duty to 'collect and cause to be collected every part of inland revenue', but the Revenue had also a general power of care and management of income tax, and a power to make special arrangements for collection of income tax from casual workers; thus the duty to collect was expressed in the vaguest terms, as one might expect.

Lord Wilberforce (with whom Lords Fraser and Roskill agreed), though in no doubt that the Revenue could be compelled at the instance of a taxpayer to fulfil its duties towards him,[26] was at pains with regard to persons other than the taxpayer to contrast the duty to assess the rateable value of property with the duty to collect taxes:

> assessments and all information regarding taxpayers' affairs are strictly confidential. There is no list or record of assessments which can be inspected by other taxpayers. Nor is there any common fund of the produce of income tax in which income taxpayers as a whole can be said to have any interest. The produce of income tax, together with that of other inland revenue taxes, is paid into the consolidated fund which is at the disposal of Parliament for any purposes that Parliament thinks fit. . . .
>
> [T]he amount of rates assessed upon ratepayers is ascertainable by the public through the valuation list. The produce of rates goes into a common fund for the benefit of the ratepayers. Thus any ratepayer has an interest, direct and sufficient, in the rates levied upon other ratepayers. . . .[27]

His Lordship went on to say that the Revenue's duties were owed to the Crown, but was not prepared to assert that, in a case of sufficient gravity, the court might not be able to hold that other taxpayers could challenge the acts or abstentions of the Revenue; indeed he stressed the importance of fairness between taxpayers, and said that

[26] Ibid. 632–3; and see *R.* v. *Income Tax Special Commissioners* (1888) 21 QBD 313; *Income Tax Special Commissioners* v. *Linsleys (Est. 1894) Ltd.* [1958] AC 569 (HL).

[27] [1982] AC 632–3; cf. *Arsenal Football Club Ltd.* v. *Ende* [1979] AC 1 (HL).

a sense of unfairness might be the beginning of a recognizable grievance.[28]

Lord Diplock, holding that the applicants had failed to show any breach of duty by the Revenue, took a different approach. Not only did he give the applicants standing, but he rejected Lord Wilberforce's reasons for confining the justiciability of the duty to collect taxes within narrow limits. The argument based on confidentiality he appeared to reject on the grounds that preferential treatment not of particular taxpayers but of a class of taxpayers was in question, and that no problem of confidentiality arose because the taxpayers concerned had not made any returns or provided any information. He also appeared to reject the idea that the Revenue's duties are owed only to the Crown, asserting that the Revenue is responsible to Parliament with regard to efficiency and policy, but to the courts with regard to legality. He also made it clear that he would have granted mandamus if the applicants had been able to show that the Revenue had breached their duty.[29]

Lord Scarman analysed the duty to collect taxes in more detail than any of their Lordships, and his analysis, taken with that of Lord Diplock, seems preferable to that of the other three judges. He specifically rejected the Revenue's argument that the statute did not involve any duty owed to the general body of taxpayers, and also rejected the idea that the duty to collect taxes is owed only to the Crown. He described the duty as one

> owed by the revenue to the general body of the taxpayers to treat taxpayers fairly; to use their discretionary powers so that, subject to the requirements of good management, discrimination between one group of taxpayers and another does not arise; to ensure that there are no favourites and no sacrificial victims. The duty has to be considered as one of several arising within the complex comprised in the care and management of a tax, every part of which it is their duty, if they can, to collect.[30]

Significantly, and in a manner which directly contradicts Lord Wilberforce's refusal to involve the courts in a management exercise, he added:

> I would not be a party to the retreat of the courts from this field of public law merely because the duties imposed upon the revenue are complex and

[28] [1982] AC 632–3; see also *R.* v. *Commissioners of Customs and Excise, ex p. Cook* [1970] 1 WLR 450.

[29] [1982] AC 637.

[30] Ibid. 651; in *Butler* v. *Cobbett* (1709) 86 ER 1023, Holt CJ said that mandamus would lie against the tax commissioners 'if they refused to tax any part'. Contrast *Ex p. Cook*, above n. 28.

call for management decisions in which discretion must play a significant role.[31]

The position as it stands after this case with regard to the justiciability of the duty to enforce fiscal laws is somewhat confused. A general and enforceable duty to treat ratepayers equally has been specifically recognized, and none of their Lordships precluded the possibility of compelling the Revenue to treat taxpayers equally. However, between Lord Wilberforce's view and Lord Scarman's view there lies a yawning chasm which is only slightly diminished by the concessions each makes to the other. It is suggested that the two main pillars of Lord Wilberforce's position have been effectively removed by Lord Scarman and Lord Diplock, at least in so far as those pillars support the non-justiciability in practice of the duty to collect taxes. It is difficult to see how Lord Wilberforce can at one and the same time say that the Revenue's duties are owed to the Crown and that the court can intervene at the instance of a taxpayer in a case of sufficient gravity; indeed if 'a case of sufficient gravity' means a case where a particularly serious breach of duty is established, his analysis is of no assistance in ascertaining precisely what is the legal extent of the duty to collect taxes. Lord Scarman has left the way open for development of this duty in future cases. In spite of the caution of some of the judges in the *National Federation* case it seems inevitable and desirable that the courts should ensure that the individual taxpayer's right not to be subjected to discrimination is observed.

If the rule of law is to be maintained, then enforcement duties must be taken seriously. In spite of the difficulties, the courts will have to look at cases of this kind carefully. There seems to be a tendency to fulminate but refuse a remedy, which is not an adequate judicial response to the problem. Enforcement agencies should be required to justify their disinclination to act by reference to genuine policy choices which are within the intention of the statute. The level of enforcement demanded depends of course on the particular circumstances, but those responsible for enforcing the law should not be allowed to feel that their actions can never in practice be subjected to serious scrutiny.

2.5 CONSTITUTIONAL DUTIES

Constitutional duties are dealt with more fully than can be attempted here in books on constitutional law, and vary in their content and justiciability from jurisdiction to jurisdiction.

[31] [1982] AC 652.

The enforcement of duties of a constitutional nature was one of the earliest and most important functions of the writ of mandamus. Many cases appear in the early reports in which local authorities or officials were compelled to appoint[1] or admit to office[2] one duly chosen or elected, to deliver up records to a successor in office,[3] to hold an election,[4] or to levy a rate.[5]

In more recent times mandamus and other remedies have been granted for similar purposes, for example to hold an election,[6] and to comply with local-authority standing orders.[7]

The dividing line between enforceable constitutional duties and purely political obligations remains obscure. In *South Australia* v. *Commonwealth*[8] thc High Court of Australia held that an agreement between the two governments concerning railway standardization was only enforceable through the political process. The legislature presumably cannot be coerced,[9] but on the other hand a statutory duty on a minister to lay a boundary commission report before the legislature must, it is suggested, be enforceable in the courts.[10]

Those constitutional duties which contain a conventional element are suitable to be enforced through the political process rather than the judicial process, for example the duty of a government which has

1 *Peart* v. *Westgarth* (1765) 97 ER 1007.

2 *R.* v. *Ward* (1730) 94 ER 716; *R.* v. *Bedford Level Corporation* (1805) 102 ER 1323; *R.* v. *Lichfield (Mayor)* [1841] 1 QB 453.

3 *Case of the Sheriff of Nottingham* (1660–1) 82 ER 951; *R.* v. *Street and Stroud* (1722) 88 ER 77; *R.* v. *Clapham* (1751) 95 ER 632; *R.* v. *Christchurch Overseers* (1857) 119 ER 1303. Cf. a modern instance, *R.* v. *Launceston City, ex p. Auditor-General* (1923) 19 Tas. LR 7.

4 *R.* v. *Evesham (Mayor)* (1733) 87 ER 1167; *R.* v. *Cambridge Corporation* (1767) 98 ER 46.

5 *R.* v. *Wilkinton (Churchwardens)* (1729) 94 ER 155; *R.* v. *St Luke's Vestry, Chelsea* (1862) 1 B. & S. 903. For a modern instance, see *R.* v. *Hackney London Borough Council, ex p. Fleming* (1987) 85 LGR 626.

6 See *R.* v. *Kibble, ex p. Bryand* (1907) QSR 31; *R.* v. *Newry Urban District Council* (1909) 43 ILT 172; *R.* v. *Edward (1912) 8 DLR 450; Re Barnes Corporaton, ex p. Hutter* [1933] 1 KB 668; *Tonkin* v. *Brand* [1962] WAR 1; *Cholod* v. *Baker* [1978] 2 SCR 484 (SCC); *Re Maharaj* (1966) 10 WIR 149; *Re Mckay and Minister of Municipal Affairs* (1973) 35 DLR (3d) 627; *Re Peters* (1978) 22 N & PEIR. 54.

7 *R. (McKee)* v. *Belfast Corporation* [1954] NI 122; *R.* v. *Hereford Corporation, ex p. Harrower* [1970] 1 WLR 1424.

8 (1962) 108 CLR 130 (HCA).

9 *Belliveau* v. *Legislative Assembly, Nova Scotia* (1978) 31 NSR (2d) 346. None the less mandamus has issued to amend a by-law: *R.* v. *Manchester Corporation* [1911] 1 KB 560.

10 *R.* v. *Secretary of State for the Home Department, ex p. McWhirter, The Times*, 21 Oct. 1969, in which the substantive point remained undecided, the minister having conceded performance but having also obtained an indemnity from Parliament.

lost the confidence of the legislature to resign, and the duty of a head of state to assent to a bill passed by the legislature, or to summon the legislature. Where enforcement of the duty involves coercion of the head of state, the duty may not be directly enforceable against him at all.[11] The content of these duties is not amenable to any very systematic treatment, because there are no general principles which are applied in deciding whether a constitutional duty has been breached. An extended discussion of this aspect, while no doubt interesting, would not shed much light on the present inquiry and would have to extend far beyond it.

There are of course public duties which are based on fundamental-rights provisions in written constitutions. However, these too will not be dealt with here because the significance of fundamental-rights provisions has already been alluded to above in relation to the implication of duties,[12] because most of such provisions are prohibitive rather than prescriptive in nature, and because little case law has been generated on this aspect of constitutional law. It can, however, be noted that fundamental-rights provisions are capable, exceptionally, of giving rise to enforceable duties as opposed to constitutionally invalid acts; one obvious example is the right to counsel, which imposes a duty on the state to provide legal representation to an accused or an arrested person, a duty which can in an appropriate case be enforced by mandamus.[13] It is possible also that a provision requiring equal protection of the law to be afforded to all citizens could give rise to enforceable duties, in the sense that in order to comply with such a provision a state organ may have to provide some good for members of a number of different groups if it intends to provide it for one group;[14] it is debateable whether equal protection could be satisfied in such a case simply by not providing the good for any group.[15] Of course most duties imposed on the state by written

[11] See below Ch. 3.2(i).

[12] See above Ch. 1.3.

[13] *Lee Mau Seng* v. *Minister of Home Affairs* [1971] 2 MLJ 137, 141; *Ooi Ah Phua* v. *Officer-in- Charge, Criminal Investigations, Kedha/Perlis* [1975] 2 MLJ 198, 201.

[14] In the United States the courts have gone a long way in founding duties to provide particular programmes on constitutional provisions, particularly the equal protection doctrine; see Driver, 'The Judge as Political Pawnbroker: Superintending Structural Change in Public Institutions' (1979), 65 *Va. LR* 43 for a survey of the case law and the extent of judicial supervision. See also Polyviou, *The Equal Protection of the Laws* (1980), pp. 331–40.

[15] In *Bachmann* v. *Government of Manitoba* [1984] 6 WWR 25 the court went so far as to impose a duty on an education authority to provide 100% school transport subsidy for French-speaking pupils, for whom it had originally decided to provide only 35% as against 100% for English-speaking pupils; the court did not leave open the possibility that the authority could validly provide less than 100%, or nothing,

constitutions are contained in directive principles or preambles, and are therefore perhaps properly regarded as duties of imperfect obligation.[16]

2.6 THE DUTY TO DISCLOSE INFORMATION

In Australia the duty to disclose information is dealt with at some length in the Freedom of Information Act 1982 (Cwlth).[1] Elsewhere, however, until the duty to disclose becomes the subject of special legislation along the lines of the Australian Act, we are left with a series of statutory provisions which impose particular duties to disclose on particular authorities for a variety of reasons. These duties are constitutional duties in the sense that most of them exist for a purpose which is particularly associated with the system of government at the national, state, or local level. The law sometimes imposes on a public authority a duty to disclose accounts, minutes, and other documents in the nature of public records[2] so that members of the electorate may be able the better to judge the performance of the authority in question, or raise particular questions for review by the authority or elsewhere. In addition to these statutory duties a common-law duty rests on members of local authorities to be acquainted with such documents as their duties as members demand, and this in turn requires the relevant officers of local authorities to disclose these documents.[3]

Whether a duty to disclose has been breached or not depends of course on whether the document concerned falls within the category of documents which the statute requires to be disclosed,[4] and is therefore purely a matter of statutory interpretation. However, most cases involving a duty to disclose are decided on the issue of standing,

for any group, even though the statute did not impose a duty but only gave a power to provide. It is submitted that the decision is correct, however. In the circumstances the authority had committed itself to provide, and should provide on a non-discriminatory basis rather than be allowed to renege in a fit of pique.

[16] See above Ch. 1.4.

[1] See Bayne, *Freedom of Information* (1984); several articles appearing in (1983–4) 14 *FLR*.

[2] See e.g. Local Government (Access to Information) Act 1985 (England and Wales).

[3] See *R.* v. *Barnes Borough Council, ex p. Conlan* [1938] 3 All ER 226, 230.

[4] See e.g. *Re McAuliffe and Metropolitan Toronto Board of Police Commissioners* (1976) 61 DLR (3d) 223; *Re Simpson and Henderson* (1977) 71 DLR (3d) 24. The first of these cases held, rather oddly, that the Board were under no duty to make available to the public, regulations dealing with the government of the police force.

which is clearly crucial because disclosure, apart from situations where documents are actually open to inspection by any member of the public, must be made to an individual, so that the real issue is whether the individual has a right to demand disclosure; occasionally, however, the issue is described as being either whether there is a duty to disclose at all, or whether the court should in its discretion grant the remedy sought.[5]

The question of standing which arises here is, however, a question which cannot be answered by referring to general principles of standing, which are discussed below,[6] because the courts have not generally referred to the position with regard to standing for mandamus, but have treated standing to compel production of a document as *sui generis*; the subject can therefore be conveniently dealt with here.

Persons who seek disclosure fall into two groups which require separate treatment: members of the public (i.e. ratepayers, electors, residents, and 'public-interest' litigants such as newspaper reporters), who can be called 'outsiders', and members of the authority itself, who can be called 'insiders'.

(i) *The right to disclosure: outsiders*

The main obstacle to securing performance of a duty to disclose is that the burden of proof lies on the applicant to show that he has a genuine interest in the matter.[7] This position at common law has been in effect reversed in Australia by the Freedom of Information Act 1982, section 11. The basic principle of the Act is that every person has a legally enforceable right of access to documents covered

[5] *R.* v. *Godstone Rural District Council* [1911] 2 KB 465, for example, is a case where it is not at all certain which of the three possible ratios indicated in the text was applied. To treat the issue as one of discretion is inadequate because it tends to relegate to a matter of discretion crucial questions of individual rights, and is also confusing in view of the general preference for seeing the issue as one of standing.

[6] See Ch. 6.

[7] *R.* v. *Wiltshire & Berkshire Canal Co.* (1835) 3 A. & E. 477. In *R.* v. *London and St Katharine's Docks Co.* (1874) 44 LJQB 4, Blackburn J. held that mandamus to compel inspection of company accounts should be refused where the applicant gave no reason for his request, even though he was a shareholder and the statute required the accounts to be made available for inspection by shareholders; see also *R.* v. *Clear* (1825) 107 ER 1293; *Ex p. Briggs* (1859) 28 LJQB 272; *R.* v. *Bank of England* (*Governor and Company*) [1891] 1 QB 785; *Bank of Bombay* v. *Suleman Samji* (1908) 99 LT 62 (PC). This position was affirmed by the House of Lords in *City of Birmingham District Council* v. *O* [1983] 1 All ER 497, 504 per Lord Brightman.

by the Act, and it is for the authority concerned to show that the document comes within one of the various statutory exceptions.[8]

As one might expect, the law has shown some affection for ratepayers and electors.[9] However, the English cases have been quite restrictive because of the superadded requirement, which has been applied to insiders as well as outsiders, that since a genuine interest is required, access will be refused if the applicant has an ulterior motive, which has invariably been the motive of gaining an advantage in legal proceedings against the authority concerned.[10] This preoccupation with motives is perhaps unfortunate; if the applicant is able to show that he is within the statutory provision his motive should be regarded as irrelevant, as indeed it is under the Australian Act. It might have been preferable if the English cases had been decided on the basis that disclosure would be a breach of legal professional privilege; this would explain the cases where the applicant sought access to a counsel's opinion obtained by the authority.[11] Otherwise, as where the document concerned is the minutes of the authority's proceedings,[12] the fact that the document is a public record required to be disclosed must override the purely incidental fact that it also benefits a party in litigation against the authority; indeed the motive in seeking disclosure cannot be logically relevant where the applicant is exercising a statutory right of access. Furthermore, it has been held in one case

[8] See ss. 4, 7, 11 and Pt. IV of the Act. For a full discussion of these provisions, see Hotop, *Principles of Australian Administrative Law*, 6th edn. (1985), pp. 406–7.

[9] Even to the extent that in *Re Simpson and Henderson*, above n. 4, access was compelled where the applicant applied as reporter but was given standing on the ground that he was a resident and elector. For ratepayers see *R.* v. *Wimbledon Urban District Council, ex p. Hatton* (1897) 77 LT 599; *R.* v. *Bedwellty Urban District Council, ex p. Price* [1934] 1 KB 333; *Re Ristimaki and Municipal Council of Timmins* (1974) 3 OR (3d) 609. For electors see *R.* v. *Bradford-on-Avon Rural District Council* (1908) 99 LT 89; *R.* v. *Godstone Rural District Council*, above n. 5.

[10] See the *Wimbledon*, *Bradford*, and *Godstone* cases, above n. 9; *R.* v. *Hampstead Borough Council, ex p. Woodward* (1917) 116 LT 213; *R.* v. *Barnes Borough Council, ex p. Conlan*, above n. 3; and cf. also *R.* v. *Staffordshire Justices* (1837) 112 ER 33. In *R.* v. *Southwold Corporation, ex p. Wrightson* (1907) 97 LT 431, 432, Lord Alverstone CJ required not merely a right to see, but also a bona fide ground for seeing. Where the applicant has mixed motives, it would appear that access will be refused if the motive is not solely that of securing the public interest: see the *Hampstead* case at p. 215. The Malaysian courts have granted mandamus to produce at the instance of an accused person documents held by the police: *Gomez* v. *Ketua Polis, Daerah Kuantan* [1977] 2 MLJ 24 (first information report); *Khoo Siew Bee* v. *Ketua Polis, Kuala Lumpur* [1979] 2 MLJ 49 (cautioned statements by accused); cf. *Re MacIntyre and R.* (1980) 110 DLR (3d) 289.

[11] See the *Bradford* and *Godstone* cases, above n. 9; the *Wimbledon* case involved council minutes, so that on the theory advanced here the decision would be wrong.

[12] As in the *Wimbledon* case; see above n. 9.

that production would not be compelled, at the instance of a councillor who was about to appear as a witness in litigation against the council, concerning a demolition order which he felt was too severe, because the applicant must be motivated solely by the public interest;[13] this is tantamount to holding that an ulterior motive can override legitimate public-interest motives where the two are present, and sits uneasily with another decision which makes it clear that the purpose of opposing the policy of the authority is a proper one.[14]

(ii) *The right to disclosure: insiders*

With regard to insiders the problems are more complex, given that the insider has an ill-defined common-law right to see documents necessary for the performance of his duties,[15] and that the exercise of this right is likely to conflict with the presumptively confidential nature of documents which are not available to outsiders. These problems have recently received the attention of the House of Lords in *City of Birmingham District Council* v. *O*.[16] W, a councillor and member of the Council's Housing Committee, in pursuance of her duties in the latter capacity, became concerned about a proposal of the Council's Social Services Committee to approve the adoption of a child in its care. She sought access to the files of the Social Services Department, and the Council, acting on the advice of its solicitor, decided to grant her access. The prospective parent of the child sought judicial review of this decision in the form of an order of prohibition to prevent access being granted. The House of Lords, reversing a split decision of the Court of Appeal, held that since ultimate responsibility for agreeing to the adoption lay with the Council itself, even though it had properly delegated its function in the matter to its Social Services Committee, W as a councillor had an interest in seeing the files; however, their Lordships made it clear that the decision was one for the authority itself, and could only be attacked if it was one to which no reasonable authority could come. It is important to note that the documents concerned in this case were very sensitive and

[13] The *Hampstead* case, above n. 10.

[14] See *R.* v. *Southwold Corporation, ex p. Wrightson* (1907) 97 LT 431.

[15] *R.* v. *Barnes Borough Council, ex p. Conlan*, above n. 3. In the *Southwold* case, above n. 14, Lord Alverstone CJ, although prepared to grant mandamus to compel access where the documents were withheld merely because the councillor seeking access disagreed with the council's policy in relation to a matter, was concerned to restrict the right of access: 'a councillor has no right to a roving commission to go and examine books or documents of a corporation because he is a councillor. Mere curiosity or desire to see and inspect documents is not sufficient' (pp. 431–2).

[16] [1983] 1 All ER 497.

could affect individuals in a particularly intimate way; while this may explain the cautious approach adopted by the House of Lords, it would be unfortunate if the very important duty of an authority to disclose information to its own members were to be watered down by making it always a matter of discretion.[17] It is true that the discretion would be exercised by elected representatives and not by officials, and that the concept of reasonableness enables the court to intervene in a clear case of abuse; none the less it is difficult to refute the proposition that a duty to disclose is not a discretion not to disclose, and the court's proper function in matters of this kind is to decide whether the duty has been breached and order its performance if it has.

The state of the common law with regard to the duty to disclose seems lamentable from many points of view; the rights of both insiders and outsiders are ill-defined and quite limited, and most of the important issues relating to freedom of information have not been discussed in the courts. Fortunately the position has been largely rectified in Australia by the Freedom of Information Act 1982, and no doubt other jurisdictions will follow suit.

[17] In *R.* v. *Clerk to Lancashire Police Committee, ex p. Hook* [1980] 2 All ER 353, a decision on a par with the *Birmingham* case, but with a different result, Lord Denning MR, dissenting, pointed (p. 358) to the difference between the decided cases and those in which the applicant is a person in actual conduct of affairs, and considered that a member has an interest in all affairs of a committee, past and present. The case concerned a sensitive report containing defamatory material. In *R.* v. *Hackney London Borough Council, ex p. Gamper* [1985] 1 WLR 1229, in which the *Birmingham* case was applied, access to meetings was put on the same footing as access to documents; see also *R.* v. *Sheffield City Council, ex p. Chadwick* (1985) 84 LGR 563.

3
Enforcement: Mandamus

3.1 INTRODUCTION: A HISTORICAL PERSPECTIVE

The precise origins of the prerogative writ of mandamus,[1] which is the principal remedy for the enforcement of public duties, are not known and its early history is shrouded in obscurity.[2] It is not possible here to shed much light on this matter, but it will be necessary to make a few assertions concerning the development of the writ in the seventeenth and early eighteenth centuries. The ensuing discussion is broached not for its antiquarian interest, but in order to discover the *raison d'être* of the writ and explain the main principles which governed, and still govern, its availability.

Mandamus is thought to originate with the judgment of Coke CJ in *Bagg*'s case in 1615. A chief burgess of Plymouth had been removed from office by the mayor and chief burgesses. At his suit the King's Bench issued a writ commanding them to restore him to office or signify the cause of his removal. The mayor's return to the writ being judged insufficient, a writ of restitution was awarded, restoring Bagg to office. Coke based his authority for the writ on a principle of majestic width:

> . . . to this court of King's Bench belongs authority, not only to correct errors in judicial proceedings, but other errors and misdemeanours extra-judicial, tending to the breach of the peace, or oppression of the subjects, or to the raising of faction, controversy, debate, or to any manner of misgovernment;

[1] The term 'prerogative writ' is strictly incorrect, at least in England, where the prerogative writs were replaced by prerogative orders: Administration of Justice (Miscellaneous Provisions) Act 1938, s. 7; see now Supreme Court Act 1981, s. 29(1). However, the word 'writ' is used throughout this discussion, because it is still in common use, is not necessarily incorrect or less appropriate in other jurisdictions, and is less likely to cause confusion.

[2] The only detailed history is contained in Henderson, *Foundations of English Administrative Law* (1963), chs. 2, 4; but see also De Smith, *Judicial Review of Administrative Action*, 4th edn. by Evans (1980), pp. 591–4; Jenks, 'The Prerogative Writs in English Law' (1923), 32 *Yale LJ* 523; and a particularly interesting article by Howell, 'An Historical Account of the Rise and Fall of Mandamus' (1985), 15 *VUWLR* 127.

so that no wrong or injury, either public or private, can be done but that it shall be (here) reformed or punished by due course of law.[3]

Legal scholars both ancient and modern have attempted to discover authorities for Coke's decision in *Bagg*'s case, but the truth of the matter seems to be that Coke sought to assert the supremacy of the law and to establish a principle of constitutional importance[4] rather than delve into the Year Books for authority, though such authority could have been found.[5] The theory implicit in this principle seems to be that the Crown always has power to give justice to its subjects, which power it has delegated to the court of King's Bench, so that the court can invoke this inherent jurisdiction to correct any wrong where the complainant has no other remedy at law.[6] For his boldness Coke earned the sharp rebuke 'he hath as much as insinuated that this Court is all-sufficient in itself to manage the state' from one of his contemporaries.[7] However, in later cases authority was supplied to support the new writ's validity and Coke's somewhat unmanageable principle was moulded into 'one of the few effective instruments of public policy in the era between the abolition of the Star Chamber in 1640 and the creation of the modern system of local government in the nineteenth century'.[8]

Two types of development can be noted: a tendency for the writ to be applied to novel situations by the familiar process of analogical reasoning and purposive generalization; and a tendency to confine the availability of the writ, within those situations, to cases of evident need and expediency.

With regard to the former tendency, mandamus was held to lie to compel admission to an office of one duly elected;[9] to compel the

[3] (1615) 77 ER 1271, 1277.

[4] Cf. *Patrick*'s case (1665–7) 83 ER 54, 56, where counsel went so far as to say that 'the judges finding the mischief, did out of the foundation of law frame that writ'.

[5] See *Middleton*'s case (1574) 73 ER 752. Henderson, above n. 2, pp. 58–76, shows that there was considerable authority of recent origin, and notes that early cases seem to have been connected with the notion of privilege. Bowen LJ even sought the origins of mandamus in Magna Carta: *Re Nathan* (1884) 12 QBD 461, 478; while Lord Mansfield was content with a case of Edward II's reign: see *R.* v. *Askew* (1768) 98 ER 139; both would appear to have been guilty of wishful thinking.

[6] This theory can be gleaned from the passage quoted and also the theory expressed in *Co. Inst.* iv. 71.

[7] *Observations on the Lord Coke's Reports* (1710?), ed. G. Paul, 11, sometimes attributed to Lord Ellesmere.

[8] Wade, *Administrative Law*, 5th edn. (1983), p. 630.

[9] *Case of South Balaunce* (*Parish*) (1619) 81 ER 973.

giving of judgment in a case properly brought;[10] to compel a probate court to appoint an executor named in the will;[11] to compel scavengers in London to perform their statutory duties;[12] to compel a sheriff to deliver his records to his successor;[13] to compel the entering of judgment as to the value of land acquired for public purposes in favour of one entitled to it;[14] to compel a statutory tribunal to exercise its jurisdiction;[15] to admit a scholar to his scholarship;[16] and to compel an election for office.[17]

With regard to the latter tendency, it was quickly realized that mandamus could not be granted in every case where a litigant was able to establish a grievance. The association of mandamus with the prerogative writs[18] had the convenient consequence that its award was discretionary, the prerogative writs being writs of grace, not of right. Thus mandamus would not be granted if it were unnecessary, ineffectual, or would lead to double vexation,[19] nor if there was another remedy available to the litigant, notably a right of appeal.[20] More importantly, perhaps, it was neither feasible, nor even proper, for the courts to sit on appeal from every questionable exercise of administrative discretion. In a group of rating cases it was held that mandamus would not lie to compel the making of an equal rate by parish officers, because they were the 'proper judges'[21] of the matter, and to allow mandamus to go in such cases would be 'of very bad

[10] *R.* v. *London* (*Mayor*), *ex rel. Crispe* (*1626–7*), unreported, see Henderson, above n. 2, p. 81.

[11] *Luskins* v. *Carver* (1646) 82 ER 488, in which the word 'mandamus' was first used.

[12] *Anon.* (1652) 82 ER 765.

[13] *Case of the Sheriff of Nottingham* (1660–1) 82 ER 951.

[14] *Amherst*'s case (1671–2) 83 ER 112.

[15] *Groenvelt* v. *Burwell* (1691) 91 ER 1202, 1212 (*obiter*).

[16] *R.* v. *St John's College, Oxford* (1694) 90 ER 1141.

[17] *R.* v. *Evesham* (*Mayor*) (1733) 87 ER 1167.

[18] The title of mandamus to be ranked with the other prerogative writs is somewhat doubtful: see De Smith and Jenks, both above n. 2. Jenks considers that the idea was invented by Mansfield and Blackstone, but in fact mandamus was called a prerogative writ much earlier (*Knipe* v. *Edwin* (1694) 87 ER 394, 395), and the analogies with *certiorari* and prohibition (see Henderson, above n. 2, chs. 3, 4) are so close and of such early origin, and the jurisdiction to award mandamus so hard to explain otherwise, that it seems churlish to deny its pedigree, even if the whole thing was accidental. The matter is now of course of purely antiquarian interest.

[19] *R.* v. *Heathcote* (1713) 88 ER 620.

[20] *R.* v. *Shepton Mallet Overseers* (1698) 87 ER 742; *Wilkins* v. *Mitchel* (1698) 91 ER 1129; *R.* v. *Baines* (1706) 92 ER 332; *Butler* v. *Cobbett* (1709) 88 ER 1023.

[21] *R.* v. *Weobly* (*Churchwardens*) (1746) 93 ER 1167.

consequence'.[22] Similarly mandamus was refused in cases where the applicant appeared to be trying to compel the exercise of discretion in his favour rather than generally according to law.[23] Thus the modern doctrine of limited judicial review is foreshadowed in these early mandamus cases. No doubt the judges were aware of the constitutional obstacles which mandamus might encounter as well as the state of the court lists.

However, this period was the heyday of mandamus and Lord Mansfield was able to sum up the case law confidently in *R.* v. *Barker* in 1762:

> A mandamus is a prerogative writ; to the aid of which the subject is intitled, upon a proper cause previously shown, to the satisfaction of the court. The original nature of the writ, and the end for which it was framed, direct upon what occasions it should be used. It was introduced, to prevent disorder from a failure of justice, and defect of police.
>
> Therefore it ought to be used upon all occasions where the law has established no specific remedy, and where in justice and good government there ought to be one. Within the last century it has been liberally interposed for the benefit of the subject and the advancement of justice. The value of the matter, or the degree of its importance to the public police, is not scrupulously weighed. If there be a right, and no other specific remedy this should not be denied.[24]

By this time all the important principles governing mandamus had been developed. Three features of mandamus are important for its future use. First, it was pre-eminently a public-law remedy and was not generally available to remedy private wrongs;[25] secondly, it was never confined only, or even mainly, to judicial functions, but applied to all kinds of administrative offices and duties; thirdly, it was not

[22] *R.* v. *Freshford* (*Churchwardens*) (1737) 95 ER 281; see also *R.* v. *Barnstaple* (*Inhabitants*) (1728) 94 ER 95. Similarly, in *R.* v. *Lichfield* (*Bishop*) (1734) 87 ER 1200, it was material whether the Bishop was acting ministerially or judicially—if the latter, then the matter was for him to 'determine one way or the other' (Lord Harwicke CJ).

[23] *Anon.* (1730) 94 ER 271; *R.* v. *Surrey Justices* (1734) 94 ER 586; the *Weobly* case, above, n. 21. In a number of cases concerning academic institutions mandamus was considered not to lie because of a right of appeal to the visitor; see Henderson, above n. 2, pp. 131–7, Wade, above n. 8, p. 631, and *R.* v. *University of Cambridge* (1723) 93 ER 698 (*Bentley*'s case).

[24] (1762) 97 ER 823, 824. See also *R.* v. *Blooer* (1760) 97 ER 697.

[25] *London*'s case (1658) 82 ER 1285 (Glyn CJ); *Middleton*'s case (1663) 82 ER 1037; *R.* v. *Wheeler* (1735) 94 ER 1123. In *R.* v. *Oxenden* (1691) 90 ER 1139, Holt CJ said: 'a mandamus lies for all offices of a publick nature, or related to the administration of justice'; cf. *R.* v. *Croydon* (*Churchwardens*) (1794) 101 ER 396. There are of course exceptions, of which *Barker*'s case itself is one.

tied to the notion of a statutory duty, but was created to put justice into effect, so that it applied to wrongful exercise of powers as well as refusal to perform duties.

Thus mandamus was admirably equipped to serve the needs of judicial control of the myriad local agencies which governed England for two hundred years before the growth of the modern administrative state and the system of local government. As Holdsworth reminds us, during this period there was a distinct lack of the modern techniques of administrative control, and the administrative separateness of these local bodies meant that judicial control was in fact the only control mechanism available, not just for the citizen, but for the central government too.[26] In a period of administrative diversity and disorder, mandamus was therefore essential for the carrying on of effective government, and if it had not existed, something very much like it would have had to be invented.

The growth of administrative control and the development of rational, responsible, and professionalized administrative practices, the increased provision of rights of appeal and other avenues of redress, and the decline of the concept of a 'freehold' public office,[27] naturally led to a decline in the importance of mandamus in the latter part of the nineteenth century. During the present renaissance of administrative law, mandamus has re-emerged, not simply as the principal means of enforcing public duties and the object of scrutiny in a number of important and controversial cases, but as an organic part of the modern system of judicial review of administrative action, namely that part which specifically compels tribunals and public authorities to comply with the substantive and procedural canons of the system.

3.2 THE SCOPE OF MANDAMUS

The public duties which are enforceable by mandamus have already been indicated in Chapter 1. The courts have not in fact always been entirely consistent in mandamus cases in awarding the remedy only against public authorities and in a number of nineteenth-century cases mandamus issued against officers of what would nowadays be thought

[26] Holdsworth, *A History of English Law*, 4th edn. (1936), pp. 155–7.

[27] See Henderson, above n. 2, pp. 76–80. Most of the early cases involve restoration to public office.

of as private bodies.[1] The modern tendency to distinguish clearly between public law and private law for the purpose of statutory judicial review[2] will ensure that these cases are not used to revive 'private-law mandamus'. Thus the scope of mandamus can be said to be determined by the nature of the duty to be enforced rather than the identity of the authority against whom it is sought. As Aronson and Franklin have it:

> Any incumbent of a public duty (even if that incumbent is in all other respects a private person) can be a respondent to a mandamus. In a sense, an incumbent of a public duty is *pro tanto* an official [amenable to mandamus]—there is no need to look for further evidence of the official status of his position.[3]

There are however three exceptions, or possible exceptions, which restrict the operation of this reasoning: mandamus is said not to lie against the Crown, against minor officials, and against superior courts of record.

(i) *The Crown*[4]

As we have seen, according to the original theory of mandamus set out by Coke, the Crown is the very *fons et origo* of mandamus. The

[1] See above Ch. 1.3. Although the courts often insisted that the duty must be a public one, even to the extent that the public nature of the duty must be apparent on the face of the writ (*R.* v. *Bank of England* (1819) 106 ER 492; *R.* v. *London Assurance Co.* (1822) 106 ER 1420; *R.* v. *Hopkins* (1841) to LJQB 63; *Holland* v. *Dickson* (1888) 37 Chanc. D. 669; cf. *R.* (*Butler*) v. *Navan Urban District Council* [1926] IR 92, 466 (SCI)), in practice the concept of a public duty was quite wide. This was no doubt due to the fact that many legally private companies in fact performed public duties; thus mandamus was granted at the instance of shareholders to compel access to company accounts or minutes: *R.* v. *Abrahams* (1843) 114 ER 857; *R.* v. *London & St Katherine's Docks Co.* (1874) 44 LJQB 4; *In re Burton and the Saddler's Co.* (1861) 31 LJQB 62; cf. *R.* v. *Severn and Wye Railway Co.* (1819) 106 ER 501. Canadian cases have shown a willingness to extend the scope of mandamus to public duties owed by private persons to public authorities; see e.g. *Re Corner Brook and Goodyear & House Ltd.* (1966) 53 MPR 305; *Smythe and Humbert* v. *Anderson* (1970) 73 WWR 536. This is not a tendency to be encouraged because it would tend to create confusion in an already confused area of the law, namely the remedies available to public authorities to enforce the law against private persons, and would also tend to blurr the distinction between public and private law.

[2] See below Ch. 5.1.

[3] *Review of Administrative Action* (1987), p. 490. With regard to Australian federal officials, the scope of mandamus is determined by the term 'officer of the Commonwealth' under the Constitution (Cwlth), s. 75(v); see ibid. 500–3, for the case law in so far as it affects the law of mandamus.

[4] The best discussions of this topic are in Aronson and Franklin, above n. 3, and *Mitchell* (*G. H.*) *& Sons* (*Australia*) *Pty. Ltd.* v. *Minister of Works* (1974) 8 SASR

Courts were therefore constrained to hold that mandamus would not lie against the Crown:

That there can be no mandamus to the Sovereign there can be no doubt, both because there would be incongruity in the Queen commanding herself to do an act, and also because disobedience to a writ of mandamus is to be enforced by attachment.[5]

Not only has this theory persisted; it has been extended also to servants of the Crown on the basis that 'where an obligation is cast upon the principal and not upon the servant, we cannot enforce it against the servant as long as he is merely acting as a servant'.[6] If it were taken seriously, it would mean wholesale governmental immunity from mandamus.[7] Fortunately, however, a third theory has been relied upon to restrict the immunity to instances in which a Crown servant acts in that capacity:

Mandamus will not lie against the Crown or aganist a Minister of the Crown when acting purely as a servant of the Crown, but it may go against [the

7, 26–30 (Zelling J.). See also De Smith, *Judicial Review of Administrative Action*, 4th edn. by Evans (1980), pp. 553–6; Wade, *Administrative Law*, 5th edn. (1983), pp. 645–7; Lucas, 'The Immunity of the Crown from Mandamus' (1909), *LQR* 290; Street, *Governmental Liability* (1953, repr. 1975), pp. 135–40; Hogg, *Liability of the Crown* (1971), pp. 12–15.

[5] *R.* v. *Powell* (1841) 113 ER 1166, per Lord Denman CJ; 'the thing is out of the question' said Cockburn CJ in *R.* v. *Treasury Lords Commissioners* (1872) LR 7 QB 387, 394. Of course similar considerations apply to the prerogative writs of *certiorari* and prohibition, but since these writs are relevant only to determinations, the immunity of the Crown is less serious; see *R.* v. *Minister of Health* [1939] 1 KB 232; *Border Cities Press Club* v. *Attorney-General of Ontario* [1955] 1 DLR 404; *Re Gooliah and Minister of Citizenship and Immigration* (1967) 63 DLR (2d) 224; *Banks* v. *Transport Regulation Co. Board* (1968) 119 CLR 222 (HCA); *FAI Insurances Ltd.* v. *Winneke* (1982) 56 ALJR 388, 391 (HCA). With regard to attachment, Lord Denman's argument is surely untenable. As De Smith says (above n. 4, p. 554), a disobedient Crown could be dealt with differently; the same considerations have not been argued in relation to declarations.

[6] (1872) LR QB 398 per Blackburn J., followed in *Armytage* v. *Wilkinson* (1878) 3 App. Cas. 355, 367 (PC). Cf. Coleridge J. in *Re De Bode* (*Baron*) (1838) 6 Dowl. 776, 792: '. . . against the servants of the Crown, as such, and merely to enforce the satisfaction of claims upon the crown, it is an established rule that mandamus will not lie.'

[7] '[W]ere the Crown immunity principle honoured in spirit as well as to the letter, it would virtually confine mandamus to the area of local government', Aronson, and Franklin, above n. 3, pp. 492–3. Lord Parker CJ observed in *R.* v. *Commissioners of Customs and Excise, ex p. Cook* [1970] 1 All ER 1068, 1072 that the principle that mandamus will not lie against the Crown or an officer or servant of the Crown was 'in this day and age as a general proposition quite untrue'.

Minister] where he is charged with the performance of some statutory duty and the applicant for mandamus is entitled to have an act done in the discharge of that duty without which he cannot enforce or enjoy some right which he possesses. As it is sometimes put, mandamus will lie against a Minister when he is acting, not simply under a duty to the Crown as its servant, but as a *persona designata*.[8]

None the less this particular species of Crown immunity, limited as it is, has caused considerable injustice, because a wide definition of 'Crown servant' has been adopted, and sometimes because the duty concerned is said to be owed to the Crown, not to the public generally, or to the particular applicant.[9] The most famous instance is *R.* v. *Secretary of State for War*,[10] in which mandamus was refused to compel the Secretary of State to make an addition to the applicant's pension due under royal warrant. Many of the cases falling into this category were decided in the nineteenth century, but some are surprisingly recent. In one case, mandamus was refused to compel payment of a pension by a workmen's compensation board because the funds it

[8] *Mitchell's* case, above n. 4, p. 14 per Bray CJ; followed in *Re Federal Commissioner of Taxation, ex p. Just Jeans Pty. Ltd.* (1986) 65 ALR 147.

[9] The argument has pre-empted judicial review in many cases where the substance of the argument might have been doubtful, but the applicant was in effect precluded from making out a case, and it is in this that the real injustice lies. See *R.* v. *Customs Commissioners* (1836) 111 ER 1209; *Ex p. Pering* (1836) 111 ER 1040; *Re De Bode (Baron)* (1838) 6 Dowl. 776; *Ex p. Napier* (1852) 118 ER 261; *Ex p. McKenzie* (1867) 6 SCR (NSW) 306; *R.* v. *Treasury Lords Commissioners*, above n. 5; *Ex p. Cox* (1876) 14 SCR (NSW) 287; *Re Nathan* (1884) 12 QBD 461; *Re Massey Manufacturing Co.* (1886) 11 OR 444; *McQueen* v. *R.* (1887) 16 SCR 1; *Awatere Road Board* v. *Colonial Treasurer* (1887) 5 NZLR 372; *R.* v. *Secretary of State for War* [1891] 2 QB 326; *R.* v. *Arndel* (1906) 3 CLR 557 (HCA); *Re Carey and Western Canada Liquor Co. Ltd.* (1920) 3 WWR 329; *Gartley* v. *Workmen's Compensation Board* (1942) 57 BCR 217; *R.* v. *Warnock* [1946] NI 171; *Ex p. Cornford, re Minister for Education* [1962] 62 SR (NSW) 220; *Kariapper* v. *Wijensinha* [1968] AC 717 (PC); *Re Lofstrom and Murphy* (1972) 22 DLR (3d) 120; *Re Central Canada Potash Co. Ltd. and Minister of Mineral Resources of Saskatchewan* (1972) 32 DLR (3d) 107 (affd. (1973) 38 DLR (3d) 317 n (SCC)); *Re Le Blanc and Board of Education of Saskatoon East* (1980) 117 DLR (3d) 600; *Alberta Mortgage and Housing Corporation* v. *Hindmarsh* (1987) 77 AR 263. Perhaps the most extraordinary case is *Ex p. Mackenzie*, in which money was appropriated by the legislature to compensate a man who had been unjustly convicted; his estate was unable to obtain mandamus to compel the Treasurer to pay the money over to it. In *Ex p. Cox* the applicant took possession of Crown land, paid rent, and was told a lease had been prepared and would be issued when executed; later the lease was cancelled without notice or reason and another lessee was put in possession, and yet mandamus would not lie to compel the issue of the lease.

[10] Above n. 9; see, however, *Council of Civil Service Unions* v. *Minister for the Civil Service* [1984] 3 All ER 935, 943, where the reasoning in this case was doubted.

held were Crown funds;[11] in another, mandamus was refused to compel a Minister to release on grounds of ill-health a prisoner on hunger strike because no writ of mandamus could issue against a Minister of the Crown.[12] A notable feature of the cases in which mandamus has been refused against Crown servants is that most of them concern claims for pensions and the like out of the general revenue,[13] and this may explain, partly at least, the reason for the persistence of the manifestly unjust immunity of the Crown from mandamus. The doctrine which emerged from a series of cases against the Treasury in England in the nineteenth century was that mandamus would only lie in respect of funds specifically voted by Parliament and held for the particular purpose in question;[14] it may therefore be that the judges were reluctant to fetter the expenditure of the revenue except in the face of clear parliamentary intention.[15]

[11] *Gartley*'s case, above n. 9. See, however, *R.*, *ex rel. Lee* v. *Workmen's Compensation Board* [1942] 2 DLR 665, where mandamus issued against the Board to continue paying a pension.

[12] *R.* v. *Warnock*, above n. 9. This decision is clearly against the trend of the authorities, for the Minister's functions were statutory.

[13] e.g. all the 19th-cent. cases and also *Carey*, *Gartley*, *Kariapper*, and *Lofstrom*, above n. 9.

[14] *R.* v. *Treasury Lords Commissioners* (1835) 111 ER 794; *R.* v. *same*, *re Hand* (1836) 111 ER 1053; *Re De Bode* (*Baron*), above n. 9; *R.* v. *Treasury Lords Commissioners*, *ex p. Brougham and Vaux* (*Lord*) (1851) 20 LJQB 305; *Re same*, *ex p. Walmsley* (1861) 121 ER 644; *R.* v. *same* (1872), above n. 5; *R.* v. *same* [1909] 2 KB 183. Similar principles were applied in cases against the Revenue for recovery of taxes paid: *R.* v. *Income Tax Special Purposes Commissioners* (1888) 21 QBD 313; *Income Tax Special Purposes Commissioners* v. *Pemsel* [1891] AC 531 (HL); *same* v. *Linsley's Ltd.* [1958] AC 569 (HL). The vigorous decision of Lord Denman CJ in the 1835 case, if followed and developed, could have avoided the problem, but it was distinguished in the 1836 case and criticized in the 1872 case. See also Campbell, 'Private Claims on Public Funds' (1969), *UTLR* 138; Whitmore and Aronson, above n. 3, pp. 365–6, esp. n. 90. The same principle has been followed in Australia in *Ex p. Krefft* (1876) 14 SCR (NSW) 446, followed in *Ex p. Miller*, *re NSW Ambulance Transport Service Board* (1950) 67 WN (NSW) 17. In this latter case Owen J.'s judgment seems to suggest that mandamus will not lie even to enforce equality of treatment by an authority of district authorities within its jurisdiction, merely because the statute did not require allocation of funds to any particular purpose or to any particular authority; however, he did not expressly decide the point. See also *R.* v. *Dickson*, *ex p. Barnes* [1947] Qld SR 133.

[15] See the arguments of Jessel S.-G. (as he then was) in *R.* v. *Treasury Lords Commissioners*, above n. 5, pp. 389–90: 'Where the legislature has constituted the Lords of the Treasury agents to do a particular act, in that case a mandamus might lie against them as mere individuals designated to do that act; but in the present case, the money is in the hands of the Crown or of the Lords of the Treasury as Ministers of the Crown; in no case can the Crown be sued even by writ of right. If the Court granted a mandamus they would be interfering with the distribution of public money; for the applicants do not show that the money is in the hands of the

Unfortunately this reasoning seems to have been carried over into areas in which it is wholly inapplicable. In *Ex p. Cornford, re Minister for Education*[16] the parents of some boys were told they would have to attend a different school after the Ministry had redrawn the boundaries of the catchment area, and applied for mandamus to compel the Minister to provide places at the original school. The case was approached on the basis that

> [F]irstly, a public duty must rest upon the Minister, the responsibility for the discharge of which is upon the Minister personally and not upon the Crown; secondly the applicant must be entitled to have an act done in the discharge of such a duty without the doing of which he cannot enforce or enjoy some right which he possesses. The presence of these conditions must appear from the statute relied upon for the establishment of the duty and the right.[17]

In spite of the fact that the relevant statutory duty, if it could be clearly established, was imposed on the Minister, it was held that the duty lay on the Crown and was not personal to the Minister, and that the parents had no right of selection of the school the boys were to attend. Similarly, in *Re Central Canada Potash Co. Ltd. and Minister of Mineral Resources of Saskatchewan*,[18] the Minister was designated by a statute, whose purpose was to secure proper utilization and conservation of minerals, as the authority responsible for carrying out the purposes and policies of the statute. When he refused a licence to the company without taking into account what the company alleged to be relevant considerations, mandamus was refused because the Minister was said to be the representative of the Crown. The general trend, however, has been to hold that a Minister acting under statutory provisions is a *persona designata*[19] and 'is called upon to do

Lords of the Treasury to be dealt with in a particular manner. . . . The effect of the annual Appropriation Act is not to give any third person a right to the money; but it is to prevent the Crown from appropriating money given for one purpose to another.' Campbell, above n. 14, p. 157, summarizes the case law, correctly it is suggested, as follows: 'mandamus may lie to compel payment from monies of the Crown when by or pursuant to legislation a public duty is imposed on the defendant to pay and the applicant has a right to payment. For the remedy to lie, payment of the money claimed must have been authorized by parliamentary appropriation, and by such officer or officers whose sanction or approval is required before the monies may be lawfully paid.'

16 (1962) 62 SR (NSW) 220.

17 Ibid. 223–4.

18 Above n. 9.

19 *In re Sooka Nand Verma* (1905) 7 WALR 225; *R.* v. *Secretary for Public Lands, ex p. Moore* [1908] Qld SR 55; *R.* v. *Watt, ex p. Slade* [1912] VLR 225; *R.* v. *Minister of Health, ex p. Rush* [1922] 2 KB 28; *Minister of Finance of British Columbia* v. *R.* (1935) 2 DLR 316 (SCC); *Re Bukit Sembawang Rubber Co. Ltd.* [1961] MLJ 269; *Bradley* v.

the act, not in his capacity of Minister, but as filling the position which the statute has conferred upon him';[20] it is suggested that the two cases referred to above are anomalies which could and should have been decided on other more relevant grounds concealed by the identification of the Minister with the Crown.

Where the *persona designata* is not a Minister, the argument that the official is a Crown servant is rarely raised and is, in modern cases, rarely successful.[21]

In a number of statutes the incumbent of the duty is the representative of the Crown, and in Australia mandamus applications have caused some difficulty in this regard, because the Governors of the States are often entrusted with statutory or constitutional powers and duties.

The immunity from mandamus of a representative of the Crown was established in *R.* v. *Governor of South Australia*.[22] The seats of three senators for South Australia became vacant under the Constitution. Elections were held, but the election of one candidate was held by the Court of Disputed Returns to be void. An elector applied for mandamus to compel the Governor to cause a writ to be issued for the election of a senator to fill the vacancy, arguing that Parliament had no power to make the choice, which was therefore a nullity, so that a new election was required. The High Court of Australia dealt with the argument that mandamus would not lie against the Governor as follows. When causing a writ to be issued, the Governor was acting as a constitutional Head of State, and if he failed in his duty, the duty was one owed to the State collectively, a duty of imperfect

Commonwealth (1973) 128 CLR 557 (HCA); *Mitchell*'s case, above n. 4. In England the point does not seem to have required decision: see e.g. *Padfield* v. *Minister of Agriculture* [1968] AC 997 (HL); *R.* v. *Secretary of State for the Environment, ex p. Bilton (Percy) Industrial Properties Ltd.* (1976) 31 P & CR 154; *R.* v. *Secretary of State for the Home Department, ex p. Phansopkar* [1976] QB 606. In the Canadian cases of *Minister of Finance* v. *R., Re Dumont and Commissioner of Provincial Police* [1940] 4 DLR 721 (affd. [1941] SCR 317 (SCC)), *R., ex rel. Lee* v. *Workmen's Compensation Board* (above n. 11), and *Central Canada Potash* (above n. 9), the position reached, based on *R.* v. *Treasury Lords Commissioners* (above n. 5), seems to be that mandamus will only issue against the Minister if he acts as the agent of the legislature, i.e. without discretion. This notion is clearly wrong. See also *Mitchell*'s case, above n. 4, p. 30, where Zelling J. criticizes this approach; and Aronson and Franklin, above n. 3, pp. 493–6.

[20] *In re Sooka Nand Verma*, above n. 19, p. 230.

[21] See the cases cited above n. 14, and also *R.* v. *Leong Ba Chai* [1954] SCR 10 (SCC), in which an immigration officer was held to be a *persona designata* rather than a Crown servant acting as such, his duties being statutory.

[22] [1907] 4 CLR 1497 (HCA); see also *Horvitz* v. *Connor* (1908) 6 CLR 38 (HCA). In *Re Maharaj* (1966) 10 WIR 149, however, the argument from immunity was not raised in relation to the Governor-General of Trinidad and Tobago.

obligation which could be reviewed only by intervention of the sovereign and not by the courts on an application for mandamus. Although the Governor was not acting here as an agent for the sovereign, the duty being imposed by the Constitution and not by delegation, it was none the less imposed upon him as Head of State and his position was therefore the same as the sovereign's. This casc was followed in *Ex p. McWilliam*,[23] in which an applicant for a Crown grant was refused mandamus, directed to the Governor of New South Wales, to issue the grant when the applicant had fulfilled all the statutory requirements, the duty was ministerial, and conditions attached on the grant were held to be unlawful.

Against these decisions one can set others made in rclation to different remedies and circumstances, in which the immunity of the Crown has been whittled down.[24] The underlying principle in these cases was well put by Mason J. in *Re Toohey*,[25] a case which concerned the reviewability of a statutory discretion vested in the Administrator of the Northern Territory, who was assumed to be the representative of the Crown:

> The foundations of the old rule [of Crown immunity] have been undermined. Procedural reforms have overcome the Sovereign's immunity from suit which in turn was the source of the principle that the king can do no wrong. Appropriate as it is that this principle should apply to personal acts of the Sovereign, it is at least questionable whether it should now apply to acts affecting the rights of citizens which, though undertaken in the name of the Sovereign or his representative, are in reality decisions of the executive government. In the exercise of the prerogative as in other matters the Sovereign and her representatives act in accordance with the advice of her Ministers. This has been one of the important elements in our constitutional development. The continued application of the Crown immunity rule to the exercise of prerogative power is a legal fiction.[26]

The High Court of Australia held in this case that the Administrator's discretion would be reviewable, but the decision does not affect non-statutory exercises of prerogative power by the government, or the exercise of powers by the Head of State personally. The High Court also expressly left intact the law established in *R.* v. *Governor of South Australia*.[27]

[23] (1947) 47 SR (NSW) 401.

[24] See e.g. *Conway* v. *Rimmer* [1968] AC 910 (HL); *Laker Airways Ltd.* v. *Department of Trade* [1977] 2 All ER 182; *Sankey* v. *Whitlam* (1978) 142 CLR 1 (HCA); *Re Toohey (Aboriginal Land Commissioner), ex p. Northern Land Council* (1981) 56 ALJR 164 (HCA).

[25] Above n. 24.

[26] Ibid. 184.

[27] Ibid. 170

Is it then only weight of authority which stands between the judges and abolition of Crown immunity from mandamus?

The first point is that the immunity of the Crown from prerogative relief does not spring from the same source as Crown immunity from suit in the general sense. It has been held that a declaration is available against the Crown, and thus the unavailability of mandamus may not be as restrictive as it might appear: a declaration is nearly always available instead.[28] This very fact, however, makes the overturning of authorities on mandamus rather difficult, because it pre-empts arguments by analogy.

Is there perhaps some good reason for maintaining Crown immunity from mandamus? Three reasons have been put forward for Crown immunity, though none of them has been mentioned in relation to mandamus[29] (and in fact no theory of mandamus has ever sought to improve on Lord Denman CJ's dictum of 1841 cited above):

1. Judges must not interfere in the administrative process.[30]
2. The secrecy of governmental deliberations.[31]
3. The doctrine of ministerial responsibility.[32]

All these reasons for Crown immunity have long since been exploded and the courts have, by degrees, reformed the law accordingly.[33] However, let us put the case of mandamus at its highest. It might be argued that mandamus stands in a different case from other remedies because it actually compels particular action to be taken, and the Crown should not be fettered in this way, particularly in the case of emergencies and the like. The short answer to this is that mandamus is in this respect the same as an injunction, and yet these considerations are not regarded as decisive in the case of an injunction against the Crown.[34] However, even if this argument is substantial, it does not

[28] *Dyson* v. *Attorney-General* [1911] 1 KB 410; *Ex p. McWilliam*, above n. 23; *Marks* v. *Commonwealth* (1964) 111 CLR 549, 565 (HCA).

[29] Save the doctrine of ministerial responsibility on one occasion: see *R.* v. *Watt, ex p. Slade*, above n. 19. After *Re Toohey*, above n. 24, it is unlikely that this justification will resurface.

[30] See *Re Toohey*, above n. 24, p. 184; Lee, *Emergency Powers* (1984), pp. 277 et seq.; Hogg, 'Judicial Review of Action by the Crown Representative' (1969), 43 *ALJ* 215.

[31] See refs. above n. 30, and *Sankey* v. *Whitlam*, above n. 24.

[32] See refs. above n. 30, and *FAI Insurances Co. Ltd.* v. *Winneke* (1982) 56 ALJR 388 (HCA).

[33] See refs. above nn. 24, 30.

[34] At least in Australia, see below Ch. 4.3; and also Street, above n. 4, p. 142; Hogg, above n. 4, pp. 22–3.

of course justify the unavailability of mandamus against the Crown in all cases.

Probably the complete abolition of Crown immunity from mandamus cannot be achieved except by the legislature, but the judges can restrict its scope considerably. There are signs that the process is already under way, and a possible avenue has been provided by Lord Diplock in *Teh Cheng Poh* v. *Public Prosecutor*.[35] That case concerned the validity of a proclamation made under Malaysian internal security laws, under which the appellant had been convicted. His argument before the Privy Council was that since the need for the proclamation in question was no longer evident, the court should treat it as having lapsed by operation of law. This argument was rejected, but Lord Diplock said *obiter*:

> Apart from annulment by resolutions of both Houses of Parliament it can be brought to an end only by revocation by the Yang di-Pertuan Agong [Head of State]. If he fails to act the court has no power itself to revoke the Proclamation in his stead. This, however, does not leave the courts powerless to grant to the citizen a remedy in cases in which it can be established that a failure to exercise his power of revocation would be an abuse of his discretion. Article 32(1) of the Constitution makes the Yang di-Pertuan Agong immune from any proceedings whatsoever in any court. So mandamus to require him to revoke the Proclamation would not lie against him; but since he is required in all executive functions to act in accordance with the advice of the Cabinet, mandamus could, in their Lordships' view be sought against the members of the Cabinet requiring them to advise the Yang di-Pertuan Agong to revoke the Proclamation.[36]

Admittedly the situation in question involved a statutory function performed by the Head of State acting on advice, as in *Re Toohey*, but it was a situation with a distinct whiff of prerogative about it. Whether or not other judges will follow Lord Diplock into this no man's land remains to be seen, but one can compare the decision in *Tonkin* v. *Brand*,[37] in which the Full Court of Western Australia granted a declaration that the State's Executive Council were under a duty to advise the Governor to issue an election proclamation. If the approach in these cases is accepted, it would mean that only the exercise of personal prerogatives by the Head of State is immune from

[35] [1979] 2 WLR 623 (PC).

[36] Ibid. 633–4.

[37] [1962] WAR 2, distinguishing *R.* v. *Governor of South Australia*, above n. 22, on the ground that there was no coercion of the Governor; 'I am unable', said Hale J. at p. 22, 'to follow the argument that to declare the true meaning of a statute is to interfere with a duty owed by a Minister to the Crown'.

mandamus, a result which might well be thought to be consonant with the present policy of the courts.

(ii) *Minor officials*

In *R.* v. *Bristow* Lord Kenyon CJ said:

> . . . it would be descending too low to grant a mandamus to inferior officers to obey [an] order; we might as well issue such a writ to a constable, or other ministerial officer, to compel him to execute a warrant directed to him . . .[38]

This dictum gave rise to a misconception that mandamus does not lie against minor officials and the remedy has been refused on this ground.[39] The cases can be explained on two grounds: first, that an alternative remedy in the form of an indictment was available;[40] and secondly, that the official concerned was not the person charged by statute with performance of the duty, but only the servant of such person.[41] The first ground no longer holds, because indictment has all but died out as a method of enforcing public duties and is not in any case an equally convenient remedy. The second ground still holds, but does not support the general proposition that minor officials *per se* are not amenable to mandamus. In the only case in this century on the question, mandamus was held not to lie against the clerk of a probate court to issue letters of administration but only against the judge, because the clerk could only act on the judge's direction.[42] The result is therefore that the cases are merely examples of Blackburn J.'s principle that the obligations of the principal cannot be enforced against the servant acting as such.[43]

(iii) *Superior courts of record*

It has long been established that superior courts of record are not subject to the prerogative writs,[44] and despite attempts by Lord

[38] (1795) 101 ER 492, 494.

[39] See *R.* v. *Treasurer of the County of Surrey* (1819) 1 Chit. R. 650; *R.* v. *Jeyes* (1835) 111 ER 471; *R.* v. *Payn* (1837) 112 ER 150; *R.* v. *Oswestry Treasurer* (1848) 116 ER 858.

[40] See esp. *R.* v. *Jeyes*, above n. 39, where Lord Denman CJ regarded it as 'some remedy, though imperfect'.

[41] In this situation, as observed in *R.* v. *Wood Ditton Highways Surveyors* (1849) 18 LJMC 218, the defendant is nominal, and the persons to whom he is responsible should be compelled; see also *R.* v. *Jeyes*, above n. 39, p. 474 per Lord Denman CJ.

[42] *Re Macdonald* [1930] 2 DLR 177.

[43] See above text and n. 6.

[44] See *The Rioters' Case* (1683) 23 ER 396 (mandamus refused against the Chief Justice of England to sign a bill of exceptions in a criminal case).

Denning MR in *Pearlman* v. *Harrow School*[45] and in *Re Racal Communications Ltd.*[46] to extend the ambit of judicial review to such courts in England, namely, in these instances, the County Court and the High Court, this extension has been firmly rejected by the House of Lords in the latter case. The Crown Court, while it is a superior court of record, is made subject to the prerogative jurisdiction by statute.[47]

Australian decisions have dealt with the problem, 'what is a superior court of record?' The courts have construed the term very narrowly even in the face of statutory provisions designating a court as a superior court of record. In this they have adopted what has been called an objective test.[48] Thus the Commonwealth Court of Conciliation and Arbitration,[49] the Industrial Court of Queensland,[50] the Family Court of Australia,[51] and the Victorian County Court[52] have all been held to be subject to the prerogative jurisdiction of the higher courts. This approach is based on sound constitutional principle: if it were possible to set up specialist courts and escape judicial review by calling them 'superior courts of record', the jurisdiction of the ordinary courts would be seriously eroded.

3.3 DEMAND AND REFUSAL

Performance of public duties has been discussed in Chapter 2. However, many cases on mandamus have been decided on the basis that, regardless of the question of performance of the duty, there has been no demand for performance followed by refusal to perform.[1]

[45] [1979] QB 56.

[46] [1981] AC 374; the Court of Appeal's judgment, s.n. *In re a Company*, is reported at [1980] 1 All ER 284. For a case in which mandamus was refused against a superior court of record, see *R.* v. *Central Criminal Court* (1883) 11 QBD 479.

[47] See Supreme Court Act 1981 (England and Wales), s. 29(3). The rule does not apply to trial on indictment. See also *R.* v. *Cardiff Crown Court, ex p. Jones* [1973] 3 WLR 497; *In re Sampson* [1987] 1 WLR 195.

[48] Sykes, Lanham, and Tracey, *General Principles of Administrative Law*, 2nd edn. (1984), p. 203; see also Aronson and Franklin, above n. 3, p. 490.

[49] *R.* v. *Commonwealth Court of Conciliation and Arbitration, ex p. Ozone Theatres (Australia) Ltd.* (1949) 78 CLR 389 (HCA).

[50] *Attorney-General of Queensland* v. *Wilkinson* (1958) 100 CLR 422 (HCA).

[51] *R.* v. *Watson, ex p. Armstrong* (1976) 9 ALR 551.

[52] *R.* v. *Martin (Judge), ex p. Attorney-General* [1973] VR 339.

[1] See *R.* v. *Brecknock & Abergavenny Canal Co.* (1835) 111 ER 395; *Ex. p. Whitmarsh* (1840) 8 Dowl. 431; *R.* v. *Cheadle Highway Trustees* (1842) 7 Jur. 373; *R.* v. *Bristol & Exeter Railway Co.* (1843) 114 ER 895; *Ex p. Thompson* (1845) ER 272; *R.* v. *Brassingbourne (Vicar)* (1845) 9 JP Jour. 83; *R.* v. *St Mary, Newington (Guardians)* (1851) 17 LTOS 163; *In re Wall* (1890) 16 VLR 686; *R.* v. *Bodmin Corporation* [1892] 2 QB

One can readily see the utility of the requirement of demand and refusal. If an unequivocal demand is met by an unequivocal refusal, the problem of non-performance becomes simpler, and the only question is whether the respondent was under a legal duty to do what was demanded. It is also fair to ensure that the respondent is clearly informed of what is expected of him and given the opportunity to decide whether to accede or test the duty in the courts.[2] A similar rule applies in the tort of detinue, in which the wrong is committed only by refusing to deliver the goods when requested to do so.[3] The significance of the rule is, however, slightly different, because whereas demand and refusal is the gist of detinue it is not the gist of failure to perform a public duty: there are many ways of failing to perform a duty other than by blank refusal. It is hardly therefore surprising that modern cases make little reference to the rule, especially as the refusal is often a highly artificial one, sometimes called a 'constructive refusal'; for example, when a tribunal takes into account a legally irrelevant consideration, it is said to have (constructively) refused jurisdiction.[4] In such cases the notions of demand and refusal have very little relevance. This is not to say, however, that the rule is now totally obsolete. Perhaps it can be regarded in this way: demand and refusal constitute one way of establishing non-performance of a public duty, and on that understanding, the old rules remain applicable;[5] it is not, however, an essential requirement, and it is open to an applicant to prove any other relevant facts which will constitute a failure to perform the duty such as to warrant the issue of mandamus.

The rule itself stems from *R.* v. *Brecknock & Abergavenny Canal Co.*, in which a statutory undertaker agreed to complete certain works if indemnified against claims resulting therefrom. Deciding in their favour Lord Denman CJ said:

> We cannot grant a mandamus unless there has been a direct refusal, and here, I think, there has not. It is not indeed necessary that the word 'refuse', or any equivalent to it, should be used; but there should be enough to show

21; *In re Gray* (1892) 8 WN (NSW) 84; *Fleming* v. *Waverley Town Board* (1914) 33 NZLR 831; *Re Dunlop and Halifax City Charter* [1944] 3 DLR 257; *Hughes* v. *Henderson* (1963) 42 DLR (2d) 743; *R.* v. *Board of Commissioners of Public Utilities, ex p. Halifax Transit Corporation* (1971) 15 DLR (3d) 720; *R.* v. *Warden at Cloncurry, ex p. Sheil* [1971] Qld 406.

[2] This justification was articulated by Coleridge J. in *R.* v. *Brecknock & Abergavenny Canal Co.*, above n. 1, p. 398.

[3] See Rogers, *Winfield and Jolowicz on Tort*, 12th edn. (1984), pp. 484–5; Trindade and Cane, *The Law of Torts in Australia* (1985), pp. 132–5. In conversion, demand and refusal were *one* way of establishing the wrong.

[4] See above Ch. 2.2.

[5] See *R.* v. *Kent County Council, ex p. Bruce*, *The Times*, 8 Feb. 1986.

that the party withholds compliance, and distinctly determines not to do what is required. The question is . . . whether the party had done what the Court distinctly sees to be equivalent to a refusal. Here I cannot perceive that in the correspondence and conduct of the company. Their answer to the last application is, that they are ready to do the works, if indemnified. That leaves the case short of the point to which it would have been brought if such an application had been made that any non-performance afterwards must have amounted to a refusal.[6]

This very passage shows that the rule was not regarded as exclusive or absolute, though it was treated as such in a number of later cases (though few modern ones), in which a good deal of irrelevant casuistry was spun as to whether the facts amounted to a demand and refusal. As a result great care was needed to couch the demand in the correct form, for a refusal to do something not strictly enjoined by the law would not be a proper refusal;[7] and to make it opportunely, since an early demand, or one met with an equivocal refusal, might be premature and a fresh demand might be needed.[8] The rule was sometimes applied with severity: if an application was dismissed on this ground and a demand and refusal secured later, a second application would not be allowed;[9] in one case even palpable disobedience to the statute and the passing of the time for performance was not allowed to displace the rule;[10] and the notion of 'constructive refusal' allowed in Lord Denman CJ's dictum was not applied with the liberality one might have expected, so that, for example, delay in replying to the demand was held not to be a refusal.[11] As a result many an application failed as a result of a pure technicality. Given the more complicated facts of modern cases, the sometimes difficult policy questions involved in them, and the extension of the notion of constructive refusal, the demand and refusal requirement is conspicuous by its absence from modern reports, and rightly so. The question to ask is not, 'has A refused to perform?', but, 'has A substantially fulfilled the duty?', or, in the case of duties to determine, 'has A decided in accordance with law?' One could of course regard all successful applications as cases in which a constructive refusal was

[6] Above n. 1, pp. 397–8. See also the remarks of the same judge in *R.* v. *Grand Western Canal Co.* (1837) 1 Jur. 53.

[7] See e.g. *Re Dunlop and Halifax City Charter*, above n. 1.

[8] *Re Aldham and United Parishes Insurance Society* (1854) 18 JP Jour. 311: and see *R.* v. *Bristol & Exeter Railway Co.*, above n. 1.

[9] *Ex p. Thompson* and *R.* v. *Bodmin Corporation*, above n. 1.

[10] *R.* v. *Bristol & Exeter Railway Co.*, above n. 1.

[11] *R.* v. *Wiltshire & Berkshire Canal Co.* (1840) 8 Dowl. 623. For the present attitude to this question, see below n. 22 and text.

established, but it is hard to see how this way of looking at them is at all helpful.

In a number of cases demand and refusal have been specifically held not to be required. In *R.* v. *Hanley Revising Barrister*[12] the barrister, who was under a duty to settle the voters' lists in court hearings, employed a clerk who inadvertently omitted to strike off some names which the barrister ordered expunged after upholding objections to their inclusion. The mistake was not discovered until the lists had been delivered to the town clerk and had come into operation, the original lists being no longer available. Mandamus was granted against the barrister and the town clerk to correct the lists, even though there had been no demand and refusal, which were held to be unnecessary because the time for performance of the duty had passed and the barrister could only act on a court order.

In relation to a duty to determine, the requirement of a demand can cause confusion. Depending on the facts there may be two duties: to hear and determine the matter according to law, and to act on the finding, for example by granting a licence if the necessary conditions are satisfied.[13] The former duty will only arise if there is an application, which should also be treated as a demand for performance of the duty to hear and determine. For the latter duty no further demand ought to be necessary, for it is implicit in the application that the applicant demands appropriate action as well as appropriate procedure, and the application will usually be couched in these terms. Yet it has, rather oddly, been held that a demand must be made when the decision is handed down, because the original petition is not a demand.[14] However, if failure to decide in accordance with law is a constructive refusal of jurisdiction, a subsequent demand must, surely, be superfluous.

For the most part it is not thought necessary to construe a demand from the applicant's conduct if there is clear refusal, because '[j]ust as the courts will *imply* a refusal to perform a duty—a *constructive* refusal—it seems to be assumed without argument that there is a "constructive demand" '.[15] If necessary, however, it can be construed from conduct: to second a motion for a poll, if one proved necessary,

[12] [1912] 3 KB 518. See also *R.* v. *Richmond City, ex p. May (E. B.) Pty. Ltd.* [1955] VLR 379, where the *Hanley* case was applied, and *R., ex rel. Mikklesen and McGaughey* v. *Highway Traffic Board* [1947] 2 DLR 373.

[13] See above Ch. 2.2.

[14] *R.* v. *Board of Commissioners of Public Utilities, ex p. Halifax Transit Corporation*, above n. 1. And see *Re Maharaj* (1966) 10 WIR 149, where the court drew a distinction between an act giving rise to a duty and a demand for its performance.

[15] Aronson and Franklin, *Review of Administrative Action* (1987), p. 484.

was held to constitute a demand when the original motion was withdrawn;[16] however, when some public employees agitated for several years to be awarded clerical status and one of them brought a test case, it was held there was no demand.[17]

Constructive refusal is also of course readily implied from conduct. In *State* (*Modern Homes Ltd.*) v. *Dublin Corporation*,[18] the corporation were under a statutory duty, on resolving to make a planning scheme, to give effect to their decision with all convenient speed. The company's business suffered as a result of the Corporation's failure to act on its decision, and on demand for performance the Corporation decided to continue with 'interim control', pending revision of the legislation. This was held by the Supreme Court of Ireland to be a refusal to perform the duty. Statements made in correspondence can of course also be construed as refusal,[19] as for example where the authority's solicitor says he will accept process.[20] Often statements made in negotiating the matter amount to a refusal to act except on conditions or an agreement to act on certain conditions. Whether such statements amount to a refusal depends on the nature of the conditions: if they are unlawful, then clearly there is a refusal to act,[21] but if they are merely unreasonable there is presumably no refusal unless the imposition of the conditions can be reviewed on other grounds. The *Brecknock* case itself was a case which should, it is suggested, have been decided on these principles. Failure to respond to an unequivocal demand must of course be treated as a refusal,[22] but cases of delay are more problematical and depend on the reasons for the delay and the nature of the duty. This is discussed above in Chapter 2 as an aspect of performance.

3.4 GROUNDS FOR REFUSING MANDAMUS

Even where there is a breach of a public duty by an authority against whom mandamus may issue, an order may still be refused where the

[16] *R.* v. *Dover Corporation* [1903] 1 KB 668. And see *Re Williams and Brampton* (1908) 17 OLR 398 (demand implied from the sending of a deputation).

[17] *In re Wall*, above n. 1.

[18] [1953] IR 202. And see *Re Williams and Brampton*, above n. 16 (refusal implied from council proceedings).

[19] See e.g. *Mitchell* (*G. H.*) *& Sons* (*Australia*) *Pty. Ltd.* v. *Minister of Works* (1974) 8 SASR 7.

[20] *R.* v. *Norwich & Brandon Railway Co.* (1845) 815 LJQB 24.

[21] *R.* v. *Lancaster* (*Inhabitants*) (1900) 64 JP 280; *R.* v. *City of Preston, ex p. Sandringham Drive-In Theatre Pty. Ltd.* [1965] VR 10.

[22] *Wijeyesekera* v. *Principal Collector of Customs, Colombo* (1951) 53 NLR 392; *City Motor Transit Co.* v. *Wijesinghe* (1961) 63 NLR 156.

applicant lacks standing, or where the court in its discretion refuses to make an order on one of the grounds discussed below.[1] Standing is discussed separately in Chapter 6.[2]

(i) *General considerations*

Before we look at the principles upon which this discretion is exercised, it would be well to see how and when the discretion arises, and how it relates to the rest of the substantive law relating to mandamus. As we have seen, mandamus is a writ of grace, not of right; it is therefore for the applicant to show why it should issue. Presumably he does this by showing that he has standing to apply for mandamus and that the respondent has failed to perform a public duty reposed in him. In some cases it has been said that the applicant must satisfy the court that its discretion should be exercised in his favour,[3] but the better view, and one more in keeping with the weight of authority, is that mandamus will issue unless the court is satisfied that, for some reason, it should not.[4]

In some cases the awarding of mandamus is a matter of right, not of discretion. It has been held that where the conditions for the granting of mandamus are provided by statute, the court has no discretion if the conditions are fulfilled, at least 'in the absence of some legal error or omission of legal form'.[5] Since this problem arises

[1] The best discussion of these grounds can be found in Aronson and Franklin, *Review of Administrative Action* (1987), pp. 517–30.

[2] Yardley, in 'Prohibition and Mandamus and the Problem of Locus Standi' (1957), 73 *LQR* 534, argues that standing is also a matter for the court's discretion. This view does not appear to have found favour and seems wrong in principle, because standing ought surely to be a matter of right, not discretion; see below Ch. 6. Considerations relevant to standing, particularly motive, may be relevant also to discretion: see e.g. *R.* v. *Peterborough Corporation* (1875) 44 LJQB 85, and text below.

[3] Dicta of this kind are numerous (see e.g. text to n. 13 below; *R.* v. *Bridgman* (1846) 2 New Sess. Cas. 232), but are best seen not as statements concerning the burden of proof, but as indications that there must be some substantial purpose or advantage in granting mandamus.

[4] The most telling statement of this view is that of Martin B. in *Rochester* (*Mayor*) v. *R.* (1858) 120 ER 794: 'Instead of being astute to discover reasons for not applying this great constitutional remedy for error and misgovernment, we think it our duty to be vigilant to apply it to every case to which, by any reasonable construction, it can be made applicable.'

[5] *R.* v. *Staines Union* (1893) 62 LJQB 540, 543 per Cave J. The view of Lord Widgery CJ in *R.* v. *Hounslow London Borough Council, ex p. Pizzey* [1977] 1 WLR 58, 62 that mandamus is less discretionary than the other prerogative orders is not borne out by the case law. *R.* v. *Staines Union* has been followed in two Victorian cases: *R.* v. *Metcalfe Shire Council, ex p. Public Health Commission* [1920] VR 578; *R.* v. *Rochester Shire Council, ex p. Public Health Commission* [1928] VLR 492.

only where a superior authority seeks to enforce a default order or ministerial direction made under statutory provisions, another rule impinges on the question, namely that the court has no discretion where the applicant for prerogative relief is the chief law officer of the Crown; this rule was applied in *R.* v. *Martin (Judge), ex p. Attorney-General.*[6]

In *Lord Advocate* v. *Glasgow Corporation*[7] the court's discretion was held to be excluded where the conditions for granting an order of specific performance (the Scottish equivalent of mandamus) of ministerial directions to a local education authority were laid down by statute, the provision made the granting of the order permissive, not obligatory, and the applicant was the chief law officer applying on behalf of the minister; it is not clear which if any of the above rules were being applied, but the leading judgment of Lord Reid seems to indicate that principles similar to those in *Julius* v. *Oxford (Bishop)*[8] were being applied, though his Lordship did say that if the House of Lords held otherwise the courts would have to 'consider and act upon wide questions of policy'.[9] It will be seen in the discussion that follows that the discretion to refuse mandamus is an important vehicle for importing public-policy considerations into mandamus cases. In view of this it is hard to see how either of the above rules can be justified. What they amount to is in reality a refusal of the courts to investigate the conduct and motives of high authorities, as applicants, which is out of keeping with modern ideas of administrative law as evidenced by, for example, *Tameside*[10] and *Re Toohey.*[11] Even if one were to proceed on the basis that high authorities can never be guilty of delay or bad faith, many of the grounds for refusing mandamus have nothing to do with the identity or conduct of the applicant. The cases can probably be easily distinguished, but if necessary these strange rules should be abandoned. As a matter of statutory interpretation, a provision which allows the granting of mandamus to an authority seeking to enforce a default order or ministerial direction cannot possibly be said to exclude the most

[6] [1973] VR 339 (a case of *certiorari* and mandamus), citing *HLE*, 3rd edn. vol. xi, para. 263, where the rule is regarded as applying to *certiorari* but not mandamus.

[7] 1973 SLT 33 (HL).

[8] (1880) 5 App. Cas. 214 (HL), for discussion of which see above Ch. 1.3.

[9] Above n. 7, p. 36; thus the *effect* of the decision is that the court has no discretion where the applicant is the chief law officer of the Crown or, it is suggested, a minister.

[10] *Secretary of State for Education and Science* v. *Tameside Metropoliton Borough Council* [1976] 3 WLR 641 (HL).

[11] *Re Toohey (Aboriginal Land Commissioner), ex p. Northern Land Council* (1981) 56 ALJR 164 (HCA).

important rules relating to the operation of that remedy; on the contrary the entire corpus of the common law relating to mandamus is imported into the statutory provision. If the court's discretion protects superior authorities from the unwarranted demands of inferior authorities, why should not inferior authorities be similarly protected from the unwarranted demands of superior authorities?[12]

In all cases other than those indicated above, the discretion to refuse mandamus comes into play once all the other requirements for issuing it have been satisfied. A judge said two hundred years ago:

> It is true that an application for a mandamus is made to the discretion of the court, but that discretion must be governed by certain principles. It is never granted merely for the asking: some reason must be assigned for it.[13]

Unfortunately the courts, while they have adhered to the second of these propositions, have not adhered very carefully to the first, and in fact it is still to a large extent unclear on what principles the discretion will be exercised. There is a vast number of cases of discretionary refusal, but none of them attempts to explain the purpose and extent of any one principle, let alone all of them. This is perhaps attributable to the fact that the court exercises a discretion in the matter and does not therefore feel obliged to explain carefully the basis for its exercise in the particular instance, or to relate its decision to others on the same point.[14] This tendency is most unfortunate because it has prevented the court's discretion from being placed on a rational and predictable footing. It has also given rise to some confusion about reviewability of the exercise of the discretion. This matter was discussed in *R.* v. *All Saints, Wigan* (*Churchwardens*),[15] in which the Queen's Bench had granted mandamus at the instance of the Public Works Loan Commissioners to the Churchwardens to levy a rate for repaying loans made by the Commissioners more than twenty years after the loans were made; the House of Lords held that, *inter alia*, the writ had to be refused on grounds of inordinate delay in applying for it. Lord Chelmsford said this:

> A writ of mandamus is a prerogative writ and not a writ of right, and it is

[12] The same arguments, it is suggested, apply in some measure to the equally strange notion that there is no discretion where the duty is ministerial: see *R.* v. *Sarum* (*Bishop*) [1916] 1 KB 466; the kind of duty involved is of course logically irrelevant to the applicant's motives and conduct or the availability of alternative remedies.

[13] (1786) 99 ER 1174, 1176 (Ashurst J.).

[14] An exception to this is the corpus of authority on alternative remedies: see below Ch. 3.4(ii).

[15] (1875-6) 1 App. Cas. 611.

in this sense in the discretion of the Court whether it shall be granted or not. The Court may refuse to grant the writ not only upon the merits, but upon some delay, or other matter, personal to the party applying for it; in this the Court exercises a discretion which cannot be questioned. So in cases where the right, in respect of which a rule for a mandamus has been granted, upon shewing cause appears to be doubtful, the court frequently grants a mandamus in order that the right may be tried upon the return; this also is a matter of discretion. But where the Judges grant a peremptory mandamus, which is a determination of the right, and not a mere dealing with the writ, they decide according to the merits of the case, and not upon their own discretion, and their judgment must be subject to review, as in every other decision in actions before them.[16]

In the context of the decision in the case itself it is suggested that this does not mean that the exercise of discretion cannot be reviewed on appeal, but in fact means the exact opposite. Certainly, whatever his Lordship may have meant, appeal courts have not been unduly inhibited in overturning decisions of the lower courts which depended on the exercise of discretion to refuse an order.[17]

(ii) *Alternative remedies*

The most important rule[18] involved in the exercise of discretion to refuse mandamus is that the existence of an alternative remedy precludes the award of mandamus. In these days of proliferation of statutory remedies this rule is obviously of great importance and deserves some close scrutiny.

The historical reason for the existence of this rule is that the jurisdiction of the courts to grant mandamus depended on the absence of any other remedy, so that the granting of mandamus where there was another remedy would have been controversial.[19] This supplementary nature of mandamus has persisted until the present day, and there are hundreds of cases in which mandamus has been refused because of an alternative remedy being available, but the *raison d'être* of the rule has changed. Although no judge has sought to

[16] Ibid. 620.

[17] The case under discussion is a good example. See, however, the views to the contrary expressed in *R.* v. *Maidenhead Corporation* (1882) 9 QBD 494, 503, 505, to the effect that it requires a strong case before the discretion of the judge at first instance will be overturned.

[18] The word 'rule' seems appropriate here because the courts have sought to justify the exclusion of mandamus on this ground by reference to precedent rather than by pointing to their discretion, as is apparent from the discussion below.

[19] For early instances of the application of the rule see above Ch. 3.1, and below n. 22.

justify the rule in general terms, it seems clear enough that it exists as a filter to enable the courts to direct particular kinds of complaint towards particular remedies. The only alternative remedies which regularly exclude mandamus are administrative appeal and administrative default powers; this shows that the purpose of the rule is to ensure that the administrative process is exhausted before the complainant comes to court—a sound limitation, generally, on the jurisdiction of the courts, at least as regards appeals. A well-known example of this policy is the refusal of the courts to interfere with the duties of universities with regard to academic and disciplinary matters on the ground that there is recourse to the university visitor.[20]

Initially the view was taken that the existence of any alternative remedy precluded mandamus. This rule worked injustice because mandamus was, in spite of its cumbrous procedure,[21] generally more efficacious than other remedies, which tended to proliferate with the growth of statute law generally. The discretionary refusal of mandamus on this ground came therefore to be restricted to situations where the alternative remedy was 'equally convenient, beneficial and effective'.[22] Apart from the two exceptions mentioned above, this development made great inroads into the rule because naturally there would be few situations in which the applicant opted for a *less* convenient, beneficial, and effective remedy; at the same time it allowed the courts to investigate the relative merits of mandamus and the other remedies in the context of the particular facts. Increasingly the alternative remedies considered came to be administrative rather than legal ones, and it became necessary to define the relationship between mandamus and administrative remedies. The opportunity to effect this definition came in *Pasmore* v. *Oswaldtwistle Urban District Council*.[23] In this case the plaintiff sought mandamus to compel the authority to fulfil its obligation under the Public Health Act 1875 (England and Wales), section 15, to 'cause such sewers to be made as may be necessary for effectually draining their district' in such a way as to provide sewerage

[20] For mandamus cases of this kind see *R.* v. *Hertford College* (1878) 3 QBD 693; *R.* v. *Dunsheath, ex p. Meredith* [1951] 1 KB 127.

[21] See Tapping, *Mandamus* (1848), *passim.*

[22] For a catalogue of formulae, see Whitmore and Aronson, above n. 1, p. 390 nn. 286–93. Lord Mansfield spoke in *R.* v. *Barker* (1762), *R.* v. *Cambridge University* (*Vice-Chancellor*) (1765) 97 ER 1027, 1032, and *R.* v. *Bank of England* (1780) 99 ER 334, 335 of 'no other specific remedy'. The formulation in the text, which represents a liberal version of Lord Mansfield's words, can be traced at least as far back as *Re Barlow* (1861) 30 LJQB 270, 271. It had previously been held that mandamus would issue where the other remedy was at all doubtful: *R.* v. *Nottingham Old Water Works Co.* (1837) 112 ER 135; *R.* v. *Rawlinson* (1844) 2 LTOS 354.

[23] [1898] AC 387.

facilities for his factory, which was affected by a nuisance alleged to result from the authority's default. The same statute in section 299 provided a typical administrative default remedy:

Where complaint is made to the Local Government Board that a local authority has made default in providing their district with sufficient sewers . . . the Local Government Board, if satisfied after due inquiry that the authority has been guilty of the alleged default, shall make an order limiting a time for the performance of their duty in the matter of such complaint. If such duty is not performed by the time limited in the order, such order may be enforced by writ of mandamus, or the Local Government Board may appoint some person to perform such duty . . .

The judge issued mandamus in the terms of section 15, but with the addition of the words 'and in particular the plaintiff's premises'. The writ was overturned on appeal and the House of Lords affirmed the decision of the Court of Appeal. Lord Halsbury LC based his decision on the principle that 'where a specific remedy is given by statute, it thereby deprives the person who insists upon a remedy of any other form of remedy than that given by the statute'.[24]

The principle, said to be very familiar and one which ran through the law, is not in fact generally manifest in mandamus cases, in which the principles were as has been indicated above, but in actions for damages or an injunction.[25] The real objections to granting mandamus

24 Ibid. 394, citing *Doe* v. *Bridges* (1831) 109 ER 1001 (not a case of mandamus).

25 See below Chs. 7.2, 4.3. In the then recent case of *R.* v. *Lewisham Union Guardians* (1897) 1 QB 498, Wright J. refused mandamus to a local sanitary authority to compel the Guardians to enforce vaccination legislation partly on the ground that the Local Government Board, a supervisory authority, had power to do all that the Guardians could do and could also compel the Guardians to act; no reason was given why these facts should exclude mandamus. However, in *R.* v. *Leicester Union Guardians* [1899] 2 QB 632 mandamus was granted, on similar facts, but at the instance of the Local Government Board itself, and without reference to *Pasmore*; see also *R.* v. *Norwich & Brandon Railway Co.* (1845) 15 LJQB 24, which is also contrary to Lord Halsbury's dictum. The introduction of the 'exclusive remedy' principle into mandamus cases seems to have caused confusion. In *R.* v. *Poplar London Borough Council, ex p. London County Council (No. 1)* [1922] 1 KB 72 the Court of Appeal seems to have regarded the 'exclusive remedy' principle as a rule *in addition* to the principle that mandamus will not be granted where there is another equally convenient remedy. Fortunately the judges were also able to point to another dictum, of Lord Macnaghten, also in *Pasmore*: 'Whether the general rule is to prevail, or an exception to the general rule is to be admitted, must depend on the scope and language of the Act which creates the obligation and on considerations of policy and convenience': [1898] AC 397. In the result the remedy of distress was held not to be equally convenient to mandamus to levy a rate for the purpose of paying a precept. But for the 'loophole' of Lord Macnaghten's dictum, Lord Halsbury's principle would exclude mandamus even where the alternative remedy was not equally convenient, provided only that it was created by the same statute as that which created the public duty.

were that the plaintiff sought not merely to enforce the duty generally, but for his own benefit, as indicated by the wording of the writ of mandamus; that the subject-matter was considered to be better judged by the Local Government Board; and that if mandamus were granted there would be a flood of wasteful litigation. These reasons prompted Lord Halsbury LC to say this:

> . . . the particular jurisdiction to call upon the whole district to reform their mode of dealing with sewage and drainage should not be in the hands, and should not be open to the litigation, of any particular individual, but should be committed to a Government department.[26]

Pasmore has been followed on numerous occasions in England, most notably in cases involving educational duties.[27] Under the Education Act 1944, section 99(1)

> [i]f the Minister is satisfied, either upon complaint by any person interested or otherwise, that any local education authority . . . have failed to discharge any duty imposed upon them by or for the purposes of this Act, the Minister may make an order declaring the authority . . . to be in default in respect of that duty, and giving such directions for the purpose of enforcing the execution thereof as appear to the Minister to be expedient; and any such directions shall be enforceable, on an application made on behalf of the Minister, by mandamus.

This provision has been held to exclude mandamus,[28] actions for an injunction,[29] and damages for breach of statutory duty.[30] However, the latest of these cases, *Meade* v. *Haringey London Borough Council*,[31]

Another peculiarity of *Pasmore* is that the judges were at pains to point out that the duty was a new one; presumably the idea was to relate the specific remedy to the duty and so reinforce the unavailability of mandamus, but of course all statutory duties are new when they are introduced, and it is puzzling that this point should have been considered at all relevant; cf. the position with regard to actions for damages, below Ch. 7.2, where the policy reasons are also discussed.

[26] [1898] AC 394.

[27] The cases have occurred in several contexts: (1) education: *Watt* v. *Kesteven County Council* [1955] 1 QB 408; *Wood* v. *Ealing London Borough Council* [1966] 3 All ER 514; *Bradbury* v. *Enfield London Borough Council* [1967] 1 WLR 1311; *Lee* v. *Enfield London Borough Council* (1967) 66 LGR 195; (2) public health: *Clark* v. *Epsom Rural District Council* [1929] 1 Chanc. 287; *R.* v. *Kensington and Chelsea London Borough Council, ex p. Birdwood* (1976) 74 LGR 424; (3) housing: *Southwark London Borough Council* v. *Williams* [1971] Chanc. 734; *Kensington and Chelsea London Borough Council* v. *Wells* (1973) 72 LGR 289; *Roberts* v. *Dorset County Council* (1976) 75 LGR 462.

[28] *Watt*, above n. 27.

[29] *Wood* and *Bradbury*, above n. 27.

[30] *Watt*, above n. 27.

[31] [1979] 1 WLR 637; for further discussion of this case, see above Ch. 2.3, and below Ch. 7.2.

sheds some new light on this question. It will be recalled that in this case the Council closed the schools in its area as a result of a caretakers' strike, and some parents sought an interlocutory mandatory injunction to compel the Council to reopen the schools, having first, it may be noted, failed to induce the Minister to put section 99(1) into operation. In the Court of Appeal Lord Denning MR, ignoring all the cases in which *Pasmore* has been followed, but referring instead to cases in which damages had been awarded for breach of an education authority's duty to maintain schools to the prescribed standard, held that section 99(1) left open '*all* the established remedies which the law provides in cases where a public authority fails to perform its statutory duty by an act of *commission or omission*';[32] he went on to link this with the idea of *ultra vires* action, concluding that in the event of an *ultra vires* breach of duty[33] an action or an injunction would lie (and presumably mandamus too). The remarks of the other two judges were entirely consistent with this view, Eveleigh LJ going so far as to say that section 99(1) was no bar because the authority were positively misusing their powers for an improper purpose.

What then is the scope of the decision in *Pasmore* with regard to alternative remedies? First of all, it has been accepted in the cases on section 99(1) that the rule in *Pasmore* only applies to cases of nonfeasance, not misfeasance, because in the case of misfeasance the alternative remedy does not apply;[34] after *Meade*'s case, however, it is quite unclear how the distinction between nonfeasance and misfeasance is to be drawn, and it may be that, if Lord Denning's view prevails, the rule in *Pasmore* will be done away with completely. Secondly, *Pasmore* appears now not to apply to cases of *ultra vires* breach of duty, an exception which, if it is different from that relating to misfeasance, makes another large hole in the rule. Thirdly, Whitmore and Aronson[35] have suggested that *Pasmore* might be interpreted as applying only where the default procedure has not first been tried. Fourthly, it might be interpreted as applying only where the default procedure includes a right of an individual to make

[32] Ibid. 646 (emphases mine).

[33] See above Ch. 1.5.

[34] *Bradbury* v. *Enfield London Borough Council* [1967] 1 WLR 1311, 1326 per Diplock LJ. The principle involved here depends on the wording of the statute; by way of contrast, in *R.* v. *Marshland, Smeeth, and Fen District Commissioners* [1920] 1 KB 155, the alternative remedy of appeal lay only in respect of anything *done* by the Commissioners in the execution of their duty to carry out drainage works, so that it applied only in cases of misfeasance: mandamus was therefore available for nonfeasance.

[35] *Review of Administrative Action* (1978), p. 393.

representations to the authority empowered to use the default procedure.[36]

The reasons which impelled the House of Lords to establish the rule in *Pasmore* are far from convincing, with the exception, possibly, of the argument that a government department may be better able to judge whether an authority is in default, given that difficult questions of, for example, educational administration or sanitary engineering may have to be determined.[37] It is no argument, however, for the courts not retaining jurisdiction to enforce the duty, either directly or else by judicial review of the exercise of the default powers themselves.[38] The nonfeasance/misfeasance distinction is a very unsatisfactory means of defining the role of the courts and is in any case very difficult to draw in particular cases. The right to make representations is an insufficiently radical basis for redrawing of the rule because experience shows that default powers are rarely invoked. The conclusion reached here is therefore that Whitmore and Aronson's suggestion has much to recommend it, but that Lord Denning's view is preferable, because there may be little point in insisting upon an attempt to invoke the default procedure where it is obviously unlikely to be effective.[39] It could be objected that a conflict of view between the courts and the administration in these matters is undesirable, and it is interesting that the judge at first instance in *Meade* was influenced by this consideration.[40] After *Tameside*[41] and *Meade*, however, this argument seems very unconvincing. The courts and the administration may well come to different conclusions, but they will consider different matters before reaching their decisions: the courts may plausibly act where the administration refuses to act, or refuse to act where the administration acts. In the uncertain state of the authorities it may well be advisable for a complainant to try the administrative remedy,

[36] This view of the case law was specifically rejected, however, by Edmund Davies LJ in *Kensington and Chelsea London Borough Council* v. *Wells*, above n. 27, p. 301.

[37] '[T]he court ought to confine itself to questions of law and should not entertain questions of sanitary science', *R.* v. *Staines Union*, above n. 5, p. 543 per Cave J.; if this reasoning were consistently adopted, mandamus would serve little purpose outside of the decision-making context, and in any case it points rather to the advantages of having alternative remedies than to the exclusion of mandamus.

[38] See below Ch. 4.2; *De Wolf* v. *Halifax City* (1979) 37 NSR (2d) 259 (no obligation on Minister to act, therefore no equally beneficial remedy).

[39] In *Wilson* v. *Independent Broadcasting Authority* 1979 SLT 279 Lord Ross held that ministerial powers of intervention did not preclude an application by an individual for an injunction to restrain a breach of statutory duty; see also *Attorney-General, ex rel. McWhirter* v. *Independent Broadcasting Authority* [1973] QB 629.

[40] [1979] 1 WLR 641; the case itself shows, however, that there are no terrors involved in such a conflict of view.

[41] Above n. 10.

but this should not be regarded as a circumstance precluding later application to the courts where necessary.

There are of course other statutory remedies to which the rule in *Pasmore* has no direct application. Many statutes imposing a duty to determine provide for an appeal to a higher administrative authority or tribunal, or a court of law, and there is a wealth of authority, especially in Australia,[42] to the effect that such avenues of appeal preclude mandamus to review the determination. These remedies are of course genuinely more convenient, from every point of view, and the position established by the courts is consistent with modern notions of ripeness for review. However, even in these cases it is apparent that the courts will look to the substance rather than the form, and the 'convenient, beneficial and effective' formula still applies; thus in *R.* v. *Stepney Corporation*[43] the Corporation deducted a portion of the compensation due to the holder of an abolished post under a misapprehension that they had no discretion in the matter, and mandamus was granted in spite of the existence of a statutory right of appeal to the Treasury, because the Corporation itself knew the circumstances better and a preliminary investigation would be useful

[42] The principle is deeply rooted in case law which goes back as far as the 18th cent. The earliest examples are quite similar to the modern cases: see *Butler* v. *Cobbett* (1709) 88 ER 1023 (appeal to tax commissioners held to exclude mandamus); *R.* v. *Barnstaple (Inhabitants)* (1728) 94 ER 95; *R.* v. *Wheeler* (1735) 94 ER 1123; and see above Ch. 3.1 n. 20. The modern cases are of course legion, and only the most important ones are cited here: *R.* v. *London Union Assessment Committee* [1907] 2 KB 764; *R.* v. *Army Council, ex p. Ravenscroft* [1917] 2 KB 504; *R.* v. *Port of London Authority, ex p. Kynoch Ltd.* [1919] 1 KB 176; *Stepney Borough Council* v. *John Walker and Son Ltd.* [1934] AC 365 (HL); *Cuming Campbell Investments Pty. Ltd.* v. *Collector of Imposts (Vic.)* (1938) 60 CLR 741 (HCA); *Ex p. Tooth & Co Ltd., re Parramatta City Council* (1954) 55 SR (NSW) 282; *Ex p. Australian Property Units (No. 2) Ltd., re Baulkham Hills Shire Council* (1962) 80 WN (NSW) 34; *Ex p. Arnold Homes, re Blacktown Municipal Council* [1963] NSWR 806; *Ex p. Catlett (S. R.) Constructions Pty. Ltd., re Baulkham Hills Shire Council* [1964-5] NSWR 1667; *McBeatty* v. *Gorman* [1975] 2 NSWR 262; *Harelkin* v. *University of Regina* [1979] 3 WWR 676 (SCC). The principle has been applied even where the applicant is out of time for his appeal: *R.* v. *West Norfolk Assessment Committee, ex p. Ward* (1930) 94 JP 201; but not where the appeal was precluded by the respondents' failure to give the grounds of decision, which was the reason for the application: *R.* v. *Thomas* [1892] 1 QB 426. For discussion of the principles involved, see Campbell, 'Judicial Review and Appeals as Alternative Remedies' (1982), 9 *Mon. ULR* 14.

[43] [1902] 1 KB 314. The decision is an isolated one and perhaps hard to justify. There was no element of the determination which the Treasury were not competent to consider, and the decision appears to rest on the fact that the Corporation had never exercised its discretion in the matter, a consideration which is not apparent in any of the other cases; surely an exercise of discretion by the appellate body is better than an exercise of discretion by the original decision-maker?

in the event of a later appeal to the Treasury; and in a South Australian case[44] mandamus was granted in spite of a right of appeal from a departmental decision to an appeal board because in a previous appeal in relation to the same matter the minister had overruled the board, making a further appeal an inferior remedy, in the circumstances, to mandamus against the Minister to decide in accordance with law: and the High Court of Australia refused to regard an appeal from the Commonwealth Court of Conciliation and Arbitration to the Full Court thereof as excluding mandamus because the latter would be more effective and less embarrassing to the parties.[45] An interesting case of this kind is *R.* v. *Paddington Valuation Officer, ex p. Peachey Property Co. Ltd.*[46] in the Court of Appeal. The applicants challenged the validity of the entire valuation list for the borough. If the list was invalid then possibly all the lists for London and even the rest of England might be affected. The majority found on the facts that the list taken as a whole was valid. There was an alternative statutory remedy by way of proposals for alteration of the list, and appeals, ultimately to the courts, where the proposals were not accepted by the valuation officer. This remedy was held not to be so convenient, beneficial, and effectual where the whole list was impugned, though it would be in the case of challenge to a particular valuation, and the reasons given by Lord Denning MR are instructive: proposals for alteration of the valuation of every one of 31,565 hereditaments on the list would be impossible; the parties might not agree which should be taken as test cases; and the procedure would be deficient because there could be no discovery against occupiers. On the facts the applicants had an adequate remedy under the statute because their attempt to invalidate the whole list failed.

In recent cases the inconvenience factor has been regarded as extending to the public interest, not simply the applicant's convenience.[47] Sometimes the statute gives the complainant himself default

[44] *Mitchell (G. H.) & Sons (Australia) Pty. Ltd.* v. *Minister of Works* (1974) 8 SASR 7; see also the division of the Full Court of Victoria in *R.* v. *Beecham (H.) & Co., ex p. Cameron (R. W.) & Co.* [1910] VLR 204; *R.* v. *Anderson* [1942] 1 DLR 58; *R.* v. *Perth Shire, ex p. Dewar and Burridge* [1968] WAR 149.

[45] *R.* v. *Foster, ex p. Commonwealth Steamship Owners' Association* (1953) 88 CLR 549; and see *R.* v. *Town Planning Committee, ex p. Skye Estate Ltd.* [1958] SASR 1 (mandamus preferred to consideration by a joint committee appointed by both Houses of Parliament).

[46] [1966] 1 QB 380; see also *Anderson* v. *Valuer-General* [1974] 1 NZLR 545 (chaos and delay, mandamus refused).

[47] See *McBeatty* v. *Gorman*, above n. 42; and the more explicit formulation of this view by Glidewell J. in *R.* v. *Huntingdon District Council, ex p. Cowan* [1984] 1 All ER 58, 63. Cf. also *Ex p. Lloyd* (1889) 5 TLR 180. A striking instance of this wide view of discretion is *Ex p. Coff's Harbour Shire Council, re Allen* (1957) 2 LGRA 243; in that

powers, and these too have been held to exclude mandamus, particularly where the complainant is a supervisory authority with powers to do that which has been neglected by the inferior authority, or powers to issue directions of the kind mentioned in the default provisions discussed above.[48] These directions are generally enforceable by mandamus.

Apart from administrative remedies there are of course also legal remedies to consider. Where a legal remedy is provided by the statute which imposes the duty, this will often exclude mandamus because the jurisdictional problems discussed above do not arise.[49] For example, many statutes give a right of direct appeal to the courts from an administrative determination, whether on the merits or on a point of law, or else a specific statutory right to judicial review.[50] The existence of a legal remedy other than appeal does not generally exclude mandamus, because it will rarely be a genuine alternative.[51] Indictment was originally considered an adequate remedy,[52] but the courts have now clearly recognized that, at least as regards private prosecution, the hazards and expense involved make it a very inadequate alternative to mandamus.[53] Indictment has in fact all but

case 450 objections to rating valuations depended on the outcome of the case, and the council were accordingly unable to prepare their estimates; for these reasons mandamus was preferred to appeal, even though on application of the usual principles mandamus would have been refused. The decision is commendable.

[48] In *R.* v. *Halifax Overseers* (1841) 10 LJMC 81 mandamus was refused to a district auditor because he had a remedy (disallowance) in his own hands; see also *R.* v. *Gamble* (1839) 113 ER 339; *R.* v. *Cotton* (1850) 117 ER 575; *In re Glenelg Shire, ex p. Sealey* (1885) 11 VLR 64; *R.* v. *Monken Hadley Overseers, ex p. Harnett* (1910) 8 LGR 363; *Ex p. Jarrett* (1946) 62 TLR 230.

[49] See, however, the *Poplar* case, discussed above n. 25.

[50] See e.g. *R.* v. *London Union Assessment Committee* and *Stepney Borough Council* v. *John Walker & Son Ltd.*, above n. 42.

[51] See the *Poplar* case, above n. 25 (distress); *R.* v. *St George's, Southwark, Vestry* (1892) 61 LJQB (obsolete remedy under local act); *Perpetual Executors and Trustees Association of Australia Ltd.* v. *Hosken* (1911) 14 CLR 286 (HCA) (summoning Registrar before Supreme Court to substantiate refusal to register transfer). A petition of right has been held to exclude mandamus: *Re Nathan* (1884) 12 QBD 461; *Ex p. McWilliam* (1947) 47 SR (NSW) 401. These decisions might, however, not be followed today because the remedy is virtually obsolete. The right to ask for a case stated excludes mandamus because it approximates to appeal: see e.g. *R.* v. *Epping and Harlow General Commissioners, ex p. Goldstraw* [1983] 3 All ER 257.

[52] *R.* v. *Bristow* (1795) 101 ER 492; *R.* v. *Jeyes* (1835) 111 ER 471. Indictment was a common remedy for non-performance of public duties, even where other remedies were available: see *R.* v. *Davis* (1754) 96 ER 839. In *R.* v. *Incorporated Law Society* [1895] 2 QB 456 mandamus was refused to compel the Law Society to investigate charges of misconduct against a firm of solicitors, because the applicants could have brought the charges directly before the court.

[53] See below Ch. 4.6.

died out as a means of enforcing public duties, except in cases of punishment for contempt for disobedience to an order of mandamus.[54]

An action for damages used to be generally regarded as an appropriate alternative remedy such as to exclude mandamus.[55] However, as Wright J put it in *R.* v. *London & North Western Railway Co.* (*1894*):

> There must be many cases in which a prerogative writ of mandamus ought to be granted for the purpose of speedy justice, or other reasons, in which it would be uncertain whether it was possible to achieve the same results in any other way, and if it were possible would take much time and difficulty.[56]

(iii) *Unfortunate consequences of mandamus*

Mandamus has sometimes been refused where the performance of the duty is impossible or would involve illegality, or would be inconvenient or futile. All these grounds relate to the consequences of making an order, the consequences, that is, not simply as between the parties, but from the point of view of the wider public interest. They therefore give the courts an opportunity to consider questions of public policy which would not come to be considered if the award of mandamus depended purely on whether the authority was in breach of its duty; this is an advantage because it does not follow from the fact that an authority has breached its duty that it ought to be commanded to rectify the breach. None the less it is of course important that the discretion be confined to refusing an order only in cases which will not set a precedent inviting or excusing dereliction of public duties.

It is said, first of all, that where the duty becomes impossible to perform, mandamus will not be granted: *lex non cogit ad impossibilia.*[57]

[54] See *R.* v. *Bedwellty Urban District Council, ex p. Price* [1934] 1 KB 333; and below Ch. 3.5.

[55] *R.* v. *Free Fishers of Whitstable* (1806) 103 ER 137; *Nicholl* v. *Allen* (1862) 31 LJQB 286; *R.* v. *Lambourn Valley Railway Co.* (1888) 22 QBD 463; *R.* v. *London & North Western Railway Co.* (1896) 65 LJQB 516. An action for damages was held to exclude mandamus in the Tasmanian case of *Butler* v. *Zeehan District Hospital* (1912) 8 Tas L. R. 91, but that was an action for reinstatement (see discussion of the *Evans* case below and n. 82), and probably lays down no general principle. An action in detinue for delivery of mail was held, *obiter*, in *R.* v. *Arndel* (1906) 3 CLR 557 (HCA) to exclude mandamus.

[56] [1894] 2 QB 512, 518; and see *R.* v. *Dymock* (*Vicar*), *ex p. Brooke* [1915] 1 KB 147; *R.* v. *Christ's Hospital* (*Governors*), *ex p. Dunn* [1917] 1 KB 19.

[57] See *Re Bristol & North Somerset Railway Co.* (1877) 3 QBD 10. Although it is true that one 'cannot put people in prison for not complying with an order when they have no means of doing so' (per Cockburn CJ at p. 12) that is of course not the only option open in proceedings for attachment; see also *R.* v. *Marshland, Smeeth, and Fen District Commissioners*, above n. 34.

In no decided case, however, has the impossibility taken the form of total physical impossibility, as distinct from extreme inconvenience. The principle has, however, been applied in cases where performance has become *legally* impossible,[58] as, for example, where the authority concerned has ceased to exist,[59] or its powers to effect performance have expired;[60] this is merely a corollary of the principle that the respondent must at least have a power to act in the matter.[61] Performance may of course be impossible because of impending duress or obstruction, and it is not inconceivable that mandamus would be refused in such circumstances; it is more likely, however, that the court would order performance and the duress or obstruction would be considered in relation to contempt for disobedience to the order.[62]

Inconvenience is more problematical. In a sense an order of mandamus always causes inconvenience of some kind, but the courts are apt to say 'fiat justitia, ruat coelum'. However, arguments based on extreme inconvenience are not always to be rejected out of hand. While it is true that if performance is generally very inconvenient the duty should be subjected to legislative review, it is also true that the inconvenience may be due to special circumstances rather than the normal hazards to be expected in the fulfilment of a duty. Thus the courts retain discretion in the matter, but that discretion is sparingly exercised. The cases do not, however, indicate that any clear principles are applied.

The most important argument based on inconvenience is the argument that the authority has no money to do what is demanded, or that it has money but performance would involve unreasonable expenditure which cannot be justified considering the advantages to be gained. There is of course a tendency for the duty itself to be construed so as not to involve unduly heavy expenditure,[63] but there are cases where the expenditure involved was not envisaged by the legislature, and restrictive construction of the duty is not possible.

[58] *R.* v. *Ambergate Railway Co.* (1853) 118 ER 475.

[59] *Ex p. Gosford District Land and Investment Co. Pty. Ltd., re Gosford Shire Council* (1951) 69 WN (NSW) 85; or the applicant has ceased to exist: *Ex p. Clatworthy* (1906) 23 WN (NSW) 193.

[60] *R.* v. *London & North Western Railway Co.* (1851) 117 ER 1113.

[61] *R.* v. *Ely (Bishop)* (1750) 95 ER 610, 96 ER 28; *R.* v. *St Saviour's, Southwark (Churchwardens)* (1834) 110 ER 1252.

[62] Cases in which duress has been considered in relation to breach of the duty are considered above Ch. 2.3. The Courts have not had to consider cases where the duress, being continuing or imminent, is set up as an argument for refusing mandamus, the breach of duty being conceded.

[63] See e.g. *Rodd* v. *Essex County Corporation* (1910) 44 SCR 137 (SCC); *R.* v. *Bristol Corporation, ex p. Hendy* [1974] 1 WLR 498 (see text below).

The authorities are uncertain. Some lean towards the attitude of the court in *R.* v. *Luton Roads Trustees*,[64] in which lack of funds was held to be no excuse, and others lean towards the more liberal attitude of the court in *Re Bristol & North Somerset Railway Co.*,[65] in which it was held mandamus would not be granted where it was shown that performance was due to lack of funds not involving default on the part of the authority. However, cases of the latter kind can often be explained on other grounds: in *R.* v. *Kings County Council*[66] performance of a duty to pay compensation for criminal damage necessitated the levying of a special rate, but it was shown that the authority had provided for the compensation in its estimates for the following year, so that there was no pressing need for a special rate, which would be both troublesome and expensive;[67] in *R.* v. *Publicover*[68] the Court refused to order the holding of a poll whose result was not likely to be of benefit to the applicant (whose standing was also doubted) and which would involve unreasonable expenditure in wartime, when there was a pressing need to curtail expenditure. The only recent case which bears on this question is *R.* v. *Bristol Corporation, ex p. Hendy*,[69] which was discussed in Chapter 2. Two appeal judges in that case held that the authority had not on the facts failed to carry out its statutory duty to provide the applicant with suitable alternative accommodation; the third judge Scarman LJ, on the other hand, proceeded on the assumption that they had failed in their duty and dealt with the case under the rubric of discretionary refusal:

In my judgment, if, in a situation such as this, there is evidence that a local authority is doing all that it honestly can to meet the statutory obligation, and that its failure, if there be failure, to meet that obligation arises really

[64] (1841) 1 QB 860.

[65] Above n. 57. See also *R.* v. *Paddington Vestry* (1829) 109 ER 170 (no mandamus where effect is to put burden of repairing private road on public).

[66] [1908] 2 IR 176.

[67] In a case in the same year *Croydon Corporation* v. *Croydon Rural District Council* [1908] 2 Chanc. 321, the plaintiffs claimed certain payments due from the defendants which the plaintiffs had overlooked for several years; although the defendants admitted liability, mandamus was refused to compel the levying of a rate for previous years. The ratio is unclear, and could rest on the illegality of a retrospective rate or on inconvenience.

[68] [1940] 4 DLR 43; cf. *Smith* v. *Commissioner of Corrective Services* [1978] 1 NSWLR 317 (declaration refused because it would entail the alteration of a prison to ensure that an accused person would not suspect breach of privacy of consultation with his lawyers). In *R.* v. *Greater London Council, ex p. Royal Borough of Kensington and Chelsea*, *The Times*, 7 Apr. 1982, the court refused declarations impugning the validity of a rate precept because to do so would deprive the precepting authority of necessary funds for public purposes.

[69] Above n. 63.

out of circumstances over which it has no control, then I would think it would be improper for the Court to make an order of mandamus compelling it to do that which either it cannot do or which it can only do at the expense of other persons not before the court who may have equal rights with the applicant and some of whom would certainly have equal moral claims.[70]

The crucial point here is that to have made an order would have put the applicant in an advantageous position as regards priority on the authority's housing list. The failure to provide, if any, was only due to circumstances beyond the control of the authority in the sense that to provide what was asked would set a precedent resulting in unreasonable expenditure, or else unfairness to others awaiting their turn in the housing queue. Seen in this light the dictum of Scarman LJ is not really authority for the admissibility of a 'lack of funds' argument, but for an 'equality of treatment' argument. In any event the argument is more properly to be regarded as relevant to the content of the duty, not to the exercise of the court's discretion, and in this respect the reasoning of the majority is to be preferred. If the cases are explained as indicated in this discussion, we find that the 'lack of funds' argument has never been clearly acceded to in any case from any jurisdiction in this century, except where it has gone to the construction of the duty.[71]

Perhaps the matter can be explained in this way. Public authorities, governments, and legislatures have very effective methods of estimating expenditure on the fulfilment of public duties. Thus lack of funds for the performance of a public duty, as opposed to the exercise of a power, is only likely to arise (1) where the authority itself overspends unreasonably on the exercise of powers, (2) where the duty in a particular instance proves more expensive than could have been envisaged, or (3) where the courts construe the duty unexpectedly widely. The last two situations are unlikely to occur, and the first puts the authority beyond the court's sympathy, and may even involve illegality.[72] When one considers in addition that the courts can hardly be expected to review the totality of an authority's expenditure in one year, which is the function of the executive and legislative branches, there seems to be only one real option open to the courts in the overwhelming majority of instances, and that is to insist on performance and leave the consequences to the executive (by way of

[70] Ibid., p. 503; cf. *R.* v. *Ryde Corporation* (1873) 28 LT 629, in which Blackburn J. refused mandamus against the corporation because they had done all they could to comply with their duty.

[71] See, however, above n. 67.

[72] See e.g. *Roberts* v. *Hopwood* [1925] AC 578 (HL); *Bromley London Borough Council* v. *Greater London Council* [1982] 2 WLR 62 (HL).

fiscal readjustment) or the legislature (by way of review of the duty). In the words of a Canadian judge who saw no difficulty in the matter:

> To hold otherwise might very well encourage public and governmental authority to disregard prudent limitations upon their expenses and then permit them to rely upon their own improvidence as an excuse for non-fulfilment of their statutory duties.[73]

It is difficult to say what other heads of inconvenience will be accepted by the courts. Many of those accepted are extremely unconvincing both as general propositions and with regard to the facts of the individual cases. These range from fear of chaos[74] or of a vast number of similar applications,[75] to the possibility of exposing the respondent to double vexation,[76] or an accused person to double jeopardy where a court dismisses charges without giving the prosecution a fair hearing.[77] The decision of the House of Lords in *Chief Constable of North Wales Police* v. *Evans*[78] puts the whole matter of inconvenience on a new footing. The Chief Constable dispensed with the services of a police officer in a summary fashion without even the pretence of observing the principles of natural justice. The House of Lords refused mandamus to reinstate the officer, which was

[73] *McLeod* v. *Salmon Arm School Trustees* (1952) 4 WWR (NS) 385, 386 per Sloan CJ; followed in *Mountain* v. *Legal Services Society* [1984] 1 WWR 113 (duty to provide legal aid executory in spite of lack of funds); see also *R.* v. *Secretary of State for Social Services, ex p. Greater London Council and Child Poverty Action Group*, *The Times*, 8 Aug. 1985.

[74] *Re Central Canada Potash Co. Ltd. and Minister of Mineral Resources of Saskatchewan* (1972) 32 DLR (3d) 107 (affd. (1973) 38 DLR (3d) 317 n.) In *Ex p. Gibson* (1881) 2 NSWLR 203 mandamus was refused to compel granting of a liquor licence in respect of railway premises because it might interfere with the duties of the railway authorities, while in *Ex p. Wallace & Co.* (1892) 13 NSWLR 1 it was refused merely because a bill was being considered which would have the effect of making the refusal to pass imported goods lawful (cf. *Eternit SA* v. *Newfoundland* (1985) 168 APR 48), and in *Ex p. Zillman* (1907) SR (NSW) 362 it was suggested that mandamus would not be granted to compel registration of a minister of religion for the purpose of running a marriage bureau, because marriage bureaux were pernicious! Naturally the fact that an unlawful policy has been operated for many years is no ground for refusing mandamus: *R.* v. *Birmingham Licensing Planning Committee, ex p. Kennedy* [1972] 2 QB 140. Of course inconvenience may also be a good reason *for* granting mandamus; see *R.* v. *Commonwealth Court of Conciliation and Arbitration ex p. Ozone Theatres (Australia) Ltd.* (1949) 78 CLR 389 (HCA).

[75] *R.* v. *Kerrier District Council, ex p. Guppys (Bridport) Ltd.* (1976) 32 P & CR 411.

[76] *R.* v. *Heathcote* (1713) 88 ER 620.

[77] *R.* v. *Birmingham Justices, ex p. Lamb* [1983] 3 All ER 23; see also, however, *R.* v. *Clerkenwell Metropolitan Stipendiary Magistrate, ex p. DPP* [1984] 2 All ER 193; *Weight* v. *Mackay* [1984] 2 All ER 673 (HL); *Harrington* v. *Roots* [1984] 2 All ER 474 (HL).

[78] [1982] 3 All ER 141.

conceded to be, from his point of view, the only remedy that would suffice,[79] and granted instead a declaration which affirmed that the officer's services had been unlawfully dispensed with but stopped short of reinstating him. This was done on the basis that to grant mandamus was 'impractical and might border on usurpation of the Chief Constable's powers'.[80] This decision is not only a lamentable denial of justice. If the reasoning adopted were regularly applied in mandamus cases it would make the remedy virtually useless. Presumably, however, it will be confined to cases of dismissal from office, which have always caused problems of selecting the appropriate remedy.

The case of *R.* v. *Paddington Valuation Officer, ex p. Peachey Property Co. Ltd.*, discussed above,[81] also dealt with the question of inconvenience. Salmon LJ, echoed by the other two judges, took a bold view of the matter:

> I would add that whatever inconvenience or chaos might be involved in allowing the appeal, the court would not be deterred from doing so if satisfied that the valuation officer had acted illegally. One of the principal functions of our courts is to protect the public from an abuse of power on the part of anyone, such as a valuation officer, entrusted with a public duty which affects the rights of ordinary citizens. If the valuation officer acted illegally and thereby produced an unjust and invalid list, this would be such an abuse of power and one which the courts would certainly redress. It could be no answer that to do so would produce inconvenience and chaos for the rating authority—otherwise the law could be flouted and injustice perpetrated with impunity.[82]

However, his Lordship might not have been so sanguine if the majority had not found the valuation list valid, and had not Lord Denning MR provided a *deus ex machina* to resolve the difficulty: mandamus would lie to prepare a new list, on completion of which the old list could be quashed by *certiorari*.[83]

One ground which has been consistently applied is that of illegality. An order will not be granted where to do so would be to compel the respondent to act illegally or even to be a party to illegality.[84] Thus

[79] Damages were not claimed, though they would have been substantial. See also *R.* v. *Saddlers' Company* (1863) 11 ER 1083 (HL).

[80] Above n. 78, p. 156.

[81] See above n. 46 and text.

[82] [1966] 1 QB 419.

[83] Ibid. 402. His Lordship was constrained to say that the old list would be voidable and not void in order to provide this solution; Salmon LJ doubted, however (pp. 418–19), whether Lord Denning MR was right on this point.

[84] *R.* v. *Conyers* (1846) 115 ER 1143; *R.* v. *Garland* (1870) LR 5 QB 269; *Manning* v. *Bergman* (1915) 32 WLR 519; cf., however, *R.* v. *Woodbury Licensing Justices, ex p. Rouse* [1960] 1 WLR 461.

mandamus was refused when a union official, applying as a private person, sought to compel a public tramway company to employ its striking workers to comply with its statutory duty to operate tramcars, when it could only do so by acceding to terms demanded by the workers which were being disputed before a tribunal and being asserted by unlawful means.[85] Similarly it has been held that mandamus will not be used to allow circumvention of the marriage laws of another country, even where there has been an error of law in the decision.[86] Recognized heads of public policy, such as the principle that a person should not benefit from his own wrong, have been treated as implied limitations on the performance of public duties, rather than occasions for the exercise of discretion to refuse an order.[87]

The rule that mandamus will not be granted where to grant it would be futile is one which overlaps to a certain extent the grounds of impossibility and illegality. Mandamus was intended to remedy injustice and the judges were vigilant in refusing the remedy where no substantial injustice had been suffered, or where mandamus would not remedy the injustice.[88] The most common application of the rule these days is in the context of decision-making, where, although an illegality has occurred, the correction of it is not likely to be to the applicant's ultimate advantage because the determination impugned will probably be reached validly on reconsideration,[89] or if reversed

[85] *McPherson* v. *Perth Electric Tramways Ltd.* (1910) 12 WALR 192.

[86] *R.* v. *Brentwood Superintendent Registrar of Marriages, ex p. Arias* [1968] 2 QB 956; but the principle would not extend to refusing an order for the purpose of complying with UN Security Council resolutions: *Bradley* v. *Commonwealth* (1973) 128 CLR 557 (HCA) (declaration and injunction sought, but the decision cannot have been different if mandamus had been sought).

[87] *R.* v. *Chief National Insurance Commissioner, ex p. Connor* [1981] QB 758; *R.* v. *Secretary of State for the Home Department, ex p. Puttick* [1981] QB 767. In *Re an Application by Derbyshire County Council* [1988] 1 All ER 385 Knox J. identified a novel head of public policy to which public duties are subject, namely the due administration of justice, when he held that a library authority could not purchase material containing confidential information contrary to an interlocutory injunction obtained in litigation to which the authority was not a party.

[88] See e.g. *R.* v. *Axbridge Corporation* (1777) 98 ER 1220; *R.* v. *Herefordshire Justices* (1819) 1 Chit. 700; *R.* v. *Griffiths* (1822) 106 ER 1358; *R.* v. *Pembrokeshire Justices* (1831) 109 ER 1188; *R.* v. *Bateman* (1833) 110 ER 563; *R.* v. *Northwich Savings Bank* (1839) 112 ER 1388.

[89] *R.* v. *West Riding of Yorkshire Justices* (1842) 114 ER 275; *R.* v. *Blackall Licensing Court, ex p. Chiconi* (1919) 13 QSR 4; *Seymour* v. *Railway Commissioners* (1919) 19 SR (NSW) 30; *Ex p. Falkiner* (1929) ALR 303; *R.* v. *Warden at Rockhampton, ex p. McPaul* [1937] QSR 96 (affd. HCA [1938] QSR 68); *R.* v. *Hudson* [1915] 1 KB 133; *Ex p. Voge* (1915) 15 SR (NSW) 345; *R.* v. *Lampe, ex p. Madolozzo* [1966] ALR 144; *R.* v. *Doncaster and Templestowe City, ex p. Mayor* (1970) 27 LGRA 193. In *Wade* v. *Burns*

by the decision-maker, overturned by another body,[90] or because of some other supervening factor.[91] This use of the rule has cut quite deeply, and yet is hard to justify. If the applicant is sufficiently convinced that he has a possibility of ultimate success to go to the trouble of applying for mandamus, why should the court deny him the opportunity he seeks (not to mention the emotional release involved in winning his case) merely because it can only assist him over one obstacle out of many? And why should not the decision-maker be subjected to a coercive order which highlights the illegality which has occurred and ensures that it will not only be corrected but will not be repeated? If the courts are to judge legality and not merits, why should they presume to judge the merits at this stage, particularly when the decision-maker himself has not yet judged the merits in accordance with law?

In the cases in which mandamus has been refused on this ground, the applicant's chances of success have varied from almost nil to probably worthy of pursuit; but no principles have been laid down which would assist in discerning which cases are appropriate for refusing an order. Many cases are explicable only on the basis that the court simply did not feel inclined to interfere with the determination, a matter of justiciability which is more suitably settled at an earlier stage of the reasoning process, or else on the basis that the court was unwilling to see the decision-maker cock a snook at the law by, in effect, ignoring the court's order with impunity, a ground which is of course also quite irrelevant to the issue of the legality of his determination. In the face of this confusion it is suggested that in the context of decision-making the 'futility rule' should be applied only in the following circumstances:

1. Where there is no real possibility at all of a result favourable to the applicant, for example where, even if the decision is remitted to the decision-maker, he cannot find in favour of the applicant because it transpires he clearly has no jurisdiction in the matter,[92] or where

(1966) 115 CLR 537 (HCA) mandamus to a mining warden to hear and determine in accordance with law was held not to be futile merely because the warden had said that if he had a discretion in the matter he would refuse the application, the High Court having decided that he did have a discretion in the matter.

[90] *R.* v. *Richmond City, ex p. May (E. B.) Pty. Ltd.* [1955] VLR 379; *Mitchell (G. H.) & Sons (Australia) Pty. Ltd.* v. *Minister of Works*, above n. 44.

[91] e.g. on restoration to office the applicant could be easily removed validly: see the cases cited above n. 88, and *R.* v. *Saddlers' Company*, above n. 79 (House of Lords split 8 to 3 in favour of mandamus to reinstate).

[92] *R.* v. *Fermanagh Justices* [1897] 2 IR 559.

the illegality involved is a breach of natural justice but it transpires the applicant has no real merits to argue.[93]

2. Where the possibility of a result favourable to the applicant is sufficiently remote not to outweigh the trouble and expense involved, for example where the order is to compel the holding of an election in which the result is unlikely to be different even if it is validly conducted.[94]

3. Where the granting of the order would affect the rights of third parties without substantial benefit to the applicant, for example where the applicant seeks reinstatement to an office which has already been filled by another person.[95]

4. Where a decision in favour of the applicant would be of no practical use to him, for example where it would involve the granting of a licence for a period which has already expired.[96]

[93] *R.* v. *West Norfolk Assessment Committee, ex p. Ward* (1930) 94 JP 201; *R.* v. *Liquor Control Commission, ex p. Dickens (S. E.) Pty. Ltd.* [1983] VR 303 (see esp. the caustic remarks of Anderson J. at pp. 312–13).

[94] *Ex p. Mawby* (1854) 118 ER 1310; *R.* v. *Newry Urban District Council* (1910) 43 ILT 172. And see *Ex p. Lucas* (1910) 10 SR (NSW) 120 (no mandamus to the mayor to call a special meeting where the Council's business was at a standstill due to quarrels, because the writ would not settle the disputes or result in the resumption of business); *Smythe and Humbert* v. *Anderson* (1970) 73 WWR 536; *R.* v. *Broadcasting Complaints Commission, ex p. Owen* [1985] 2 WLR 1025.

[95] See the first four cases at n. 88 above; *Re Barnes Corporation, ex p. Hutter* [1933] KB 668; *R.* v. *Central Railway Appeal Board, ex p. Sparks* [1941] QSR 73; *R.* v. *National Dock Labour Board, ex p. National Amalgamated Stevedores and Dockers* [1964] 2 Ll. Rep. 420. It is otherwise of course where the position has not yet been filled: *R.* v. *Public Service Board of Victoria, ex p. McDonald* [1948] VLR 310; see also *Ex p. Duncan, re Mooney* (1942) 59 WN (NSW) 25. In *Re Cheyne and Shire of East Loddon* (1899) 24 VLR 900 mandamus was sought to reopen obstructed highways; the applicant included in his application some highways in respect of which he had no real interest, and with regard to these mandamus was refused, as it could be of no benefit to him but might cause harm to others; cf. *R.* v. *Epsom Urban District Council, ex p. Course* (1912) 10 LGR 609.

[96] *R.* v. *Birmingham Recorder* (1855) 24 LTOS 256; *Ex p. Montgomery, re Blue Mountains City Council* (1955) 20 LGR (NSW) 174; *O'Brien* v. *Fagg* [1972] QSR 559; cf., however, *Ex p. Wright, re Concord Municipality* (1925) 7 LGR (NSW) 79 and *Vic Restaurant Inc.* v. *Montreal* [1959] SCR 58 (SCC) (mandamus used to declare rights even where no effective order could be made, the issue having expired due to lapse of time and sale of business). An interesting case of this kind is *Ex p. Northern Rivers Rutile Pty. Ltd., re Claye* (1968) 72 SR (NSW) 165. The applicants sought to compel the upholding of their complaint against their business competitors of breach of labour conditions under a mining lease. An order was refused partly on the ground that even if the complaint was upheld the competitors' lease was unlikely to be forfeited, and even the chance of the applicants obtaining the lease was flimsy. These considerations also impelled the court to hold that they had no standing: cf. *R.* v. *Harlock, ex p. Stanford & Atkinson Pty. Ltd.* [1974] WAR 101. In *R.* v. *Aston University Senate, ex p. Roffey* [1969] 2 QB 538 a student was refused mandamus to reconsider

The case of *R.* v. *Anderson, ex p. Ipec-Air Pty. Ltd.*[97] raises an interesting difficulty with regard to the last of these principles. In that case the applicants obtained mandamus against the Director-General of the Civil Aviation Department to issue a charter licence. In the High Court of Australia the two dissenting judges would, however, have refused mandamus because the applicants had tried and failed to obtain the aircraft which they needed in order to operate under the licence. The majority were not persuaded that mandamus should be withheld for that reason.[98] The correct position in such a situation is, one would have thought, that the use a person makes of a licence is his own business provided it is lawful, and it was perfectly lawful for the applicants to hold the licence on the chance that the aircraft might be obtained. Had the High Court refused mandamus on this ground, and had the applicants then obtained the aircraft, they would have been unable to use them pending a subsequent, unnecessary, and probably unsuccessful application for a licence, and then judicial review of its refusal. For these reasons the majority view is much to be preferred.

The courts have considered other matters under the head of futility which do not necessarily relate to decision-making. In some cases the respondent has sought to evade an order by pointing to the impending repeal of the duty, but this has not been regarded as a good ground for refusing an order.[99] It is not for the courts to anticipate the outcome of legislative proceedings, and in any case the issue may relate to a breach of duty which ought to be corrected even if the duty is to be repealed. However, in a case where the duty is really obsolete and its revival would involve the setting up of machinery for implementation of the duty which may well have to be dismantled

his exclusion from the university partly on the rather dubious ground that he had obtained a place in a polytechnic. There is also some authority for the view that where the matter involved has been executed, e.g. an order obeyed (*R.* v. *Hopkins* (1900) 64 JP 454) or a warrant executed (*Coles* v. *Wood* [1981] 1 NSWR 723), mandamus will be refused. Hypothetical issues will not be determined: *R.* v. *Blackwall Railway* (1841) 9 Dowl. 558; *R.* v. *Workmen's Compensation Board, ex p. Kuzyk* (1968) 69 DLR (2d) 291.

[97] (1965) 113 CLR 177.

[98] e.g. ibid. 203 per Windeyer J.

[99] In *State* (*Modern Homes Ltd.*) v. *Dublin Corporation* [1953] IR 202 (SCI) it appears that the corporation were merely playing for time; having resolved to make a planning scheme, they then decided to continue 'interim control', i.e. to maintain the status quo, pending revision of the legislation; mandamus was granted partly because this refusal to act was causing the applicants financial damage; cf. *R.* v. *Rathmines Urban District Council* [1928] IR 260 (SCI); and see *R.* v. *Whiteway, ex p. Stephenson* [1961] VR 168.

on repeal of the duty, the court could postpone the operation of the order and make it conditional on continuance of the duty.[100]

Mandamus may be futile because it may be impossible for the courts to supervise obedience to the order.[101] This is undoubtedly a shortcoming of the remedy. It tends to be evident in cases outside the decision-making context, where compliance with the order may involve detailed and debatable questions of fact. In some cases it has been overcome by indicating carefully the principles applicable and leaving compliance to negotiation between the parties,[102] a course which is somewhat uncertain, though it may be effective where the parties are both responsible public bodies.

Naturally where the respondent complies with his duty before the hearing,[103] or where he indicates at the hearing his intention to do so,[104] mandamus is not normally issued.

(iv) *Conduct of the applicant*

The last group of considerations relevant to the exercise of the court's discretion to refuse mandamus relates to the applicant's conduct: mandamus is ordinarily refused where the applicant is guilty of delay in making the application, or where it is made in bad faith.[105]

Where the application is by way of judicial review of a determination, the factor of delay is important because the interests of persons other than the applicant, or the general interest of the public, may hinge on the validity of the determination, which cannot be left in doubt indefinitely. For this reason statutory time-limits have generally been imposed on mandamus applications.[106] However, it is not clear

[100] Cf. the *Peachey Property* case, above n. 46 and text.

[101] *R.* v. *Peak Park Joint Planning Board, ex p. Jackson* (1976) 74 LGR 376, 380 (*certiorari* granted instead, however); cf. *Chief Constable of North Wales Police* v. *Evans*, above n. 78.

[102] As in some declaration cases: see below Ch. 4.4.

[103] *R.* v. *Metropolitan Police Commissioner, ex p. Blackburn* [1968] 2 QB 118; *R.* v. *Kensington and Chelsea London Borough Council, ex p. Birdwood* (1973) 72 LGR 289.

[104] There is a tendency not to issue mandamus to a court or tribunal where it is clear that it will reconsider the matter in accordance with law: see *R.* v. *Commonwealth Court of Conciliation and Arbitration, ex p. Ellis* (1954) 90 CLR 55 (HCA); *R.* v. *Marlborough Street Stipendiary Magistrate, ex p. Bouchereau* (1977) 1 WLR 414; *R.* v. *Police Complaints Board, ex p. Madden* [1983] 2 All ER 353; or where the matter is best left to the co-operation of the parties: *R.* v. *Devon and Cornwall Chief Constable, ex p. Central Electricity Generating Board* [1981] 3 WLR 967 (see above Ch. 2.4); *Liddle* v. *Sunderland Borough Council* (1983) 13 Fam. Law 250.

[105] For special considerations concerning motives in applying for mandamus to produce documents, see above Ch. 2.6.

[106] See below Ch. 5.1.

why delay should be a factor in simple cases of failure to perform a duty other than the duty to determine, where it may be that there is nothing for anyone to rely on and the granting of mandamus would simply mean a change in policy;[107] the only principle which operates in these cases is that the law will only help those who help themselves, but that is a poor answer to the argument that the longer the delay the more the breach of duty needs to be corrected, especially where the public are affected, not just the applicant.[108] The courts should still be able to refuse mandamus on grounds of delay in such cases, but should not be tied to arbitrary time-limits, because the time factor is only one of many factors to be considered in deciding whether the delay should be a reason for refusing a remedy. It may in fact be difficult in these cases to say when the delay commenced; ordinarily the period will begin on refusal to perform following a demand for performance, but as we have seen this is not always present or necessary in mandamus cases.[109]

It is not clear what factors excuse delay, but the courts proceed on the basis that the delay must be unwarrantable;[110] in practice this means that it is for the applicant to show reason for the delay, and the courts inquire quite closely into the circumstances.[111] What the courts generally look for is merely lack of persistence in the assertion

[107] In *Cocks* v. *Thanet District Council* [1982] 2 All ER 1135, 1139 Lord Bridge said, in relation to a housing duty, that undue delay in seeking a remedy was 'perhaps not often likely to present a problem'. It may be that the courts will be lenient where the delay does not adversely affect the administration.

[108] See *R.* v. *All Saints, Wigan (Churchwardens)*, above n. 15, p. 630 per Lord O'Hagan.

[109] See above Ch. 3.3.

[110] The expression used in the Supreme Court Act 1981 (England and Wales), s. 31(6), is 'undue delay', which may be considered at the leave stage or the discretion stage, depending, presumably on the difficulty of determining whether, on the facts, the delay is undue. See *In re Wall* (1890) 16 VLR 686; *Broughton* v. *Commissioner of Stamp Duties (NSW)* [1899] AC 251 (PC); *R.* v. *Hanley Revising Barrister* [1912] 3 KB 518; *R.* v. *Hanlon, ex p. Kenny* [1933] QSR 213 (10 years' delay not explained by alleged bias of incumbent minister); *R.* v. *Transport Regulation Board, ex p. Maine Carrying Co. Pty. Ltd.* [1940] VLR 19; *Ex p. Walker, re Goodfellow* (1944) 45 SR (NSW) 103; *Ex p. McDonald, re Lake Macquarie Shire Council* [1961] NSWR 451; *R.* v. *O'Sullivan, ex p. Clarke* [1967] WAR 168; *Ee Kim Kin* v. *Collector of Land Revenue* [1967] 2 MLJ 89; *Ex p. Commissioner for Railways, re Locke* (1968) 87 WN (Pt. 1) (NSW) 430. The periods of delay held to be unwarrantable vary from a few weeks to 65 years (*R.* v. *Leeds & Liverpool Canal Co.* (1840) 113 ER 435)!

[111] In *Re Australian Broadcasting Tribunal, ex p. Fowler* (1980) 54 ALJR 549 the High Court of Australia applied this principle with some rigour even though no other remedy than mandamus was available.

of the applicant's rights, even where it is due to lack of means:[112] any reasonable and genuine excuse will probably suffice.

How far should the motives of the applicant affect his right to mandamus? The position taken by the courts here is that since the granting of mandamus is discretionary, not a matter of right, the applicant's motives are relevant.[113] While this rule undoubtedly enables the court to look more broadly at the circumstances of the case,[114] so that, for example, an order will not be granted where it would serve the ends of a scheme to defeat the purpose of the legislation,[115] it is not precisely clear what motives are considered sufficiently unworthy to disentitle the applicant to an order. The courts have looked rather too carefully at motives of a commercial nature;[116] after all, as De Smith says, 'litigation need not be inspired by public-spiritedness or altruism'.[117] Collusive applications have not been granted, because they are an abuse of the court's process.[118] In *Ex p. Mullen, re Wigley*[119] one gathers that the motive of the applicant[120]

[112] *Ex p. Anlezark, re Manufacturers' Mutual Insurance Ltd.* (1930) 31 SR (NSW) 53.

[113] *R.* v. *London Corporation* (1787) 100 ER 96; *R.* v. *Liverpool, Manchester, & Newcastle-upon-Tyne-Railway Co.* (1852) 21 LJQB 284; *Quinlivan* v. *McCracken's City Brewery Ltd.* (1899) 5 Arg. L. R. 263; *Coastal District Society of Engineers* v. *Arbitration Court (WA)* (1912) 23 WALR 87. The suggestion in *Frankel* v. *Winnipeg* (1912) 8 DLR 219 that where a legal right is asserted, motive is irrelevant, cannot be supported on the authorities: see esp. *Rodd* v. *Essex County Corporation* (1910) 44 SCR 137 (SCC); *Re Burgin and Township of King* (1973) 36 DLR (3d) 198. In *R.* v. *Peterborough Corporation* (1875) 44 LJQB 85 the applicant was refused mandamus against the corporation to hold a poll in respect of a decision to oppose a private bill because the application was in reality made on behalf of a solicitor proposing the bill, which would have given powers to a company in which the solicitor was interested. If of course the authority is guilty of bad faith, this may hasten the exercise of discretion in favour of the applicant: *Re Smith and Municipality of Vanier* (1972) 30 DLR (3d) 386.

[114] Including the history of the relationship between the parties: *Ex p. Trans Estates Pty. Ltd., re Hornsby Shire Council* (1964) 83 WN (Pt. 1) (NSW) 15. The statement in *O'Connor* v. *Jackson* [1943] 4 DLR 682, that discretion should be exercised as a result of something related to the right itself and not extraneous thereto, is patently false.

[115] *Re Burgin and Township of King*, above n. 113; *Re Stinson (George) Construction Inc. and Township of Ameliasburgh* (1977) 15 OR (2d) 547.

[116] See *R.* v. *Whiteway, ex p. Stephenson* [1961] VR 168; *R.* v. *Commissioners of Customs and Excise, ex p. Cook* [1970] 1 WLR 450. In both cases, however, it is arguable that the motives of the applicant went to standing and not discretion. The applicant cannot be allowed to profit from another's wrong: see *R.* v. *Pellizzon and Borough of Etobicoke* (1970) 10 DLR (3d) 313.

[117] *Judicial Review of Administrative Action*, 4th edn. by Evans (1980), p. 559.

[118] See *R.* v. *Blackwall Railway* (1841) 9 Dowl. 558 (issue hypothetical); cf. *Re Charlottetown Mayoralty, re Farmer* [1952] 1 DLR 778.

[119] [1970] 2 NSWR 297; cf. *Ex p. Hughes, re Moulden* (1947) 47 SR (NSW) 91.

[120] The judgments are remarkably coy.

was to protest against the national-service laws at the time of Australia's involvement in the Vietnam War; he had mounted a (presumably collusive) private prosecution against N, who had been convicted of inciting breaches of those laws; N did not oppose the application for mandamus to the magistrate for her committal for non-payment of the fine, and the application was accordingly refused. In a case where the order is to issue a licence and the authority has no discretion in the matter, the court's discretion in refusing an order extends to bad faith or fraud on the part of the applicant in obtaining his entitlement to the licence.[121]

3.5 ENFORCEMENT OF MANDAMUS

When an order of mandamus is issued it must be complied with by the date fixed in the order for compliance.[1] If no date is fixed, the order must be complied with immediately, or the respondent is in contempt.[2] In *Re Ample Investments Ltd. and City of Toronto*[3] mandamus was granted against the authority on 27 June to issue a demolition permit to the applicant, who, to the authority's knowledge, urgently required to commence demolition on the site in question. The officer concerned, who knew of the court's order on 27 June, allowed the permit to be issued in the normal manner, so that it was not received until 8 August, by which time the statute had been amended in a manner which rendered the permit invalid. When the applicant attempted to commence demolition, the authority prevented him. The applicant then proceeded for attachment against the authority and its officer. The court held that the order should have been complied with immediately and that no reasonable excuse had been given for the delay, and ordered the authority to purge its contempt by issuing a valid, unconditional permit within two weeks and consenting to the demolition, with a threat of a very heavy penalty if this order was disobeyed, and further ordered the officer to submit a written apology to the court within two weeks under threat of a $500 fine if he failed to do so.

[121] *R.* v. *Medical Board of Queensland, ex p. Horton* (1897) 7 Qld LJR 122; *In re Batchelor* [1905] VLR 579.

[1] See below n. 2.

[2] *HLE*, vol. ix, para. 69; Borrie and Lowe, *Law of Contempt*, 2nd edn. (1983), ch. 13; RSC (England and Wales), O. 45, rr. 5, 7, 8. For enforcement of court orders against public authorities, see Harlow, 'Administrative Reaction to Judicial Review' [1976], PL 116.

[3] (1974) 54 DLR (3d) 18.

Disobedience is a question of fact and must be established by evidence against the authority and any individual officers or members proceeded against. It is important, bearing in mind the possibility of committal for contempt, that natural justice be observed so that any respondent has an opportunity of exculpating himself.

The leading authority is *R.* v. *Poplar Borough Council*, *ex p. London County Council* (*No. 1*)[4] a case which was part of a political *cause célèbre* in England in the 1920s known as 'Poplarism'.[5] The case arose out of a refusal of the Poplar Borough Council [Poplar] to pay sums due to the London County Council [LCC] and another authority in respect of a rating precept. Poplar's case was that the precept was unfair in that the burden was distributed equally among the London boroughs and did not depend on the rateable value of the borough, Poplar's being exceptionally low; that Poplar itself was overdrawn on its bank account and unable to pay; and that to levy a rate would be useless because the ratepayers, being indigent or unemployed, would be unable to pay. The LCC applied for mandamus, and the writ was resisted on the ground that they had an alternative remedy of distress. In the event the writ was issued and Poplar's appeal to the Court of Appeal was dismissed.[6] Poplar failed to comply with the writ and the LCC proceeded for a writ of attachment against Poplar and thirty of the forty-one members served with the writ of mandamus. Of those councillors proceeded against, twenty-seven signed an affidavit, and seventeen of those addressed the court, to the effect that they refused to obey the writ of mandamus for the reasons indicated above, and because a writ of attachment would cause more chaos than their refusal to pay the precept. The writ of attachment was granted against Poplar and the twenty-seven councillors before the court. Before the Court of Appeal, counsel for Poplar again left aside the substantive issues,[7] and relied on a procedural point, namely that the individual councillors, as opposed to Poplar itself, had not been served with affidavits alleging the nature of their contempt.

The Court of Appeal directed itself first to the question of motive, which had not been argued:

Unless and until the time comes when the law of this country is that a person may disobey an order of the Court or the laws as much as he likes if he does

[4] [1922] 1 KB 95.

[5] The story is of great interest: see Branson, *Poplarism 1919–1925: George Lansbury and the Councillors' Revolt* (1979).

[6] [1922] 1 KB 72; and see above Ch. 3.4 at n. 25.

[7] Applying the principles discussed above Ch. 3.4(iii), it is possible that a frontal attack might succeed today, but it may be that counsel considered the sympathies of the judges to be against his radical socialist clients.

it conscientiously the question of motive is immaterial. That is not the law at present. There have been instances in which by statute conscientious motives have been stated to be, and have been constituted, sufficient reason for not obeying the law. But this is not a case that comes within any of those exceptional statutes. The fact that such exceptional statutes were necessary to make conscientious motives a justification of disobeying the law shows that the general law is not to that effect. The question therefore whether the members of the defendant council here acted conscientiously or not, has, in my opinion, nothing to do with the matter. They deliberately intend to continue to disobey the orders of the Court, as they have done up till now.[8]

It was accepted by the Court that where mandamus issues to a corporate body, which can only act by the persons who compose it, those persons can be attached for disobedience,[9] but the difficulty lay in the procedure to be followed. Warrington LJ laid down the following principles: (1) the writ of mandamus must be shown to have been brought to the knowledge of the person sought to be attached; (2) any individual to be treated as responsible for the disobedience to the writ must be named in the originating process; (3) the originating process must be served upon such individual; (4) copies of the affidavits proposed to be used on the hearing of the application for attachment must be served upon such individual; and (5) there must be sufficient evidence of disobedience.[10] In the instant case the second requirement had not been complied with, and though the fourth had been, the affidavits did not establish disobedience on the part of the councillors individually. The court held, however, that those irregularities had been waived by the affidavit the councillors had sworn, and accordingly the writ of attachment issued and the councillors were committed.

The decision is problematical because it produces a very peculiar result: by ignoring the writ of attachment the councillors would have escaped completely, and yet by volunteering evidence which the court agreed was given in good faith, they exposed themselves to imprisonment. However, the principles established in the case are surely correct. If the ultimate sanction for disobedience to mandamus is ineffective, the entire system of enforcement of public duties may be ineffective in cases where it is of the greatest practical value. The

[8] Above n. 4, p. 103.

[9] The conclusion is inevitable because a public authority cannot be committed and a fine would not only be inadequate, but would be in effect paid by the innocent ratepayers or taxpayers. With regard to officers of an authority subjected to mandamus, notwithstanding the case cited above n. 3, one would imagine that the responsibility lies with the members, who have the remedy in their own hands if an officer refuses to comply with their instructions to comply with the court's order.

[10] Above n. 4, pp. 112–13.

fact that there are very few reported cases[11] of attachment for disobedience to mandamus is immaterial, because it is only the seriousness with which society regards disobedience to a mandatory court order which compels obedience to such order. Proceedings for attachment cannot be allowed to become a forum for pursuing issues relevant to the issue of mandamus, and the only issues which are properly raised at this stage are the validity of the order, the observance of natural justice, and whether there is sufficient excuse for the disobedience, as opposed to the failure to perform the duty. Although motive should be regarded as irrelevant, there must be some element of *mens rea*. This element has not been defined in any of the cases, but presumably there is only disobedience to an order where the respondent deliberately or neglectfully fails to comply with it, and presumably also the non-observance of the order is sufficient evidence of guilt, assuming that a period of time has elapsed in which one might reasonably expect observance of the order to be effected. The general principle should be that mandamus should be complied with as soon as is reasonably practicable.

One limitation in contempt proceedings is that the Crown is not liable to committal.[12] One supposes that the same principles apply here as apply to mandamus writs themselves,[13] so that the limitation may not be a serious one.

These days, disobedience to court orders by public authorities is by no means unknown. In situations such as that in the *Poplar* case which are highly charged politically, the requirements of natural justice should be given equal weight with the need to secure observance of the law. Ultimately these situations should be resolved through the political process, as the judges in the *Poplar* cases clearly indicated.[14]

[11] The only other cases since the early 19th cent. discovered by the writer are *R.* v. *Ledgard* (1841) 113 ER 1268, where the writ was held to be bad, and *R.* v. *Worcester Corporation* [1903] 68 JP 130.

[12] For obvious reasons; see *HLE*, vol. ix, para. 58.

[13] See above Ch. 3.2.

[14] 'If the existing law presses too hardly upon the borough of Poplar, it is Parliament alone who can reduce the pressure. If those who control the expenditure of the borough lay a greater burden upon the ratepayers than they can bear the ratepayers have the remedy in their own hands', per Parkes LJ [1922] 1 KB 84.

4
Enforcement: Other Remedies

4.1 INTRODUCTION

Since the seventeenth century mandamus has been the principal remedy for enforcing public duties, and it, or something very much like it, will probably remain so. Indeed almost the entire law of public duties has grown out of mandamus. None the less there are many other remedies available for enforcing public duties, although of these only default powers can be regarded as a remedy specifically for enforcing public duties, the others having various functions to perform. In this chapter each remedy will be taken in turn and compared with mandamus for its advantages and disadvantages. Standing, however, is considered separately in Chapter 6.

In the context of comparison of remedies it naturally becomes important to consider administrative as well as legal remedies. Not all of these are effectively available to individuals, but from the viewpoint of society generally, these controls are the principal means by which public duties are in fact enforced on a day-to-day basis, although, again, they have other purposes.[1]

Legislatures are responsible for the creation of the overwhelming majority of public duties, and those which exist otherwise, exist by the sufferance of legislatures. Ultimately public authorities are in theory responsible to the legislature and the electorate for the fulfilment of their duties. However, except in extreme cases which happen to become political issues, the performance of public duties is not frequently discussed in legislatures, and it is rare for a legislature to amend a public duty in the light of experience. It is in reality the departments of government which determine the detailed content of public duties by means of the administrative machinery and practices which they create or promote. In particular the departmental circular determines, in the first instance, what the duty is envisaged to involve, and in the majority of instances will in practice be the authoritative pronouncement of the content and also the manner of performance of the duty. In the case of the duty to determine, however, less is left

[1] See Griffith, *Central Departments and Local Authorities* (1966), pp. 54–62: Davies, *Local Government Law* (1983), ch. 9; Loughlin, *Local Government in the Modern Constitution* (1985).

to administrative discretion and the content of the duty can be found in statutory instruments, procedural rules of practice, and decisions of the decision-maker or tribunal and the courts. In general the departmental circular is probably the best control mechanism for public duties because it is highly flexible and sensitive to changing social and economic conditions. Many statutes allow a minister or other high authority to give an authority 'directions' or 'guidance' as to the performance of its duties; such directions or guidance have the force of law and can be enforced by mandamus unless shown, on a challenge by the authority directed, to be invalid.[2] In some cases the control mechanism is negative rather than positive in that the initiative lies with the authority entrusted with the duty, but the approval of the minister is required before proposals for the performance of the duty, or a particular scheme or order, can be effected.[3] In some fields, notably public protection services and some social services, there are inspectorates which ensure a standard and high degree of compliance. It may be possible in some instances for a public official to be subjected to disciplinary procedures for failure to perform his public duty.[4] The duties of government departments themselves are subject to controls internally, but externally, apart from judicial review, they are controlled only by the legislature through the concept of ministerial responsibility, and the ombudsman investigation. This latter control mechanism is particularly useful for enforcing public duties where the default is not seen as an unlawful one, because it has the great advantage that it is available to individuals, no formal standing requirements apply, and *ex gratia* compensation can be recommended. Financial controls can also be useful. In particular, audit procedures can remedy default involving financial loss; a good example is the Local Government Act 1972 (England and Wales), section 161.[5]

[2] See e.g. *Secretary of State for Education and Science* v. *Tameside Metropolitan Borough Council* [1976] 3 WLR 641 (HL).

[3] See Griffith, above n. 1.

[4] See *Miles* v. *Wakefield Metropolitan District Council* [1987] 2 WLR 795 (HL).

[5] S. 161(4) enables a district auditor to recover sums which a person has failed to bring into account which should have been included, and losses incurred or deficiencies caused by the wilful misconduct of any person, from the person concerned; s. 165 allows an extraordinary audit, which was the remedy adopted in *Asher* v. *Lacey* [1973] 1 WLR 1412 (see also *Asher* v. *Secretary of State for the Environment* [1974] Chanc. 208), in which local councillors were surcharged a sum equal to the losses incurred to the authority by their refusal to implement their duties under the Housing Finance Act 1972 (England and Wales). Cf. Local Government Act 1958 (Vic.), ss. 494–501; Local Government Act 1919 (NSW), ss. 210–14. See also Davies, above n. 1, ch. 7; *Lloyd* v. *McMahon* [1987] 2 WLR 821 (HL).

The most important remedy of an administrative kind for enforcing public duties is, however, default powers, and to these we now turn.

4.2 DEFAULT POWERS

The term 'default powers' covers a wide variety of statutory provisions having in common the same basic structure: on the fulfilment of certain conditions, generally to the effect that an authority has failed to perform its duty, a named higher authority, often a minister, by following a certain procedure, may declare the authority to be in default, which declaration has the consequence that the higher authority may take specified action to remedy the default. It is difficult to classify these default powers as strictly legal or strictly administrative remedies—they are in fact hybrid. They are not generally available to an individual member of the public, except to the extent that he can succeed in persuading the higher authority to act,[1] and in fact they will only come before the courts (1) where the defaulting authority challenges the legality of their exercise,[2] (2) where the higher authority seeks an order (usually mandamus) requiring the defaulting authority to comply with directions or other requirements imposed in pursuance of the default provisions,[3] or (3) where an individual member of the public challenges the legality of the higher authority's refusal to act.[4] Although powers of this kind have been in existence at least since the Public Health Act 1875 (England and Wales), they have only recently been considered by the courts apart from the context of their being an alternative remedy to mandamus; this aspect of default powers has been discussed above in Chapter 3. It is likely, given the acutely controversial policy, and indeed constitutional, issues which these powers raise, and the current tensions

[1] Such attempts usually fail: for recent examples see *Meade* v. *Haringey London Borough Council* [1979] 2 All ER 1016, where the Minister refused to intervene even when all education in the borough was suspended for weeks because of a caretakers' strike, and *R.* v. *Secretary of State for the Environment, ex p. Ward* [1984] 2 All ER 556.

[2] See e.g. *Durayappah* v. *Fernando* [1967] 2 AC 337 (PC).

[3] See e.g. *Secretary of State for Education and Science* v. *Tameside Metropolitan Borough Council* [1976] 3 WLR 641 (HL). The legality of the exercise of the powers can of course be challenged on such an application, as in the *Tameside* case itself.

[4] See e.g. *Ex p. Ward*, above n. 1. Exceptionally, a statute may give a member of the public a default remedy, e.g. Highways Act 1959 (England and Wales), s. 59 (court order requiring proper repair of highway), invoked in *Worcestershire County Council* v. *Newman* [1975] 1 WLR 901.

in central-local relations, especially in Britain,[5] that they will be considered by the courts more frequently, and they therefore deserve some attention.

An interesting feature to note about default powers is that they are virtually exclusively directed towards duties to provide, and tend to concern such important services as education, housing, and public health. They are not generally, as one might have expected, in the nature of emergency reserve powers,[6] but are rather powers which are designed to bolster the introduction of a new statutory duty by enabling central government to coerce local authorities where they 'go slow' or 'go on strike' because of their dislike of having to implement a policy which possibly emanates from a government of a different political colour, or sometimes where they implement the policy in a manner which is unacceptable to the government.[7] In situations of this kind mandamus may not be the useful weapon of central government which it once was[8] for ensuring effective continuation of normal services; this is because mandamus does not enable any actual transfer of functions, but has only the law of contempt as its ultimate sanction, does not enable detailed and continuing supervision of administrative action by the administration itself, and is somewhat cumbrous in the sense that a successful application may take some time to achieve, and there may be many authorities involved. Although mandamus is now shorn of many of its difficulties, and the courts can decide successive appeals quickly where necessary,[9] mandamus cannot match the administrative process in speed, sensitivity, and comprehensiveness.

Default powers are, however, in the nature of ultimate sanctions to be considered where all other available administrative and political methods of securing performance of the duty have failed, and they tend to be drastic and controversial, involving in some cases the actual transfer of functions from the defaulting authority to another body or even to the minister himself. Default powers can thus result in a

[5] See Loughlin, *Local Government in the Modern Constitution* (1985); and for default powers in this context, id., *Local Government in the Modern State* (1986), pp. 164 et seq.

[6] For an exception, see National Health Service Act 1977 (England and Wales), s. 17, unsuccessfully invoked, because of the absence of an emergency, in *Lambeth London Borough Council* v. *Secretary of State for Social Services* (1980) 79 LGR 61.

[7] As in e.g. the *Tameside* case, above n. 3, and *Norwich City Council* v. *Secretary of State for the Environment* [1982] 1 All ER 737.

[8] See above Ch. 3.1.

[9] As in e.g. the *Tameside* case, above n. 3, and see the discussion of that case below.

temporary constitutional deviation in some extreme cases.[10] As a result this drastic nature of default powers is their weakness as well as their strength: like any dangerous weapon they are very rarely used, and when they are used, provoke extreme reaction. They are thus better suited to major political showdowns than securing the rights of individuals. None the less they are of great importance in ensuring the enforcement of public duties at the general, if not the particular, level, if only because of their mere existence.[11]

The conditions under which a default power can be invoked depend of course on the wording of the provision in question. Default provisions generally use such phrases as 'has failed to discharge its duty'.[12] It may be a moot point in particular instances whether the provision is aimed at wilful refusal to act, or mere neglect or inability to act; the answer can only be gleaned from the precise terms of the provision and its context, but the courts have tended on the whole to adopt a restrictive approach, assisted by the general tenor of the provisions, which suggests fault as the crucial factor.[13] Words such as

[10] 'Local self-government is such an important part of our constitution that, to my mind, the courts should be vigilant to see that this power of the central government is not exceeded or abused', per Lord Denning MR in *Norwich City Council* v. *Secretary of State for the Environment*, above n. 7, p. 745. For powers of state governments with respect to local authorities in Australia see Local Government Act 1919 (NSW), ss. 86, 219; Local Government Act 1958 (Vic.), ss. 13–15; Local Authorities (Amendment) Act 1910 (Qld), s. 7; and also *Ex p. R., ex rel. Warringah Shire Council, re Barnett* (1967) 70 SR (NSW) 69.

[11] In *R.* v. *Secretary of State for the Environment, ex p. Halton District Council*, *The Times*, 14 July 1983, it appeared that the policy of the Secretary of State with regard to his powers under the Caravan Sites Act 1968 (England and Wales) (see below n. 24) was to interfere only where issues other than local ones were raised. For infrequency of use of default powers, see Griffith, *Central Departments and Local Authorities* (1966), pp. 57–8.

[12] Other phrases are e.g. 'has made default in providing' (Public Health Act 1936 (England and Wales) s. 322); and 'persistently makes default' (Municipal Ordinance 1947 (Ceylon), s. 277(1)): see below. Some modern provisions in England are not strictly default powers because they do not depend on fault (see below, n. 13).

[13] See e.g. *R.* v. *Kent County Council, ex p. Bruce*, *The Times*, 8 Feb. 1986 ('manifest failure'). A borderline instance is the provision discussed in *Durayappah* v. *Fernando*, above n. 2 (see text below). The Privy Council rejected (at p. 350) the opinion expressed by the Supreme Court of Ceylon in *Sugathadasa* v. *Jayasinghe* (1958) 59 NLR 457, 475 that 'not competent' meant 'not able to undertake'. On the other hand the Housing Act 1980 (England and Wales), s. 23, discussed in *Norwich City Council* v. *Secretary of State for the Environment*, above n. 7, clearly does not envisage any fault on the part of the authority, although in that case the authority were undoubtedly found to be in default. Arguably the same is true of the Caravan Sites Act 1968 (England and Wales), s. 23 (see below). See also the *Warringah* case, above n. 10 ('advisable', natural justice not implied). Some provisions merely use the word

'default' and 'failure' have tended, further, to be restricted to cases of nonfeasance rather than misfeasance.[14] Some provisions, however, are clearly aimed at misfeasance or *ultra vires* breach of duty;[15] a good example is the Education Act 1944 (England and Wales), section 68:

> If the Secretary of State is satisfied, either on complaint by any person or otherwise, that any local education authority . . . [has] acted or [is] proposing to act unreasonably with respect to the exercise of any power conferred or the performance of any duty imposed by or under this Act, he may . . . give such directions as to the exercise of the power or the performance of the duty as appear to him to be expedient.

Others again refer to failure to exercise powers or functions, and have the effect of imposing a duty on the defaulting authority, because failure to obey any directions imposed by the higher authority may result in an application by the higher authority for mandamus or other coercive measures.[16]

The matter of default is a matter for the judgment, in the first instance, of the higher authority, and is of course subject to the normal principles of judicial review, like any other discretionary power.[17] This is so whether or not the matter arises on complaint, whether or not the statute provides for due inquiry, and whether or not it is couched in subjective or objective terms. The best-known example is the *Tameside*[18] case, which was an application by the Secretary of State for Education and Science for mandamus to compel a local education authority to comply with directions given by her under the provision of the Education Act 1944 cited above. A Labour-controlled

'expediency', which certainly implies no fault; see Town and Country Planning Act 1971 (England and Wales), s. 276. Education Act 1976 (England and Wales), ss. 1–3, gives powers to require submission of proposals where it appears to the Secretary of State that progress or further progress is required: see *North Yorkshire County Council* v. *Secretary of State for Education and Science*, *The Times*, 20 Oct. 1978.

[14] See above Ch. 3.4(ii).

[15] National Health Service Act 1977 (England and Wales), s. 85; Housing Act 1957 (England and Wales), ss. 171–7; Control of Pollution Act 1974 (England and Wales), s. 97.

[16] See the *Tameside* case, above n. 3; *Norwich City Council* v. *Secretary of State* above n. 7.

[17] See e.g. *Ex p. Ward*, above n. 1, and the *Tameside* case, discussed below.

[18] Above n. 3. In the *Norwich* case the Secretary of State was held to have acted reasonably in interfering under the Housing Act 1980, s. 23, because the issue was merely whether tenants were having or might have difficulty in exercising their right to buy council houses effectively and expeditiously, which, on the facts, was clearly the case. The authority's case that they had to consider other priorities and the tenants would lose nothing by waiting was rejected on the facts, but was clearly irrelevant in view of the conditions for invoking s. 23.

authority had put forward a scheme for comprehensive (i.e. non-selective entry) education, which was approved by the (Labour) Secretary of State. In May 1976 the authority became Conservative-controlled after an election. In June it informed the Secretary of State that it would continue selective entry. The Secretary of State notified the authority that she was satisfied that this later change of plan was unreasonable because it would give rise to considerable difficulties in implementation for the new school year commencing in September, issued a direction under section 68, and then sought mandamus to enforce that direction when the authority refused to comply with it. The case was heard by the Divisional Court, the Court of Appeal, and the House of Lords in July and the House of Lords decided unanimously that the Secretary of State was not entitled to mandamus because she could not be reasonably satisfied on the facts that the authority were proposing to act unreasonably.

The matter of default was held in *Durayappah* v. *Fernando*[19] to attract also the principles of natural justice. In this case the Minister of Local Government of Ceylon acted under a draconian provision which read as follows:

> If at any time, upon representation made or otherwise, it appears to the Minister that a municipal council is not competent to perform, or persistently makes default in the performance of, any duty or duties imposed on it, or persistently refuses or neglects to comply with any provision of law, the Minister, by order . . . may direct that the council shall be dissolved and superseded . . .[20]

Following complaints against Jaffna Municipal Council the Minister dissolved the Council without giving it a hearing. The Mayor of the dissolved Council sought *certiorari* to quash the order. The Privy Council held that he had no standing to challenge the order, but that the principles of natural justice were applicable:

> In their Lordship's opinion there are three matters which must always be borne in mind when considering whether [natural justice] should be applied or not. These three matters are: first, what is the nature of the property or the office held, status enjoyed or services to be performed by the complainant of injustice. Secondly, in what circumstances or upon what occasions is the person claiming to be entitled to exercise the measure of control entitled to intervene. Thirdly, when a right to intervene is proved, what sanctions in fact is the latter entitled to impose upon the other. It is only upon a

[19] Above n. 2. See also the *Norwich* case, above n. 7, per Lord Denning MR at p. 745; the *Warringah* case, above n. 10 (*Durayappah* distinguished).

[20] Municipal Ordinance 1947 (Ceylon), s. 277(1).

consideration of all these matters that the question of the application of the principle can properly be determined.[21]

On the facts, in relation to all three matters, there was no doubt that natural justice was applicable, and it is hard to imagine any default provision,[22] let alone the provision in question in *Durayappah*, which would not be treated as attracting the principles of natural justice. With regard to the third matter, the type of sanction applicable should not, one would have thought, be material, so long as the normal operation of the authority's functions is being in any way interfered with.

Can judicial review be invoked so as to *compel* the higher authority to act at the instance of a member of the public? This was recently considered by Woolf J. in *R.* v. *Secretary of State for the Environment, ex p. Ward*.[23] A local authority was under a duty to exercise its powers of land acquisition to provide caravan sites 'so far as may be necessary to provide adequate accommodation for gipsies residing in or resorting to their area'. The authority obtained the lease of a site, but due to pollution it proved unsuitable for human habitation. Accordingly they decided to close the site on expiry of the lease, though not to evict gypsies then living there, and not to provide an alternative site. W, a gypsy living on the site, asked the Secretary of State to exercise his default power under the following provisions:

> The Secretary of State may, if at any time it appears to him to be necessary so to do, give directions to any local authority to which [the Act] applies requiring them to provide, pursuant to [the Act], such sites or additional sites, for the accommodation of such numbers of caravans, as may be specified in the directions; and any such directions shall be enforceable, on the application of the Secretary of State, by mandamus.[24]

The Secretary of State refused to act because there was no immediate danger of the gypsies being evicted or of services to the site being cut off. W sought judicial review of the authority's decision not to provide a site, and the Secretary of State's refusal to intervene. It was held that the Secretary of State had reasonably concluded that his intervention was not necessary, but that the authority's decision should

[21] [1967] 2 AC 349. *R.* v. *Secretary of State for the Environment, ex p. Halton District Council*, above n. 11, goes further: *certiorari* and mandamus were granted in that case to compel the Secretary of State to consider the applicants' objections before directing the county council to proceed with proposals for a gypsy caravan site in the applicants' area.

[22] See, however, the *Warringah* case, above n. 10, which could be doubted.

[23] Above n. 1.

[24] Caravan Sites Act 1968 (England and Wales), s. 9; and see ss. 6(1), 8(3), 23.

be quashed and reconsidered because, on the facts, it had not appreciated that its decision involved abdication of its duty under the Act. However, it is not beyond the courts to intervene in such a case where it appears necessary, though it seems likely that they will prefer to coerce the defaulting authority rather than the higher authority, except where it would be useless to do so.[25]

Once the legality of the higher authority's action or inaction is determined, is there any possibility of judicial review of the application of a particular sanction? Default powers usually specify a particular sanction or sanctions of various degrees of severity; in addition the facts may warrant, especially after a finding that an authority is in default, other action not specified by the default provision itself or even the statute in which it appears. In these circumstances the choice of sanction may be controversial. One can infer from *Durayappah* that there is a right to a hearing with regard to the sanction as well as the finding of default, and the choice of sanction would appear to be a matter for discretion, which can be reviewed in the normal manner, but the higher authority is not confined to the sanctions listed in the default provision. In *Asher* v. *Secretary of State for the Environment*[26] a local authority refused to increase public-sector rents as enjoined by statute; although the statute allowed the Secretary of State to appoint a housing commissioner to assume the authority's functions under the statute, and to reduce the authority's housing subsidy, he instead gave the authority sixteen days to remedy their default, and then ordered an extraordinary audit under the Local Government Act 1972 (England and Wales), section 236. The district auditor surcharged the authority's losses due to the councillors' default to the councillors,[27] who thereby became disqualified for a period of five years, and the Secretary of State, nearly a year later, appointed a housing commissioner. The councillors then applied for declarations that the direction under section 236 was unlawful and that the district auditor had no power to surcharge them. The judge upheld the Secretary of State's application to strike out the statement of claim on the ground that it disclosed no reasonable cause of action and was vexatious, which decision was upheld by the Court of Appeal. The substance of the councillors' claim was that the audit had been directed for punitive purposes and that in other similar cases the Secretary of State had appointed a housing commissioner without an audit; however, these arguments were rejected on the basis that the audit was a reasonable

[25] [1984] 2 All ER 569 per Woolf J., and above Ch. 3.4(iii).
[26] [1974] Chanc. 208.
[27] See *Asher* v. *Lacey* [1973] 1 WLR 1412.

sanction to adopt because the other available remedies might well be inadequate in the circumstances: mandamus would merely assist the councillors to become martyrs; a housing commissioner would find it difficult to secure their co-operation; and reduction of the housing subsidy would merely punish the ratepayers. The sanctions following from a default order or finding are thus regarded as cumulative with any other remedies which may be available, provided there is no abuse of discretion on the part of the higher authority in selecting the appropriate sanction or sanctions. However, it may be plausibly maintained that if a higher authority acts in pursuance of a particular provision, it should be confined to the sanctions envisaged by that provision.

This brief review of default powers indicates their limited usefulness as a remedy for enforcing public duties. There are, no doubt, situations in which they are necessary and ought to be used in the wider public interest; these situations are limited to flagrant and widespread derelictions of duty at the general level. As a remedy for individuals at the specific level they are not easily invoked and may in fact be an obstacle to the availability of other and better remedies because of their exclusive effect.

4.3 MANDATORY INJUNCTION

The injunction,[1] an equitable remedy, can be used to compel or restrain the performance of an act, and in the former instance is called a mandatory (as opposed to prohibitory) injunction. Nothing turns on the distinction between mandatory and prohibitory wording, and an injunction having mandatory effect can be worded prohibitorily.[2] Its availability generally depends on the plaintiff having a right of action.[3]

The mandatory injunction has not been frequently used for enforcing public duties,[4] but now that courts have more freedom to award an

[1] See De Smith, *Judicial Review of Administrative Action*, 4th edn. by Evans (1980), ch. 9; Aronson and Franklin, *Review of Administrative Action* (1987), ch. 19, for exhaustive discussions of the injunction in public law.

[2] See e.g. *Dowty Boulton Paul Ltd.* v. *Wolverhampton Corporation* [1971] 1 WLR 204; *Glynn* v. *Keele University* [1971] WLR 487; *Winward* v. *Cheshire County Council* (1978) 77 LGR 172; and of course vice versa: see *Buckoke* v. *Greater London Council* [1971] Chanc. 655.

[3] See below Chs. 6.5, 7.2.

[4] It is almost never used in connection with the duty to determine, except in relation to private tribunals, where mandamus is unavailable. Nearly all the cases concern duties to provide.

injunction even where it is not specifically sought,[5] it may become more fashionable, at least for the enforcement of public duties to provide,[6] though it is unlikely to challenge the primacy of mandamus in this field.

The injunction is a remedy of considerable scope and flexibility. It may be used against any defendant, regardless of whether he is a public or private officer or body, regardless of whether the case arises in public or private law, and regardless of whether the function involved is judicial or administrative.[7] The only serious limitations on the availability of an injunction are first, that the rules of standing are restrictive compared with mandamus,[8] and second, that, like mandamus, it is not available against the Crown.[9] Potentially, therefore, the injunction covers almost the whole field of mandamus.

(i) *Procedure*

The main advantages of the mandatory injunction over mandamus are procedural rather than substantive:

1. Leave of court is not required for an application for an injunction, but is for an application for mandamus.[10]

2. An injunction can be, and frequently is, sought together with a declaration or damages[11] and can be the subject of a counterclaim,[12] though it cannot be sought as an adjunct to prerogative relief; mandamus on the other hand cannot be sought together with any

[5] See below, Ch. 5.

[6] See above n. 4.

[7] For these reasons the injunction may well be appropriate in cases on the borderline between public law and private law, except in England: see *O'Reilly* v. *Mackman* [1982] 3 WLR 1096 (HL), in which an action for a declaration in a public law case not commenced under O. 53 was dismissed as an abuse of the court's process; as a result, an action for an injunction where the plaintiff is unsure whether the case is one in public law will be dismissed if it turns out to be such, unless brought under O. 53; see below Ch. 5.2.

[8] See below Ch. 6.5.

[9] Except in Australia; see below n. 44.

[10] See, however, Supreme Court Act 1981 (England and Wales), s. 31(3).

[11] The conjunction of these remedies is particularly convenient because the rules for standing are somewhat similar; see below Chs. 6.5, 7.2.

[12] In *Kensington and Chelsea London Borough Council* v. *Wells* (1973) 72 LGR 289 a mandatory injunction to compel the authority to fulfil its duty to provide caravan sites under the Caravan Sites Act 1968 (England and Wales) was sought in a counterclaim to the authority's claim for possession of a site unlawfully occupied by the defendant; the counterclaim, however, failed. Since in instances of this kind a public-law issue arises collaterally, the rule in *O'Reilly* v. *Mackman*, above n. 7, has no application; see below Ch. 5.2; *Wandsworth London Borough Council* v. *Winder* [1985] AC 461 (HL).

other non-prerogative remedy and cannot be sought in a counterclaim.

3. An injunction has some elasticity of form and content;[13] in particular it can be expressed to take effect at some future date (to facilitate compliance), or else leave may be granted for a reapplication in the event of no action being taken by the defendant.[14]

4. Interlocutory relief is available in an action for an injunction, but not in an application for mandamus.[15]

Of these advantages only the last is really substantial, and is one of the principal motives for suing for an injunction rather than applying for mandamus. In many cases of urgency the award of, or refusal of, an interlocutory mandatory injunction effectively decides the case because the party aggrieved will not hazard to take the matter further. It is therefore important to discover in what circumstances an interlocutory mandatory injunction will be granted against a public authority.

The most important recent statement of the rules concerning the granting of an interlocutory injunction is to be found in a patent infringement case, *American Cyanamid Co.* v. *Ethicon Ltd.*[16] In that case the House of Lords considered the question of the 'balance of convenience', but also made an important innovation in deciding that the question of balance of convenience arises when the court is satisfied that the claim is not frivolous or vexatious, or, in other words, that there is a serious question to be tried. In so deciding their Lordships scotched the notion held previously that the plaintiff had to establish a prima-facie case.[17] The reason for this was that:

[i]n those cases where the legal rights of the parties depend upon facts that are in dispute between them, the evidence available to the court at the

[13] It is uncertain whether this is really an advantage over mandamus. One of the main defects of mandamus is that it will not enable supervision of a continuing series of acts; this is also, however, true of a mandatory injunction: see e.g. *Attorney-General* v. *Staffordshire County Council* [1905] 1 Chanc. 336; *Attorney-General* v. *Colchester Corporation* [1955] 2 QB 207; cf., however, *Gravesham Borough Council* v. *British Railways Board* [1978] Chanc. 379, 403, where Slade J. suggests that in exceptional cases an injunction will be granted to require a person to do a series of acts over a number of years, a flexible approach which could be of considerable value if accepted generally; see, further, *Fairfax (John) Ltd.* v. *Australian Telecommunications Commission* [1977] NSWLR 400, 405 per Moffitt P.

[14] However, in *R.* v. *Paddington Valuation Officer, ex p. Peachey Property Co. Ltd.* [1966] 1 QB 380, 402–3, Lord Denning MR suggested (Salmon LJ *dubitante*, pp. 418–19) that an order of *certiorari* could be postponed pending the preparation of a new valuation list in obedience to an order of mandamus, and the same is probably true of an order of mandamus.

[15] For declarations, see below Ch. 4.4(iii).

[16] [1975] AC 396.

[17] See Lord Diplock's discussion of the authorities at [1975] AC 406–8.

hearing or the application for an interlocutory injunction is incomplete. It is given on affidavit and has not been tested by oral cross-examination. The purpose sought to be achieved by giving to the court discretion to grant such injunctions would be stultified if the discretion were clogged by a technical rule forbidding its exercise if upon that incomplete untested evidence the court evaluated the chances of the plaintiff's ultimate success in the action at 50 per cent, or less, but permitting its exercise if the court evaluated his chances at more than 50 per cent.[18]

This liberalizing of the rules is certainly to be welcomed, but its effect in the present context is limited by the fact that cases involving public duties rarely involve any dispute as to the facts. More important is the matter of the balance of convenience, the rules relating to which were dealt with by Lord Diplock in the *Cyanamid* case as follows:

As to [the balance of convenience] the governing principle is that the court should first consider whether, if the plaintiff were to succeed at the trial in establishing his right to a permanent injunction, he would be adequately compensated by an award of damages for the loss he would have sustained as a result of the defendant's continuing to do [or, for our purposes, continuing to fail to do] what was sought to be enjoined between the time of the application and the time of the trial. If damages in the measure recoverable at common law would be adequate remedy and the defendant would be in a financial position to pay them, no interlocutory injunction should normally be granted, however strong the plaintiff's claim appeared to be at that stage. If, on the other hand, damages would not provide an adequate remedy for the plaintiff in the event of his succeeding at the trial, the court should then consider whether, on the contrary hypothesis that the defendant were to succeed at the trial in establishing his right to do that which was sought to be enjoined, he would be adequately compensated under the plaintiff's undertaking as to damages for the loss he would have sustained by being prevented from doing so between the time of the application and the time of the trial. If damages in the measure recoverable under such an undertaking would be an adequate remedy and the plaintiff would be in a financial position to pay them, there would be no reason upon this ground to refuse an interlocutory injunction.

It is where there is doubt as to the adequacy of the respective remedies in damages available to either party or to both, that the question of balance of convenience arises. It would be unwise to attempt even to list all the various matters which may need to be taken into consideration in deciding where the balance lies, let alone to suggest the relative weight to be attached to them. These will vary from case to case.[19]

His Lordship went on to point out that generally it is prudent in

[18] Ibid. 406, per Lord Diplock.
[19] Ibid. 408.

considering the balance of convenience to maintain the status quo, but that the relative strengths of the parties' cases may be considered, as well as 'special factors' occurring in the particular circumstances of individual cases.

How does this apply to a mandatory interlocutory injunction sought against a public authority? Lord Diplock's remarks concerning the status quo were directed towards prohibitory injunctions:

> Where other factors appear to be evenly balanced it is a counsel of prudence to take such measures as are calculated to preserve the status quo. If the defendant is enjoined temporarily from doing something that he has not done before, the only effect of the interlocutory injunction in the event of his succeeding at the trial is to postpone the date at which he is able to embark upon a course of action which he has not previously found it necessary to undertake; whereas to interrupt him in the conduct of an established enterprise would cause much greater inconvenience to him since he would have to start again to establish it in the event of his succeeding at the trial.[20]

A mandatory injunction cannot be considered in this way because it is designed to compel action which may be causing injustice to the plaintiff. Whether the performance of the acts demanded, followed by the lifting of the injunction and a return to the status quo, would cause great inconvenience, depends on the nature of the duty which the plaintiff asserts the defendant owes him, and the circumstances in which its performance is demanded. If the duty involves the setting up of machinery which will have to be dismantled if the plaintiff ultimately fails, then clearly the balance lies in favour of the status quo; but if it involves the making of some relatively simple provision for the plaintiff which can easily be withdrawn if necessary, or some provision which is being temporarily suspended or withheld, the balance may be in favour of granting an interlocutory injunction; if of course withdrawal of the provision may occur as a result of the plaintiff ultimately failing to obtain a permanent injunction, and the plaintiff would thereby have to pay damages to the defendant,[21] the court may well be unwilling (and it indeed may not be in the plaintiff's best interests) to grant interlocutory relief. In short, the matter really depends on the strength of the plaintiff's case as much as on the nature of the duty and the circumstances of the case.

As regards public authorities, the rules set out in *Cyanamid* seem to

[20] Ibid.

[21] An undertaking to pay damages may be required: see text to n. 19 above, and *Hoffmann-la Roche & Co. AG* v. *Secretary of State for Trade and Industry* [1974] 2 All ER 1128 (HL).

have received severe modification by the Court of Appeal. In *Smith* v. *Inner London Education Authority*[22] two judges held that an interlocutory injunction should not be granted against a public authority unless there is a 'real prospect' that the plaintiff would succeed in a claim for a permanent injunction,[23] though the third uttered *obiter dicta* which are consistent with the 'serious case to be tried' test in *Cyanamid*; [24] all three judges, however, were of the opinion that the 'balance of convenience' factor, when applied to public authorities, means the balance of convenience generally, including consideration of the public interest, not merely the balance of convenience as between the parties, this matter of the public interest being one of the 'special factors' referred to by Lord Diplock in *Cyanamid*.[25] The later case of *De Falco v. Crawley Borough Council*,[26] however, applied the pre-*Cyanamid* 'strong prima-facie case' test. In *Meade* v. *Haringey London Borough Council*[27] the *Cyanamid* rules were not discussed, but two judges refused to follow Lord Denning MR in his willingness to grant a mandatory interlocutory injunction against the authority to open schools closed because of a caretakers' strike, because the case involved disputed facts and the courts should be slow to grant such relief in a case involving an industrial dispute.[28]

The balance of English authority is thus against applying either the spirit or the letter of the *Cyanamid* rules to public authorities, and this seems to be so whether a prohibitory or mandatory interlocutory injunction is sought. While the decisions referred to are no doubt right in emphasizing the public-interest factor, the restricting of consideration of the balance of convenience to cases where there is a strong prima-facie case seems to inhibit unduly the courts' discretion in the matter. The strength of the plaintiff's case is certainly a relevant consideration, but should not necessarily preclude a consideration of the balance of convenience.

(ii) *Alternative remedies*

Apart from the problem of standing, the factor which has most inhibited the development of the mandatory injunction as an altern-

[22] [1978] 1 All ER 411 (prohibitory injunction).

[23] Ibid. 418, per Lord Denning MR.

[24] Ibid. 422–3, per Browne LJ.

[25] Ibid. 422. See also *Harold Stephen & Co. Ltd.* v. *Post Office* [1977] 1 WLR 1172.

[26] [1980] QB 460 (mandatory injunction); see also *Thornton* v. *Kirklees Metropolitan Borough Council* [1979] 3 WLR 1 and *Woodcock* v. *South Western Electricity Board* [1975] 1 WLR 983, in which mandatory interlocutory injunctions were refused on the merits.

[27] [1979] 1 WLR 637; and see above Chs. 2.3, 3.4(ii).

[28] See above n. 27, p. 658.

ative to mandamus is the tendency to regard the two remedies as mutually exclusive. This is most apparent in nineteenth-century cases; in *Glossop* v. *Heston and Isleworth Local Board*[29] a mandatory injunction to enforce the Board's public duty to provide adequate drainage sought by one who suffered a nuisance from their failure to perform it, was refused on the ground that mandamus was the proper remedy; in a subsequent case even the Attorney-General was refused an injunction on the same ground.[30] These decisions represent perhaps a corollary of *Holland* v. *Dickson*,[31] which held that an injunction, rather than mandamus, was appropriate where the duty was a private one. The position established by these cases is strangely restrictive when one considers that the distinction between private and public law was not generally drawn at that period.

It is suggested, however, that the mandatory injunction and mandamus are not mutually exclusive remedies, but are in fact both available in public-law cases, subject to the differing rules of standing applicable to the two remedies. Although there is no decision precisely to that effect, the availability of mandamus has not been referred to in a number of modern cases[32] where a mandatory injunction has been sought because of the convenience of linking it with a claim for a declaration or damages. These cases involve public duties to provide education and public housing, which are undoubtedly enforceable by mandamus. Although the old distinction between mandamus and mandatory injunction was accepted in *Fairfax (John) Ltd.* v. *Australian Telecommunication Commission*,[33] a case involving a duty of a public authority held to be private in character, it seems unlikely that in a case depending on the point the courts would these days refuse a mandatory injunction on such a narrow ground.

The fact that a declaration is sought, clearly does not exclude the availability of an injunction, and the existence of a right of action for damages is in most cases a precondition for the granting of an injunction;[34] however, an injunction will be refused where damages are an adequate remedy in the circumstances.[35] As far as default

[29] (1879) 12 Chanc. D. 102.

[30] *Attorney-General* v. *Clerkenwell Vestry* [1891] 3 Chanc. 527.

[31] (1888) 37 Chanc. D. 669.

[32] *Woodcock* v. *South Western Electricity Board*, above n. 26; *Winward* v. *Cheshire County Council* (1978) 77 LGR 172; *Meade*, above n. 27; *De Falco*, above n. 26; *Kensington and Chelsea London Borough Council* v. *Wells* (1973) 72 LGR 289.

[33] [1977] 1 NSWLR 400, 405 per Moffitt P.; *Jeanneret* v. *Hixson* (1890) 11 LR (NSW) Eq. 1 is to similar effect, and these two cases could be an obstacle to Australian courts in accepting the position proposed in the text.

[34] See below Ch. 6.5.

[35] See e.g. *Morton* v. *Eltham Borough* [1961] NZLR 1.

powers are concerned, in considering the application of the rule in *Pasmore* v. *Oswaldtwistle Urban District Council*,[36] the courts have not distinguished between mandamus[37] and mandatory injunction, so that the rule, in so far as it is still applicable at all, applies to both in the same manner.

(iii) *Discretionary refusal*

Since it is an equitable remedy, the award of an injunction is in the discretion of the court. As one might have expected, the discretion is exercised in a manner broadly similar to that apparent in mandamus cases. It would indeed be strange if the choice of remedy conferred any advantage in this respect.

The courts are prepared to refuse an injunction where extreme inconvenience would result: in *Attorney-General* v. *Colchester Corporation*[38] a mandatory injunction to compel the Corporation to operate a ferry service was refused, partly on the ground that the service could only be operated at a loss. Delay is also a ground for refusing an injunction.[39] In the case of an action by the Attorney-General the same problem arises as in mandamus cases: is the court's discretion ousted?[40] The better view, as with mandamus, is that it is not; however, as De Smith[41] points out, the courts should be mindful of the fact that an action by the Attorney-General may be the only remedy available. There are relatively few cases involving discretionary refusal of an injunction against a public authority, particularly a mandatory injunction, but the principles applicable in equity[42] presumably apply to public-law cases; thus the maxim 'he who comes to equity must come with clean hands' can be used to import all the case law relating to discretionary refusal of mandamus because of the applicant's conduct.[43]

[36] [1898] AC 387 (HL); see also *Clark* v. *Epsom Rural District Council* [1929] 1 Chanc. 287.

[37] See above Ch. 3.4(ii).

[38] [1955] 2 QB 207; applied in *Morton* v. *Eltham Borough*, above n. 35; cf. *Gravesham Borough Council* v. *British Railways Board* [1978] Chanc. 379.

[39] In *Legg* v. *Inner London Education Authority* [1972] 1 WLR 1245 a delay of twelve weeks was held not to be fatal to the application, which involved a number of plaintiffs.

[40] See above Ch. 3.4(ii).

[41] Above n. 1, pp. 438–9.

[42] See Sharpe, *Injunctions and Specific Performance* (1983), pp. 103–12.

[43] See above Ch. 3.4(iv); and the *per curiam* remarks of Roskill LJ in *Malone* v. *Metropolitan Police Commissioner* [1980] QB 49, 71 where it was suggested that a mandatory injunction would not lie against the police to return money seized when no charges resulted because the money, even if not stolen, was held illegally.

(iv) *The Crown*

Where an injunction is available against the Crown there is a clear advantage over mandamus. The Crown can be enjoined in Australia (with the exception of Tasmania)[44] but not in England[45] or New Zealand.[46] Two questions arise. First, where the Crown can be enjoined, is there any vestige of the old immunity at common law in relation to the Crown's representative? Secondly, where the Crown cannot be enjoined, is the position the same as for mandamus?[47]

As mentioned earlier[48] these questions are not of crucial importance when a declaration is almost always available in lieu of an injunction,[49] but they are worth considering because they affect the plaintiff's choice of remedy.

With regard to the first question, it would seem that the Crown's representative, in Australia the Governor-General, cannot be subjected to an injunction. Direct authority for this is lacking, but the intention of statutes removing the Crown's immunity from suit and not preserving its immunity from an injunction is clearly directed at the government, not the sovereign in person or the sovereign's representative.[50] However, as one commentator has put it:

> This should cause no undue difficulty, where it is sought to challenge an act of the Queen's representative by injunction; a person seeking to rely on the validity of the vice-regal act would normally be an appropriate defendant. Where no more appropriate defendant can be found, it would seem that proceedings could be taken against the Attorney-General.[51]

With regard to the second question the answer is less certain. The Crown Proceedings Act 1947 (England and Wales), section 21(2) says this:

[44] For a list of the statutory provisions giving effect to this position, see Whitmore and Aronson, *Review of Administrative Action* (1978), p. 322 n. 33; Aronson and Franklin, above n. 1, pp. 608–9. See also *McLean* v. *Rowe* (1925) 25 SR (NSW) 390; Hogg, *Liability of the Crown* (1971), p. 22; Sykes, Lanham, and Tracey, *General Principles of Administrative Law*, 2nd edn. (1984), pp. 230–2.

[45] See De Smith, above n. 1, pp. 445–9; Crown Proceedings Act 1947 (England and Wales), s. 21.

[46] See Crown Proceedings Act 1950 (NZ), s. 17.

[47] See above Ch. 3.2.

[48] Ibid.

[49] See e.g. Crown Proceedings Act 1947 (England and Wales), s. 21(1).

[50] Sykes, Lanham, and Tracey, above n. 44, p. 231, derive the proposition from the case of *FAI Insurances Co. Ltd.* v. *Winneke* (1982) 56 ALJR 388 (HCA), where this position was established in relation to declarations; *a fortiori* proceedings for injunction should not be brought against the Crown's representative.

[51] Sykes, Lanham, and Tracey, above n. 44.

The court shall not in any civil proceedings grant any injunction or make any order against an officer of the Crown if the effect of granting the injunction or making the order would be to give any relief against the Crown which could not have been obtained in proceedings against the Crown.

The only important decision on this section is unfortunately most unhelpful. In *Merricks* v. *Heathcoat-Amory*[52] the plaintiff sought a mandatory injunction against the Minister of Agriculture to compel him to withdraw a draft statutory scheme he had laid before Parliament on the ground that it was invalid, and contended that the Minister was acting not as a representative of the Crown but as a person designated in an official capacity or else as an individual. Upjohn J., refusing the injunction, found it 'very difficult to conceive of a middle classification [i.e. person designated in an official capacity]',[53] in spite of the fact that this very classification has not only been conceived of but applied in many mandamus cases. The decision is *per incuriam*, or else better explained on the basis of the unwillingness of English courts to interfere in parliamentary proceedings. A much better decision to the opposite effect is that of Lerner J. in *McLean* v. *Liquor Licence Board of Ontario*[54] in relation to the Ontario equivalent of section 21(2).[55]

The conclusion drawn is that since policy demands that the Crown's immunity be restricted, as far as possible, to personal exercise of prerogative powers by the sovereign or the sovereign's representative, an injunction should at least be available against a *persona designata*, as in mandamus law; where the Crown is not immune from an injunction, this demand of policy has in fact been achieved.

(v) *Injunctions in aid of public rights*

An injunction can only be sought by a private individual where some private right of his own is interfered with, or where he suffers special damage from interference with a public right: this is the substance of the principles established in *Boyce* v. *Paddington Borough Council*,[56] which are pursued in Chapter 6. If a plaintiff has no standing the difficulty can be circumvented by asking the Attorney-General to proceed *ex proprio motu* for an injunction in aid of public rights, or else to proceed at the plaintiff's relation, regardless of the plaintiff's standing, or lack

[52] [1955] Chanc. 567; see also *Underhill* v. *Ministry of Food* [1950] 1 All ER 591; *Maxwell* v. *Department of Trade* [1974] QB 523, 542 per Lawton LJ.

[53] Above n. 52, p. 576.

[54] (1976) 61 DLR (3d) 237.

[55] Proceedings Against the Crown Act 1970 (Ont.), s. 18(1).

[56] [1903] 1 Chanc. 109.

of it, in the matter.[57] However, these forms of action have very rarely been used to enforce public duties. In theory this should not be so, because they could have plugged an awkward gap in the remedies for enforcing public duties. This ideal has not been achieved, one can surmise, for the following reasons: first, the Attorney-General does not act against government departments;[58] secondly, his decision not to proceed with a relator action cannot be reviewed by the courts and the plaintiff cannot proceed himself even where the Attorney-General refuses;[59] thirdly, it is not clear what principles apply to injunctions of this kind to enforce public duties, because the only cases of any importance, *Attorney-General* v. *Dorking Union Guardians*,[60] and *Attorney-General* v. *Clerkenwell Vestry*,[61] proceeded on the basis that the standing rules are the same as for cases brought by the plaintiff himself.

This represents a situation which, but for the developments in the law of mandamus, would be utterly deplorable. The case for reforming this area of the law seems, however, to have been overtaken by the new statutory procedures for judicial review, and the injunction in aid of public rights will remain a dormant, if not moribund, remedy for enforcing public duties until such time as the courts decide that the Attorney-General's discretion in the matter can be reviewed.

4.4 DECLARATION

The declaration[1] is not, strictly speaking, a remedy for *enforcing* public duties at all, for a declaration merely declares the law applicable to the case, and cannot in itself actually force anyone to do anything.

[57] See e.g. *Attorney-General* v. *Howard United Reformed Church Trustees* [1976] AC 363; De Smith, above n. 1, p. 449 n. 37.

[58] Obviously this would raise difficult political and constitutional questions and would place the Attorney-General in an invidious position. The problem has been noticed judicially: *Inland Revenue Commissioners* v. *National Federation of Self-Employed and Small Businesses Ltd.* [1982] AC 617, 644 (HL).

[59] *Gouriet* v. *Union of Post Office Workers* [1978] AC 435 (HL).

[60] (1882) 20 Chanc. D. 595.

[61] [1891] 3 Chanc. 527. Both cases followed *Glossop* v. *Heston and Isleworth Local Board* (1879) 12 Chanc. D. 102. See also the relator actions in *Attorney-General* v. *Colchester Corporation* [1955] 2 QB 207; *Attorney-General* v. *St Ives Rural District Council* [1961] 1 QB 366.

[1] For detailed discussion of this remedy see De Smith, *Judicial Review of Administrative Action*, 4th edn. by Evans (1980), ch. 10; Wade, *Administrative Law*, 5th edn. (1983), pp. 522 et seq.; Zamir, *The Declaratory Judgment* (1962); Young, *Declaratory Orders*, 2nd edn. (1984).

For all that, a declaration is not without any coercive effect, and in fact is regularly resorted to in cases involving enforcement of public duties of all kinds, a fact which is not surprising because the declaration is a remedy which has a number of significant advantages. Its only disadvantages are its lack of coercive effect and its somewhat restrictive standing rules, which latter are discussed below in Chapter 6.

(i) *Non-coercive effect*

The non-coercive effect of a declaration might be thought to be a very serious limitation on the usefulness of this remedy for enforcing public duties, but in practice it is not. Public authorities very rarely defy a decision of the courts and so a declaration as to the existence or extent of a public duty is generally, in practice, equal in coercive effect to mandamus. None the less the case of *Webster* v. *Southwark London Borough Council*[2] is a salutary reminder that a declaration is of no use in itself in the face of outright disobedience. The plaintiff, a parliamentary candidate for the National Front, obtained in interlocutory proceedings a declaration that the defendant authority (Labour-controlled) was under a statutory duty to provide him with a meeting room for an election meeting. Counsel for the authority assured the court that the authority would comply with the duty as declared by the court, but in fact those instructing him had deliberately misled the court, knowing that the authority had determined to defy it. In the event the authority refused to comply with the declaration and the plaintiff proceeded against the two councillors responsible for committal for contempt and for leave to issue a writ of sequestration. The latter was granted shortly before the meeting was due to be held, and as a result the authority complied with their duty. On resumption of the hearing Forbes J. held that since a declaratory order is one which carries with it no penal notice, the councillors could not be committed for contempt,[3] but that leave to issue the writ of sequestration was properly given because of the urgency of the case and because the court could not stand by and confess it was powerless; he also doubted whether the courts should continue the practice of assuming that a public authority would obey the law when that

[2] [1983] QB 698; see also Harlow, 'Administrative Reaction to Judicial Review' [1976], *PL* 116.

[3] The decision under discussion does not explain why, if there is no contempt of court in disobeying a declaration, a writ of sequestration can issue, but there can be no committal; the only explanation for the distinction might be that the former does not involve imprisonment, but that is a poor explanation; the assertion that otherwise the court would be powerless is also inadequate to explain the distinction.

practice was open to such abuse. However, such situations are of course rare and the courts should certainly continue their practice. Where there is doubt as to whether the authority will perform the duty, either the plaintiff can seek coercive relief in addition to a declaration, or the court can grant leave to apply for coercive relief if the bare declaration is not obeyed.[4]

On the whole the non-coercive effect of a declaration is a strength rather than a weakness, because it helps to defuse the tension inherent in public-duty litigation. A declaration is often used for determining the precise legal content of a public duty to the advantage of both parties; a good example of this is the recent English case of *R.* v. *Hereford and Worcester Local Education Authority, ex p. Jones*,[5] in which the court granted a declaration that the authority's duty to provide education free of charge did not allow it to charge for music tuition provided during school hours, because music tuition was part of the educational curriculum, but refused to grant an order of mandamus to compel the authority to provide music tuition, because it was entitled to refuse to provide it if it felt its resources were too scarce. The beneficial effect of decisions of this kind is that the legal content of the duty is clarified; in the case of local-government duties the decision may result in rationality and consistency in performance of the duty over the whole jurisdiction. A declaration also avoids the necessity of government or a pressure group appearing to set up a particular authority as a target, and encourages the yielding of reasonable concessions. It has been found useful also in the definition of ongoing duties.[6]

(ii) *Conditions for granting a declaration*

A declaration can be made against any legal person, including of course any public authority or tribunal exercising administrative or

[4] *Fischer* v. *Secretary for India* (1898) LR 16 IA 16, 29 (PC); *Attorney-General for Commonwealth* v. *Colonial Sugar Refining Co. Ltd.* [1974] AC 237, 257 (PC); *Royal Insurance Co. Ltd.* v. *Mylius* (1926) 38 CLR 477, 497 (HCA); and see Zamir, above n. 1, pp. 317–19. It seems that in Australia statutory mandamus will be granted to enforce a declaration: *Mudge* v. *Attorney-General of Victoria* [1960] VR 43, 54; *Dickinson* v. *Perrignon* [1973] 1 NSWLR 72, 84, but the procedure is more cumbersome than that of granting leave to apply.

[5] [1981] 1 WLR 768; see also the almost friendly case of *North Yorkshire County Council* v. *Secretary of State for Education and Science*, *The Times*, 28 Oct. 1978; and *Re an Application by Derbyshire County Council* [1988] 1 All ER 385.

[6] See e.g. *Williams* v. *Manchester* (*Mayor*) (1897) 13 TLR 299; *Field* v. *Poplar Borough Council* [1929] 1 KB 750; *Surrey County Council* v. *Ministry of Education* [1953] 1 WLR 516.

judicial powers,[7] and even against the Crown. The limits of the availability, established in *Dyson* v. *Attorney-General*,[8] of a declaration against the Crown, are probably set by two Australian cases which fall either side of the line: in *Tonkin* v. *Brand*[9] a declaration was awarded against the Executive Council of Western Australia that they were under a duty to advise the Governor to issue an election proclamation; in *FAI Insurances Co. Ltd.* v. *Winneke*,[10] however, the High Court held that a declaration would not lie against the Governor of Victoria except where he acts under a statutory provision. The position thus established is consistent with the views expressed above in Chapter 3 concerning the availability of mandamus against the Crown.

The essence of a declaration being a statement of the law applicable to the case, it can, in the context of public duties, perform what Zamir has called an 'original' or 'supervisory' role, so that in addition to declaring the extent of an authority's duty (supervisory role), it can declare also the rights of the parties in relation thereto (original role). Given this flexibility the declaration can be used in any of the public-duty contexts described in this book. Some examples of the use of the declaration in those contexts will indicate the general utility of the remedy, whether or not it is actually granted.

1. Proceedings for a declaration can be used to clarify the extent of a duty to provide. In *R.* v. *Exeter City Council, ex p. Gliddon*[11] a

[7] The declaration is regularly used to challenge decisions of private tribunals, but not regularly those of public tribunals: see e.g. *Lee* v. *Showmen's Guild of Great Britain* [1952] 2 QB 329; *McKinnon* v. *Grogan* [1974] 1 NSWLR 295. A declaration cannot, however, be granted against a superior court of record: see Aronson and Franklin, *Review of Administrative Action* (1987), ch. 16.

[8] [1911] 1 KB 400.

[9] [1962] WAR 2.

[10] (1982) 56 ALJR 388; see also *Marks* v. *Commonwealth* (1964) 111 CLR 549, 565 per Isaacs J. (HCA). It would appear from *FAI Insurances* (per Gibbs CJ at p. 391) that the proper respondent in such a case is the Attorney-General.

[11] [1985] 1 All ER 493; see also *Surrey County Council* v. *Ministry of Education* [1953] 1 WLR 516; *Watt* v. *Kesteven County Council* [1955] 1 QB 408; *Cumings* v. *Birkenhead Corporation* [1972] Chanc. 12; *Hey & Croft Ltd.* v. *Lexden and Winstree Rural District Council* (1972) 70 LGR 531; *Bradley* v. *Commonwealth* (1973) 128 CLR 557 (HCA); *Cherwell District Council* v. *Thames Water Board* [1975] 1 WLR 448 (HL); *Roberts* v. *Dorset County Council* (1976) 75 LGR 462; *Fairfax (John) Ltd.* v. *Australian Postal Commission* [1977] 2 NSWLR 124; *Smith* v. *Commissioner of Corrective Services* [1978] 1 NSWLR 317; *Booth & Co. (International) Ltd.* v. *National Enterprise Board* [1978] 3 All ER 624; *R.* v. *Hereford and Worcester Local Education Authority, ex p. Jones*, above n. 5; *Webster* v. *Southwark London Borough Council*, above n. 2. In the *Surrey County Council* case the action was brought by the authority itself, anticipating that the Minister would uphold an appeal by a parent that the authority's provision of school transport for his child was unlawfully inadequate.

declaration that the authority had failed to perform its duty to provide him with temporary accommodation because the premises provided were in poor condition was refused, the judge finding that in all the circumstances surrounding the duty, in particular its temporary effect, the duty was only to provide premises which were fit for human habitation.

2. A declaration as to an individual's entitlement or status can similarly be used to define an authority's duty to determine. In *Mills* v. *Avon and Dorset River Board*[12] the plaintiffs, owners of fisheries, who had normally been given fishing licences by the defendants or their predecessors, obtained a declaration that they were entitled to licences, which had been refused on the sole ground that other licencees had commercialized their fishing rights to the defendants' loss.

3. Declaratory proceedings are often a convenient way of delineating the duties of public authorities *inter se*. In *Monmouthshire County Council* v. *British Transport Commission*[13] a declaration that the defendants rather than the plaintiffs were under a duty to maintain a road and an embankment was refused. Occasionally such a declaration may be granted at the suit of an individual who is unsure which of two authorities is charged with the relevant duty.[14]

4. Declarations are also sought in cases where the scope and effect of a procedural duty are in doubt. In *Grunwick Processing Laboratories Ltd.* v. *Advisory, Conciliation, and Arbitration Service*[15] the House of Lords held that a declaration should be granted that a duty imposed on ACAS to ascertain the opinion of workers to whom a recognition issue related, when inquiring into that issue, was mandatory, and failure to fulfil it rendered its subsequent recommendation void.

[12] [1955] Chanc. 341; see also *Riverina Transport Pty. Ltd.* v. *Victoria* (1937) 57 CLR 327 (HCA); *Stevens (G.E.) (High Wycombe) Ltd.* v. *High Wycombe Corporation* [1961] 3 WLR 228; *Patel* v. *University of Bradford Senate* [1979] 1 WLR 1066; *Mutasa* v. *Attorney-General* [1980] QB 114; *Cicutti* v. *Suffolk County Council* [1980] 3 All ER 689; *Chief Constable of North Wales Police* v. *Evans* [1982] 3 All ER 141 (HL); *Greater London Council* v. *Holmes* [1984] 1 WLR 1307.

[13] [1957] 1 WLR 1146; see also *Gillow* v. *Durham County Council* [1913] AC 54 (HL); *Gateshead Union Guardians* v. *Durham County Council* [1918] 1 Chanc. 46; *Monmouthshire County Council* v. *British Transport Commission* [1957] 1 WLR 1146. *Litherland Urban District Council* v. *Liverpool Corporation* [1958] 1 WLR 913; *London County Council* v. *Central Land Board* [1958] 1 WLR 1296; *South Australia* v. *Commonwealth* (1962) 108 CLR 130 (HCA); *North Western Gas Board* v. *Manchester Corporation* [1964] 1 WLR 64; *North Yorkshire County Council* v. *Secretary of State for Education and Science*, above n. 5.

[14] *Attorney-General* v. *St Ives Rural District Council* [1961] 1 QB 366; cf. *Parramatta City Council* v. *Sandall* [1973] 1 NSWLR 151.

[15] [1978] AC 655; see also *Barber* v. *Manchester Regional Hospital Board* [1958] 1 WLR 181; *Wood* v. *Ealing London Borough Council* [1966] 3 All ER 514; *Lee* v. *Department*

There are of course some limitations with regard to the circumstances in which a declaration can be granted. The most important of these is the principle that the court will not make a declaration in relation to a hypothetical issue. This rather vague proposition has been said to mean any of the following: that there is no dispute in existence; that there is a dispute but it is not attached to facts; that the dispute is based on hypothetical facts; that the dispute has ceased to be of practical significance; or that the declaration sought can be of no practical consequence.[16] The reasons for this principle lie in the courts' traditional distrust of giving advisory opinions, which is based on their fear that their independence will thereby be eroded, the fear of unnecessary litigation, and the fact that a declaration might be granted without full argument or the benefit of considering the law in relation to actual facts. An Australian judge, Else-Mitchell J., has advanced a further reason:

> . . . I am firmly of the opinion that it is not the function of the established courts to entertain applications which are designed solely or primarily as a means of obtaining legal advice for potential litigants, and that the courts should, so far as possible, avoid making determinations of hypothetical questions.[17]

This requirement of ripeness therefore restricts the circumstances in which declaratory proceedings can be used by a public authority, whether as plaintiff or defendant, as a means of clarifying its duties. The limits of the principle, and indeed the usefulness of the declaration, can be judged by reference to *Surrey County Council* v. *Ministry of Education*.[18] The plaintiff in that case was a local education authority charged under the Education Act 1944 (England and Wales) with a duty to 'make such arrangements for the provision of transport and otherwise as they consider necessary for the purpose of facilitating the attendance of pupils at school . . . and any transport provided in pursuance of such arrangements shall be provided free of charge'; they also had power to pay the whole or any part, as they thought

of Education and Science (1967) 66 LGR 211; *Ricegrowers Co-operative Mills Ltd.* v. *Bannerman and Trade Practices Commission* (1981) 38 ALR 535.

[16] See Zamir, above n. 1, pp. 50–69. Interestingly, in *Meade* v. *Haringey London Borough Council* [1979] 1 WLR 637 it was not sought to argue that the issue was hypothetical even though the granting of a declaration would have been of no consequence, the defendants having complied with their duty by the time the case was heard in the Court of Appeal; cf. *University of New South Wales* v. *Moorhouse* (1975) 133 CLR 1 (HCA).

[17] *Ku-ring-gai Municipal Council* v. *Suburban Centres Pty. Ltd.* [1971] 2 NSWLR 335, 339.

[18] Above n. 11.

fit, of the reasonable travelling expenses of any pupil in attendance at any school for whom transport arrangements were made in pursuance of that duty.[19] The authority had arranged for school transport for all children living more than three miles from school, but to cut expenditure they proposed instead to pay travelling expenses for secondary pupils in such a way that no child had further than three miles to travel to school unaided, i.e. they proposed to pay for each secondary pupil to travel, where necessary, to a point within the three-mile limit, but not the full distance to school. The Minister informed the authority that she was advised these new arrangements were unlawful, and that she might uphold an appeal by an aggrieved parent arguing that the authority had failed in its duty. The authority accordingly proceeded for a declaration that the arrangements were lawful. In the result, that declaration was refused, though it was not argued that the question was not ripe; it is suggested, however, that it was only ripe because a conflict of view between the authority and the Minister existed, and the exercise of default powers by the Minister was probably, in the circumstances, inevitable; but for these facts the issue would have been 'hypothetical'. It is interesting also to notice that the remedy of mandamus would have resolved the difficulty only if a parent had proceeded against the authority, and mandamus might well have been refused on the ground of availability of an alternative remedy, namely default powers,[20] while the exercise of default powers would only have resulted in a definitive ruling as to the extent of the duty if the authority had refused to comply with any directions the Minister might have made as to its performance; even in that event the courts' decision as to the validity of such directions would not be directly dependent on the matter of performance of the duty, but on the satisfaction, in law, of the Minister as to the authority's default. Both these methods of deciding the question would have been lengthy, unnecessarily adversarial, and possibly inconclusive; furthermore, the outcome might well have depended on the fortitude of the parents affected by the arrangements in bringing expensive and hazardous litigation against the authority. The courts should not of course be too strict in relation to the question of ripeness, and it is suggested that a likely impending conflict in the form of action or proceedings of some relevant kind should be sufficient to establish ripeness, otherwise the obvious advantages of a declaration in the resolution of doubts concerning the scope of public duties might be considerably reduced.

[19] Education Act 1944 (England and Wales), ss. 55, 39.
[20] See above Ch. 3.4(ii).

A further limitation is that the issue must be one which is justiciable, i.e. cognizable by the court in which proceedings are taken.[21] All this really means is that a declaration cannot be used in effect to enlarge the court's jurisdiction. For an action for a declaration there must of course also be a suitable plaintiff and a suitable defendant, both with an interest in the matter. This relates to standing, which is discussed in Chapter 6.

As in the case of mandamus and injunction, the remedy of a declaration is granted according to the court's discretion; this has sometimes been called, incorrectly, the only limitation on the courts' jurisdiction to grant a declaration.[22] Since the basis of the jurisdiction to grant a declaration is now statutory,[23] it is not certain whether the remedy has shed the discretionary elements attributable to its equitable origins;[24] the question is probably unimportant because the manner in which the courts' discretion is exercised closely resembles the exercise of discretion in mandamus and injunction cases. The ground of lack of ripeness, discussed above, will be seen to overlap somewhat the 'futility' rule applicable to mandamus;[25] there is authority also for the refusal of a declaration on the grounds of delay,[26] abuse of the court's process,[27] and even inconvenience or impossibility;[28] in *Attorney-General* v. *Colchester Corporation*[29] for example, a declaration was refused that the defendants were under a duty to maintain ferry services which were uneconomical.

The position with regard to alternative remedies is, however, somewhat different from that which applies in mandamus cases.[30] The main reason for this is that the declaration developed as a remedy which could be sought only where other relief was available, but was

[21] See De Smith, above n. 1, pp. 500–3.

[22] See e.g. *Barnard* v. *National Dock Labour Board* [1953] 2 QB 18, 41 per Denning LJ; *Hanson* v. *Radcliffe Urban District Council* [1922] 2 Chanc. 490, 507 per Lord Sterndale MR. It is often said that the discretion of the court in declaration cases is wider than in cases involving prerogative relief; in practice this wider discretion is not really apparent and it is surely only right that discretion should be exercised on the same principles regardless of the remedy sought.

[23] See RSC (England and Wales), O. 15, r. 16; HCR (Cwlth), O. 26, r. 19; and for the State Supreme Courts in Australia see Aronson and Franklin, above n. 7, p. 454.

[24] See Zamir, above n. 1, pp. 187–91.

[25] See above Ch. 3.4(iii).

[26] *Coney* v. *Choyce* [1975] 1 WLR 422.

[27] *Asher* v. *Secretary of State for the Environment* [1974] Chanc. 208.

[28] See De Smith, above n. 1, p. 513.

[29] [1955] 2 QB 207.

[30] See, generally, Zamir, above n. 1, pp. 225–44.

inadequate in the circumstances.[31] The result is that the declaration is much more liberal in this regard than mandamus, whose justification depended, traditionally, on the *absence* rather than the availability of alternative relief.[32]

The availability of prerogative relief is not regarded as a bar to the granting of a declaration.[33] Obviously the great utility of the declaration as a public-law remedy would be greatly reduced if that were not the case. An injunction, so far from excluding a declaration, is frequently sought in tandem with it and in order to give it coercive effect;[34] a bare declaration can, on the other hand, be sought.[35] The liberalization of the rules concerning alternative remedies in mandamus cases has tended to reduce the differences between the two remedies in this regard, apart from the matter of cumulation of remedies. With regard to administrative remedies, the policy of the courts has resulted in the exclusion of mandamus in some cases and has operated to similar effect in cases where a declaration was sought; thus the rule in *Pasmore* v. *Oswaldtwistle Urban District Council*[36] was applied first in *Clark* v. *Epsom Urban District Council*[37] (which, like *Pasmore* itself, concerned the Public Health Act 1875 (England and Wales), section 299) and then, in relation the Education Act 1944 (England and Wales), in *Watt* v. *Kesteven County Council*[38] and other cases.[39]

(iii) *Comparison with mandamus*

Provided only that the plaintiff has standing for a declaration to enforce the duty, and that the non-coercive effect of a declaration is not a problem in the circumstances of the case, the remedy of a declaration is the only one of those under discussion which can be seen as a genuine and viable alternative to mandamus in public-duty

[31] Ibid., ch. 2.

[32] See above Ch. 3.1.

[33] See Zamir, above n. 1, pp. 229–30; and the analysis in Aronson and Franklin, *Review of Administrative Action* (1987), pp. 455–6, of the case of *Toowoomba Foundry Pty. Ltd.* v. *Commonwealth* (1945) 71 CLR 545 (HCA), where it was held that the availability of prohibition excluded a declaration in the circumstances of the case; see also *Pyx Granite Co. Ltd.* v. *Ministry of Housing and Local Government* [1960] AC 260, 290 (HL).

[34] See e.g. *De Falco* v. *Crawley Borough Council* [1980] QB 460. A declaration may sometimes be granted in lieu of an injunction: see De Smith, above n. 1, p. 482.

[35] *Dyson* v. *Attorney-General* [1911].

[36] [1898] AC 387; see above Ch. 3.4(iii).

[37] [1929] 1 Chanc. 287.

[38] Above n. 11.

[39] *Wood* v. *Ealing London Borough Council*, above n. 15; *Cumings* v. *Birkenhead Corporation* and *Roberts* v. *Dorset County Council*, above n. 11.

cases. The choice of remedy may well depend on what other relief, if any, is sought. If the duty is a duty to determine, then if *certiorari* is sought, only an application for mandamus, not for a declaration, can be joined with it; if on the other hand an injunction or other non-prerogative relief, such as a claim for damages, is sought, then only a declaration, not mandamus, will do.[40]

Apart from these considerations a declaration has the following advantages over mandamus:

1. A declaration lies with regard to private as well as public duties, and is therefore appropriate where there is doubt as to whether the duty is public or private.
2. A declaration is available against the Crown, so that in cases of doubt as to whether the defendant can be properly identified with the Crown, a declaration is appropriate.
3. A declaration is also useful where interlocutory relief is sought.[41]

4.5 STATUTORY ACTION FOR MANDAMUS

The statutory action for mandamus has its origins in the Common Law Procedure Act 1854 (England and Wales), sections 68–73. The purpose of those sections was to introduce a remedy in the nature of mandamus applicable to the enforcement of private duties.[1] They were repealed, but the action for mandamus survived, in a simplified form, the fusion of law and equity,[2] and found its way into the statute

[40] In *R.* v. *Marlborough Street Stipendiary Magistrate, ex p. Bouchereau* [1977] 1 WLR 414 a declaration was granted on an application for mandamus; the jurisdiction to grant a declaration in these circumstances was not discussed and the decision seems to be *per incuriam* on this point.

[41] This is not to say that an interlocutory *declaration* can be granted; the prevailing view is that a declaration is inherently final and cannot be granted interlocutorily; see *International General Electric Co. of New York Ltd.* v. *Customs and Excise Commissioners* [1962] Chanc. 784, 789; *Wallersteiner* v. *Moir* [1974] 1 WLR 991, 1093; cf., however, *Clarke* v. *Chadburn* [1985] 1 All ER 211 (final declaration granted in interlocutory proceedings). The principle was applied also in *Meade*'s case, above n. 16, p. 648. A striking instance of the use of interlocutory proceedings (discovery in this case) to advantage is *Anisminic Ltd.* v. *Foreign Compensation Commission* [1969] 2 AC 147 (HL).

[1] See the 'Second Report' of the Commissioners appointed 'to inquire into the process, practice, and system of pleading in the Superior Courts of common law' (1853), *House of Commons Parliamentary Papers*, vol. 40, p. 693.

[2] See Supreme Court of Judicature Act 1873 (England and Wales), s. 25(8); and now Supreme Court of Judicature (Consolidation) Act 1925 (England and Wales), s. 45; RSC (England and Wales), O. 53 (prior to 1977 amendment) and O. 29, r. 1.

law of the states of Australia.[3] It has also survived the introduction of the new statutory review procedures.[4] The substantive effect of the statutory provisions is, briefly, to give a plaintiff a right to apply in an *action* for a writ of mandamus which commands the defendant to fulfil any duty in which the plaintiff is personally interested and by the performance of which he is sustaining, or may sustain, damage.[5] In England the action for mandamus, although still available,[6] is obsolete, and no case in which it has been sought has been reported since 1966.[7] In Australia, however, it has been resuscitated by a combination of judicial decision and legislative act,[8] and although relatively few cases[9] have been reported since the seminal decision in *Mudge* v. *Attorney-General of Victoria*[10] in 1960, it is considered by Australian commentators to have some potential as a public-law remedy.[11]

The action for mandamus lies somewhere between the prerogative writ of mandamus and the mandatory injunction. Unfortunately many of the principles involved are quite uncertain, and this no doubt is the reason why the action for mandamus has proved less interesting to litigants than to the textbook writers. Whitmore and Aronson[12] have analysed the English and Australian cases with great thoroughness, and

[3] See Common Law Procedure Acts 1857 and 1899 (NSW), ss. 34–41 and (NSW), ss. 65–75 respectively; Common Law Procedure Statute 1865 (Vic.), ss. 229–36; Common Law Procedure Act 1867 (Qld), ss. 34–41. The relevant provisions now are Supreme Court Act 1970 (NSW), s. 65; Supreme Court Act 1958 (Vic.), s. 62(2) and RSC (Vic.), O. 3, r. 1A and O. 53; Interdict Act 1867 (Qld), ss. 44–51, 57, Judicature Act 1876 (Qld), s. 5(8), and RSC (Qld), O. 57. For further details see Aronson and Franklin, *Review of Administrative Action* (1987), pp. 535–7.

[4] See below Ch. 5.

[5] Common Law Procedure Act 1854 (England and Wales), ss. 68–9.

[6] See above n. 2.

[7] *Thorne* v. *University of London* [1966] QB 237; the only other cases of any importance in recent times are *Watt* v. *Kesteven County Council* [1955] 1 QB 148 and *Dinzulu* v. *Attorney-General* [1958] 1 WLR 1252.

[8] See below n. 10, and above n. 3.

[9] *Mutual Acceptance Ltd.* v. *Commonwealth* (1972) 29 LGRA 123; *Dickinson* v. *Perrignon* [1973] 1 NSWLR 172; *P. & C. Cantarella Pty. Ltd.* v. *Egg Marketing Board (NSW)* [1973] 2 NSWLR 366; *Apex Development Pty. Ltd.* v. *Holroyd Municipal Council* (1974) 29 LGRA 218; *Bilbao* v. *Farquhar* [1974] 1 NSWLR 377; *Donges* v. *Ratcliffe* [1975] 1 NSWLR 501. For older Australian cases, see Aronson and Franklin, above n. 3, pp. 537–8.

[10] [1960] VR 43.

[11] Aronson and Franklin, above n. 3, p. 539, Sykes, Lanham, and Tracey, *General Principles of Administrative Law*, 2nd edn. (1984), pp. 169–70.

[12] Above n. 3, pp. 394–402; see also De Smith, *Judicial Review of Administrative Action*, 1st edn. (1959), pp. 425–8.

it is not proposed here to embark on a similar analysis. The cases in fact lay down few incontrovertible principles, and the very purpose of the remedy is somewhat evanescent.

The first problem is whether the action for mandamus is a private-law or public-law remedy or both. In spite of the original purpose of the 1854 Act provisions, those provisions did not suggest that only private duties were enforceable under them: in fact one section strongly suggests that the action for mandamus and the prerogative writ overlapped.[13] None the less Lord Campbell CJ, in the first case[14] that was decided on the new remedy, held, rather oddly, that the former was confined to duties enforceable by the latter, a position which would thwart the main reason for having the remedy at all. Interestingly enough the judges gradually moved, over the latter half of the nineteenth century, to a diametrically opposite view, that the action for mandamus was confined to enforcing private duties.[15]

Intimately related to this problem is the question whether the availability of the prerogative writ as an alternative remedy excludes the action for mandamus. The cases[16] which held that the action for mandamus was available to enforce public duties culminated in the somewhat startling decision in *R.* v. *Lambourn Railway Co.*[17] that the action for mandamus excluded, rather than was excluded by, the availability of the prerogative writ. It was as a reaction to this decision that the action for mandamus came to be seen as a purely private-law remedy,[18] and, as a corollary, that the prerogative writ came to be regarded as a purely public-law remedy. This latter principle was of

[13] S. 75: 'Nothing herein contained shall take away the Jurisdiction of the Court of Queen's Bench to grant Writs of Mandamus; nor shall any Writ of Mandamus issued out of that Court be invalid by reason of the Right of the Prosecutor to proceed by Action for Mandamus under this Act.'

[14] *Benson* v. *Paull* (1856) 119 ER 865. This decision may be explicable by the fact that the duty involved in that case was a contractual one, and the action for mandamus should not, according to Lord Campbell CJ, be used to supplant the order for specific performance.

[15] See *Norris* v. *Irish Land Co.* (1857) 120 ER 191; *Fotheray* v. *Metropolitan Railway Co.* (1866) LR 1 CP 188; *Morgan* v. *same* (1868) LR 4 CP 97; *R.* v. *Lambourn Valley Railway Co.* (1888) 22 QBD 463; *Baxter* v. *London County Council* (1890) 63 LT 767; *R.* v. *St George's Southwark, Vestry* (1892) 61 LJQB 398; *R.* v. *London & North Western Railway Co.* [1894] 2 QB 512; *R.* v. *same* (1896) 65 LJQB 516; *Smith* v. *Chorley District Council* [1897] 1 QB 572.

[16] See above n. 14 and the first three cases cited above n. 15.

[17] Above n. 15. The decision is almost certainly incorrect: see e.g. *Smith*, above n. 15.

[18] See *Baxter* and *Smith*, ibid.

course already well established, but had been applied to duties which would more sensibly be regarded as private.[19]

A third problem, again related to the first and second, is whether the 1854 Act created a new cause of action, or whether it merely created a form of ancillary relief. Cases at the end of the nineteenth century seemed to confirm the ancillary nature of the remedy.[20] However, in the latest English case it was granted where no other relief was claimed and the duty would clearly have been enforceable by the prerogative writ.[21]

In all respects other than those mentioned, the action for mandamus resembles the prerogative writ. Demand and refusal are required,[22] and the court has a discretion to refuse an order on grounds of delay[23] or the availability of an alternative remedy.[24] However, a further incident of statutory mandamus being claimable in an action is that interlocutory relief is available.[25]

The decline of the action of mandamus in England was attributed by De Smith[26] to the non-occurrence of the fact situations which generally gave rise to litigation in the old cases, the uncertainty of the public element in the duty, the uncertain relation to the prerogative writ, and the general availability of the mandatory injunction;[27] the only occasion where he considered the remedy might be useful was where the availability of a mandatory injunction was doubtful and a plaintiff might usefully combine claims for damages or a declaration and mandamus in the one action. Now that the Supreme Court Act 1981 (UK) has made damages, an injunction, or a declaration available in an application for judicial review,[28] the action for mandamus has lost its last hope of resuscitation in England.

The Australian decisions reveal that the action for mandamus is seen as pre-eminently a public-law remedy, though it has not been

[19] See above Ch. 3.2.

[20] *Baxter*, above n. 15; *R.* v. *St Giles, Camberwell, Vestry* (1897) 66 LJQB 337; *Ex p. Pager* (1897) 14 TLR 61.

[21] *Dinzulu* v. *Attorney-General*, above n. 7.

[22] The Common Law Procedure Act 1854 (England and Wales), s. 69 states as much, but see above Ch. 3.3.

[23] See the cases cited in Whitmore and Aronson, above n. 3, p. 398 n. 363.

[24] *Bush* v. *Beavan* (1862) 1 H. & C. 500; *Peebles* v. *Oswaldtwistle Urban District Council* [1893] 1 QB 384. For further instances of exercise of discretion along the lines of the prerogative writ, see De Smith, above n. 12, p. 426, esp. nn. 41–3.

[25] Supreme Court of Judicature (Consolidation) Act 1925 (England and Wales), s. 45; for the statutory provisions to similar effect in Australia, see Whitmore and Aronson, above n. 3, p. 399 n. 375.

[26] Above n. 12, p. 428.

[27] See e.g. *Davies* v. *Gas, Light, & Coke Co.* [1909] 1 Chanc. 708.

[28] See below Ch. 5.2.

confined to public law. However, the really striking development has been its employment in situations traditionally covered by the prerogative writ. Although it has never been suggested in any English case that the action for mandamus covers the *entire field* of public duties,[29] the Australian cases heavily suggest precisely that, because most of them concern duties to determine questions affecting individual rights and interests,[30] duties which are ordinarily enforced by mandamus, but, in the cases referred to, were sought to be enforced by a declaration, or other remedy, in addition to an order of mandamus. The Australian courts seem also to have decided that an action for mandamus can only be brought where other relief is claimed,[31] so that no separate cause of action is created, though in Victoria a plaintiff can claim mandamus 'in an action commenced by writ of summons, and either with or without other relief, and either as principal relief or as ancillary or interlocutory relief'.[32]

It can also be inferred from the cases referred to above that the availability of prerogative relief is no bar to an action for mandamus.

The advantages of the action for mandamus over the prerogative writ are clear. Since the remedy is (at least) an ancillary one, it can be claimed as an adjunct to a claim for damages or a declaration, and interlocutory relief is also available. In these respects it resembles a mandatory injunction, but the mandatory injunction has not been used as a means of reviewing decisions of public bodies and tribunals. These facts alone explain the re-emergence of statutory mandamus in Australia. Another possible advantage is that an action for mandamus, *qua* action, may be available against the Crown,[33] and a possible advantage over the mandatory injunction might be that the rules for standing are somewhat more liberal, approximating to those for mandamus.[34]

[29] As indicated above, the English cases suggest either that only private duties are involved, or that there is an overlap with the duties enforced by the prerogative writ. Even Lord Campbell CJ in *Benson* v. *Paull*, above n. 14, did not say that *all* public duties are enforceable by mandamus, and the case itself concerned a private duty.

[30] *Mudge*, above n. 10, and *Dickinson*, *Cantarella*, *Apex*, *Bilbao*, and *Donges*, above n. 9, all fall into this category. On the other hand, the duty in *Mutual Acceptance Ltd.* v. *Commonwealth* (1972) 29 LGRA 123 could be argued to be private in character; the same is true of *Royal Insurance Co. Ltd.* v. *Mylius* (1926) 38 CLR 477 (HCA).

[31] *Mudge*, above n. 10, following *R.* v. *Shire of Winchelsea* (1896) 22 VLR 171.

[32] RSC (Vic.), O. 3, r. 1A, passed in consequence of the decision in *Mudge*, above n. 10.

[33] Aronson and Franklin, above n. 3, p. 540.

[34] The 1854 Act required the plaintiff to be 'personally interested' in the performance of the duty. This formula survives in the Australian provisions, and can readily be equated with the current view of standing for the prerogative writ; see

What then should be done about the action for mandamus? This writer does not share the optimistic view of the Australian commentators.[35] There is no purpose in adding to the motley array of remedies available for enforcing public duties unless some tangible advantage is to be gained from doing so; in fact it may be counterproductive to do this, because the courts may well hold one remedy to exclude another, creating a hidden pitfall for the litigant rather than an embarrassment of riches: they are not after all likely to develop three remedies to cover substantially similar ground. The only tangible advantages of the action for mandamus arise from the fact that by it a mandatory order can be sought with other relief in an action; it cannot be used to develop a cause of action which does not already exist, and cannot therefore be used to develop an independent action for damages for breach of a public duty. Thus the advantages of the action for mandamus disappear if the simple reform is effected of putting the remedies of declaration, injunction, and claim for damages at the disposal of courts hearing applications for judicial review; this reform has been achieved in England and, for some purposes at least, in Australia.[36] The trend towards the kind of 'portmanteau' application envisaged by the new statutory review procedures will probably therefore render the statutory action for mandamus obsolete, at least as regards public law. This result would be desirable because the reform of procedure in public law is essentially a matter of rationalizing the existing remedies, not resuscitating old remedies to fill the gaps. Eventually therefore the action for mandamus may have to be repealed and confined to books on legal history as a 'Victorian' (in both senses) curiosity.

4.6 INDICTMENT

In the eighteenth century indictment was a common method of enforcing public duties. At common law it was

> a good general ground that wherever a statute prohibits a matter of public

below Ch. 6.5. It may account for some of the confusion concerning the use of the action for mandamus as a private law remedy.

[35] See above n. 11. According to Whitmore and Aronson, *Review of Administrative Action* (1978, p. 394) it 'could well supplant the order in the nature of the prerogative writ of mandamus completely'.

[36] See below Ch. 5.3. If it is desired to make statutory review an exclusive procedure, the existence of the action for mandamus could even be used to circumvent the law: cf. *O'Reilly* v. *Mackman* [1982] 3 WLR 1096.

grievance to the liberty and security of a subject, or *commands a matter of public convenience*, as the repairing of the common streets of a town, an offender against such statute is punishable, not only at the suit of the party aggrieved, but also by way of indictment for his contempt of the statute, unless such methods of proceeding do manifestly appear to be excluded by it . . . [emphasis added][1]

An overseer of the poor, who refused to receive a pauper in accordance with statute, committed an offence;[2] and so did a magistrate who unreasonably refused to exercise his powers to quell a riot.[3] It was even held that a remedy by way of indictment excluded the availability of mandamus.[4]

Nowadays indictment for failure to perform a public duty is obsolete as a general remedy, but a particular statute may provide that a failure to perform a duty imposed by it is a criminal offence; an example of this is wilful refusal by a local-government official to produce for inspection documents properly requested, such as council minutes.[5]

The recent decision of Lloyd J. in *R.* v. *Horseferry Road Magistrates' Court, ex p. Independent Broadcasting Authority*[6] clarifies the position somewhat. M alleged that the IBA had allowed a subliminal image of his head superimposed on a picture of a naked woman to be flashed across the screens of viewers during a television programme. This would be a breach of its statutory duty under the Broadcasting Act 1981 (England and Wales), section 4(3), to satisfy itself that the programmes broadcast by it 'do not include . . . any technical device which, by using images of very brief duration . . . exploits the possibility of conveying a message to, or otherwise influencing the minds of, members of an audience without their being aware, or fully

[1] *Hawkins' Pleas of the Crown*, 8th edn. by Curwood J. (1824), vol. ii, ch. 25, s. 4, adopted in *R.* v. *Hall* [1891] 1 QB 747, and the *IBA* case cited below n. 6.

[2] *R.* v. *Davis* (1754) 96 ER 839.

[3] *R.* v. *Kennett* (1781) 712 ER 976; *R.* v. *Pinney* (1832) 110 ER 349; and see *R.* v. *Price* (1840) 113 ER 590 (refusal to register birth of child). Penalties were imposed on public utilities in the 19th cent.: *Atkinson* v. *Newcastle and Gateshead Waterworks Co.* (1877) 2 Ex. D. 441; *Sheffield Waterworks Co.* v. *Cater* (1882) 8 QBD 632; *Re Richmond Gas Co.* v. *Richmond Borough Council* [1893] 1 QB 56.

[4] See above Ch. 3.4(iii).

[5] See *Wilson* v. *Evans* [1962] 2 QB 383 (acquittal of officer of London County Council on charge of refusing inspection of council minutes to an elector: London Government Act 1939, s. 193); *Hillingdon London Borough Council* v. *Paulssen* [1977] JPL 518 is a good example of the impropriety of using criminal proceedings in this kind of case. However, the offence at common law is not entirely defunct; in *R.* v. *Dytham* [1979] QB 722 a policeman was convicted when he failed to intervene when a man was beaten to death in the street 30 yards from him.

[6] [1986] 2 All ER 666.

aware, of what has been done'. M took out a summons against the IBA for contempt of statute. The IBA succeeded in an application to strike out the summons. The judge was not prepared to hold that 'cessante ratione legis, cessat lex ipsa', but proceeded on the basis that clear words were necessary to create a criminal offence, and laid down the following factors to be considered: (1) whether the provision is mandatory or prohibitory, (2) whether the statute is ancient or modern, and (3) whether there are other means of enforcement.

The decision is clearly correct, and it is now for legislatures to remove the remaining uncertainty by abolishing the offence of contempt of statute completely, as has been recommended by the English Law Commission.[7] De Smith's statement that 'the prosecution and conviction of a public authority may occasionally have a salutary effect in reminding it of its social responsibilities and in reaffirming the principle of equality before the law'[8] is one which might have some application to general obligations imposed on all landowners, employers, and the like, such as those relating to public health and safety, but is not applicable to the breach of purely public duties. While there may be a case for enforcing particular provisions by this drastic means (and it is hard to think of a cruder method), the offence at common law is so open-ended that it runs counter to the principle that a person should not be punished except for a distinct breach of the law; moreover, the existence of such liability could be used as a means of harrassing public officials who act in good faith. Even where the offence is quite clear, it seems wrong to expose a public official to criminal liability when mandamus is an adequate alternative. Imprisonment and fining of a public official should be confined to the situation where he disobeys an order of the court.[9] The remedy in cases such as the *IBA* case is to seek damages or mandamus.

4.7 *CERTIORARI*, PROHIBITION, AND HABEAS CORPUS

For completeness it is perhaps necessary to say something about the prerogative writs other than mandamus, in so far as they are relevant to the enforcement of public duties.

[7] *Law Commission Report*, No. 76 (1976), paras. 61–5. In Australia the offence would appear to apply except where the common law is excluded by statute, i.e. in the Criminal Codes of Queensland and Western Australia: see Howard, *Australian Criminal Law*, 4th edn. (1982), pp. 1–3.

[8] *Judicial Review of Administrative Action*, 4th edn. by Evans (1980), p. 526.

[9] See above Ch. 3.5.

Mandamus, as is clear from the discussion in Chapter 3, is overwhelmingly more important for public duties than the other prerogative writs, because it was specifically designed for enforcing public duties, and because the distinction between administrative and judicial functions has never been relevant to the award of mandamus. *Certiorari* and prohibition, on the other hand, have always been used for quashing and preventing, respectively, unlawful determinations, and were generally, though not nowadays, confined to judicial functions.

The result of these considerations is that, to use the terminology adopted in Chapter 2, while mandamus relates to all public duties, *certiorari* and prohibition relate only to the duty to determine. The function of mandamus in relation to the duty to determine is to compel the decision-maker to determine according to law. For this reason mandamus is often sought in this context along with *certiorari*, and the public nature of the duty to determine must be the same in the case of each remedy, as has been seen in Chapter 1. It has also been seen in Chapter 2 that there is a very close relation between excess of jurisdiction and failure or refusal of jurisdiction, and that the distinction between jurisdictional and non-jurisdictional errors has been all but obliterated. The question of the precise scope of *certiorari* and mandamus in relation to determinations must therefore be answered in such a way as to reflect these considerations. If all errors of law go to jurisdiction, then they do so for the purposes of mandamus as well as *certiorari* and prohibition. Thus the enforcement of a public duty to determine is achieved by having a particular theory of jurisdiction (at present a theory which says that all errors of law go to jurisdiction) and using whichever prerogative writ or writs may be appropriate to correct the error. In the case of a determination which leads to loss of liberty, the remedy will be habeas corpus. In this way the prerogative writs can be seen as a cluster of remedies which, between them, have the effect of rendering the duty to determine enforceable.

5
Enforcement: Statutory Judicial Review

5.1 INTRODUCTION

The purpose of this chapter is to inquire into the effect of the new statutory procedures for judicial review in England and Australia on the enforcement of public duties. Undoubtedly the new procedures under the Supreme Court Act 1981 (England and Wales) and the Administrative Decisions (Judicial Review) Act 1977 (Cwlth) will have considerable effect on the availability of remedies generally; in particular, however, it is likely that the necessity of unifying the rules as to standing in respect of the various remedies will be used as a means of reforming those rules, and it is this aspect which is of the greatest interest with regard to the enforcement of public duties. Standing, however, will be discussed separately in Chapter 6. The general effect of the reforms contained in these two Acts is to simplify and facilitate applications for judicial review rather than to reform the substantive law. As will be seen, however, these reforms have created some new problems of their own which have some bearing on the question of defining public duties.

5.2 SUPREME COURT ACT 1981 [ENGLAND AND WALES]

(i) O'Reilly *and public duties*

The main purpose of the new Rules of the Supreme Court (England and Wales), Order 53 and its statutory ratification in the Supreme Court Act 1981 (England and Wales)[1] is to make all the remedies used in judicial review available on a single 'application for judicial

[1] See SI 1955/1977, SI 2000/1980; *Law Commission Reports*, Nos. 20 (1969), and 73 (1976), and *Working Paper* No. 40 (1970). The best discussions of the new procedure are to be found in Craig, *Administrative Law* (1983), pp. 496–517; Aldous and Alder, *Applications for Judicial Review: Law and Practice* (1985); Gordon, *Judicial Review: Law and Procedure* (1985); but see also Blom-Cooper, 'The New Face of Judicial Review: Administrative Changes in Order 53' [1982], *PL* 250; Wade, 'Procedure and Prerogative in Public Law' (1985), 101 *LQR* 180; Woolf, 'Public Law—Private Law: Why the Divide?' [1986], *PL* 220; Beatson, ' "Public" and "Private" in English Administrative Law' (1987), 103 *LQR* 34.

review', and to abolish the unnecessary distinctions between those remedies, subjecting them to a similar regimen; in particular the standing rules are the same for each remedy, and the requirements of leave to bring the application and a time-limit are imposed whatever remedy is sought or granted.

The nub of these reforms is contained in section 31 of the Act.[2] They have, however, been given considerable clarification and impetus by the House of Lords in *O'Reilly* v. *Mackman*,[3] in which it was held that an attempt to avoid Order 53 by using the 'old' procedure[4] in an action for a declaration in a 'public-law' case is an abuse of the court's process; and it is clear that the reasoning also applies to an action for an injunction.[5]

The point of greatest importance for our present purpose is that as a result of *O'Reilly* a clear distinction must be drawn between public and private law, and hence between public and private duties: an attempt to discover how this distinction might be drawn was made in Chapter 1. It is, however, questionable whether the drawing of the distinction for the purposes of Order 53 procedure is worth the effort bearing in mind that it was not of crucial importance under the old procedures for judicial review. This aspect of *O'Reilly* is discussed below in the context of public duties, but let us first consider the precise consequences for public-duty enforcement of the new procedure and the public-law/private-law distinction.

The distinction between public law and private law is not of course a new phenomenon; as we saw in Chapter 1 it is inherent in the law relating to the prerogative writs, and is manifest in the law of mandamus in the distinction between public and private duties. What is new in *O'Reilly* is the *consequences* of the distinction, which may be

[2] For full text see App.

[3] [1982] 3 All ER 1124.

[4] The term 'old' procedure is used in this section to refer to the various procedures available for judicial review prior to the new O. 53. These of course still apply in the case of private-law duties: *R.* v. *British Broadcasting Corporation, ex p. Lavelle* [1983] 1 WLR 23; *Law* v. *National Greyhound Racing Club* [1983] 3 All ER 300; *R.* v. *East Berkshire Health Authority, ex p. Walsh* [1984] 3 All ER 425; *R.* v. *Secretary of State for the Home Department, ex p. Benwell* [1984] 3 All ER 854 (HL) (*Walsh* distinguished, duty held to be public). For the public-law/private-law distinction, see Harlow, 'Public and Private Law: Definition without Distinction' (1980), 43 *MLR* 241; Samuel, 'Public and Private Law: A Private Lawyer's Response' (1983), 46 *MLR* 558; Cripps, 'Jurisdiction, Remedies and Judicial Review' (1984), 42 *CLJ* 214; Forsyth, 'Beyond *O'Reilly* v. *Mackman*: the Foundations and Nature of Procedural Exclusivity' [1985], *CLJ* 415; Cane, 'Public Law and Private Law: A Study of the Analysis and Use of a Legal Concept', ch. 3 of Eekelaar and Bell, eds., *Oxford Essays in Jurisprudence*, 3rd Series (1987).

[5] *O'Reilly* v. *Mackman*, above n. 3, pp. 1133–4 per Lord Diplock.

serious for a litigant in the field of public duties; although in most cases the application of the distinction is not problematical, there are some areas of doubt, and these may increase if the courts wish to expand the notion of a public duty, which, for reasons discussed in Chapter 1, it is suggested they should; this difficulty is compounded by the fact that a case may raise issues relating to public law *and* private law, though the extent to which this can occur depends of course on how one defines the two concepts. This is a problem which did not exist under the old procedures, because wherever there was any doubt as to whether the case was one in public law, one could simply commence an action for a declaration and/or injunction, in which the public-law/private-law distinction was immaterial. What *O'Reilly* has done is to make this escape route impossible; the litigant must now confront the distinction squarely: if he chooses the old procedure mistakenly, his action may be struck out on the defendant's application;[6] if he chooses the Order 53 procedure mistakenly, either leave to proceed will not be granted, or else, much worse, his case will be dismissed after a full hearing on a question which in no way relates to the merits of the case. This is why it is a particularly moot point whether the advantages of the Order 53 procedure outweigh this serious disadvantage. The remedy, as Wade puts it, may be worse than the disease.[7]

The most important advantage is that it is no longer necessary for the applicant to choose his remedy and stand or fall by its application, sometimes fraught with technicalities, to the facts of his case. Under the new procedure, although it would appear from section 31 that the substantive law of the prerogative writs still determines the situations in which relief will be granted, a declaration or an injunction can be granted instead of, or in addition to, prerogative relief; in other words the procedural law of the declaration and the injunction has been added to the substantive and procedural law of the prerogative writs. The main beneficial consequences of this are that a declaration or an injunction can be granted even where an applicant would not previously have had standing for either remedy, or where, regardless of standing, such remedy is for some reason more appropriate in the circumstances of the case.[8] A claim for damages can also be added to the application under section 31(4). Thus the problems

[6] It is, however, possible for the application to continue as an ordinary action: O. 55, r. 9(5); transfer *to* O. 53 procedure is not possible: *O'Reilly*, above n. 3, p. 1133; however, as Woolf points out, above n. 1, p. 232, the High Court can none the less grant leave and treat the case as one under O. 53.

[7] Above n. 1, p. 182.

[8] See e.g. *R.* v. *Pentonville Prison* (*Governor*), *ex p. Herbage*, *The Times*, 21 May 1986.

of selection and cumulation of remedies are resolved simply and satisfactorily. While these are not the only problems involved, they are of particular importance in public duty cases, as has been seen in Chapter 4.

(ii) *Declarations and duties of the Crown*

One consequence of the new rule under section 31(2), however, remains unclear. If it means that a declaration can be granted *if and only if* prerogative relief may be granted, it is possible that a declaration cannot be granted under section 31(2) in respect of a duty of the Crown, and since an application for a declaration cannot now be made under the old procedure in respect of a public duty, it would follow that there is no method at all of compelling the Crown to perform its theoretically justiciable duties.[9] However, such a restrictive interpretation of section 31 cannot be justified, particularly in view of the absurd consequences for duties of the Crown. The availability of the prerogative orders, in this case mandamus, is only one matter to which the court must have regard in deciding whether the making of a declaration is just and convenient, the other matters being the nature of the persons and bodies against whom relief may be granted by such orders, and all the circumstances of the case. Consideration of the identity of the body concerned might conceivably restrict the making of a declaration to those bodies against whom a prerogative order could be granted, on the basis that the proper limits of substantive public law are set by those orders; however, the third matter, all the circumstances of the case, indicates that the first two are not to be considered decisive, and affords a convenient loophole for justifying the grant of a declaration against the Crown. This broader approach seems more consistent with the reasoning in *O'Reilly*, and it can be noted that under section 31(1) it is perfectly proper for an application for judicial review to be framed as, in effect, an application for a declaration.[10]

[9] See Cane, 'Standing, Legality and the Limits of Public Law' [1981], *PL* 322, 326. The argument rests of course on the unavailability of mandamus against the Crown, for which see above Ch. 3.2(i).

[10] It is also worthy of note that the Crown Proceedings Act 1947 (England and Wales), s. 21(2), specifically envisages a declaration being made against the Crown, and the courts are not likely to adopt an interpretation of s. 31(2) which would impliedly repeal that provision; see also *R.* v. *Bromley London Borough Council, ex p. Lambeth London Borough Council*, *The Times*, 16 June 1984, where it was said that a declaration could be granted where the subject-matter was the sort of thing to which a prerogative order could apply, even if one was not available on the facts; *Lavelle*, above n. 4, pp. 30–1; *R.* v. *Secretary of State for the Home Department, ex p. Herbage* [1986] 3 WLR 504, where it was held that an interim mandatory injunction can be

(iii) *Section 31: a critique from the viewpoint of public duties*

We can now assess the effect of the new procedures on the enforcement of public duties. The main effect is simply that flexibility in relation to cumulation and substitution of remedies has been purchased at the price of having to distinguish between cases involving public duties and cases involving private duties. While one applauds the remedial flexibility achieved by the reforms, one wonders whether it was necessary to pay such a large price. It is not, after all, a necessary consequence of having a statutory judicial-review procedure that other avenues of judicial review should be closed off; the Administrative Decisions (Judicial Review) Act 1977 (Cwlth), considered below, is a good example of such a non-exclusive procedure. Let us therefore look at the reasons for the exclusiveness of the new procedures, as established in *O'Reilly*, from the viewpoint of public duty enforcement.

It will be appreciated that the two reasons advanced in *O'Reilly* were (1) the need to require the applicant to obtain the leave of the court before proceeding, and (2) the need to impose time-limits on applications; both these needs are held not to apply to purely private disputes; one might add a third possible reason which was not canvassed in *O'Reilly*, although it may have been in their Lordships' minds, that (3) the procedure under section 31 and Order 53 is a kind of code for public-law cases and it might be desirable to ensure that all public-law cases are considered under this code before judges of the High Court with the relevant expertise.[11]

The requirement of leave is a double-edged sword. It can be used to protect a public authority from a vexatious litigant without the necessity of their applying to strike out, or else to protect a bona fide litigant who really has no case against a waste of his money on a fruitless hearing; it might be said also to protect the courts from wasting their time and resources on litigation from which no benefit can be derived. As far as public-duty cases are concerned the vexatious litigant is conspicuous by his total absence, unless one includes (quite misleadingly) within that category persons who do not have a sufficient interest in the matter.[12] The bona fide litigant with no case is of course in evidence in public-duty cases as well as elsewhere in the law, but perhaps deserves our sympathy less than the bona fide litigant who has a case but finds he has proceeded in the wrong court. With regard to the protection of public authorities, it is interesting to note

granted against a crown officer under O. 53 procedure; *R.* v. *Secretary of State for the Environment, ex p. Nottinghamshire County Council*, *The Times*, 10 Nov. 1986.

[11] See e.g. *Re Tillmire Common, Heslington* [1982] 2 All ER 615.

[12] See Craig, above n. 1, p. 502; and below Ch. 6.6.

that although applications for leave are made ex-parte, the respondent authority is frequently heard, and the argument does not differ significantly from that on an application to strike out. What is really lacking in all these arguments, whatever their factual bases, is some explanation why they apply to public-law cases with more cogency than to private-law cases. The leave requirement is thus an unsure foundation on which to erect an edifice of public law.

The time-limit requirement under Order 53, rule 4 is three months from the date when grounds for the application first arose, unless the court considers that there is good reason for extending the period; under section 31(6) the court may refuse leave or relief on grounds of 'undue delay'; section 31(6) is, however, expressed in section 31(7) to be subject to rules of court. The purpose of this requirement is clearly to prevent prolonged uncertainty over the validity of administrative action, in the interests of the authority, and others, who may be relying on it. In public-duty cases, however, time-limits are only important where the duty is a duty to determine; in relation to other duties there is no question of validity of a decision involved, and therefore, generally, nothing for anyone to rely on; furthermore, in relation to most public duties, at least where nonfeasance is alleged, there is a real difficulty in knowing *from* when to calculate the period of 'undue delay'.[13] The new procedure also allows only exceptionally for discovery and cross-examination, whereas these are always available under the old procedure.[14]

The third reason advanced above may have more substance, and may indeed be a highly desirable development. However, even if it is decisive it should be remembered that not all issues concerning public duties can be contained within the new procedures; the important exception of collateral issues is discussed below. The benefit of having judges with expertise in public law hear public-duty cases may be at the expense of inhibiting those very judges from developing meaningfully the concept of a public duty, when such development may be at the cost of litigants faced with an uncertain public-duty/private-duty distinction.

The conclusion drawn is therefore that as far as public-duty cases are concerned the reforms are not the best possible result. It might have been better to make the procedures non-exclusive, but allow the courts to transfer those cases more suitable, because of the issues

[13] Presumably it should generally be calculated from the time when the duty ought to have been performed, but in many cases such guidance will not be very helpful. S. 31(6) strongly suggests that the time-limit would be extended in such cases.

[14] See O. 54, r. 8; O. 24.

raised, for consideration elsewhere. Wade's suggestion[15] that the problems involved in public-law cases can be satisfactorily resolved in the context of an ordinary action has much to recommend it. The price to be paid in this event is simply that of requiring the public authority to apply to strike out in the case of an obviously unmeritorious application, or one in which the applicant is guilty of delay.

(iv) *Collateral issues*

As was mentioned above, the principle of exclusiveness is subject to exceptions which were mentioned in *O'Reilly*. The first is that the old procedure can be used where the parties agree.[16] The second is where a public-law issue arises collaterally in an action concerning a private-law right.[17] This second exception is highly material to this inquiry, as can be seen from the House of Lords' decision in *Cocks* v. *Thanet District Council*,[18] which was decided on the same day as *O'Reilly*. The plaintiff brought an action under the old procedure in the county court for a declaration that the authority were in breach of their duty under the Housing (Homeless Persons) Act 1977 (England and Wales) to provide him with permanent accommodation under the Act, and consequential relief in the form of mandatory injunctions and damages. By consent the case was removed into the High Court for determination of a preliminary issue, namely whether the plaintiff was entitled to proceed in the county court or should instead proceed under Order 53. The judge held that he could proceed in the county court, but the authority's appeal direct to the House of Lords was upheld. The reasoning adopted was as follows. Under the Act the authority were under a duty to provide the plaintiff with permanent accommodation only if they were satisfied that he had a priority need and had not become homeless intentionally; their duty to provide depended on the outcome of their duty to inquire into and determine the relevant questions. If, in pursuance of their public-law functions, the authority reached a decision giving rise to a duty to provide the accommodation, then rights and obligations were immediately created in private law, and the duty was enforceable by injunction and gave rise to liability in damages. The reasoning in *O'Reilly* was equally applicable, and for the same reasons, where 'the decision of the public authority which

[15] Above n. 1, pp. 189–90; see, however, the contrary views of Woolf above n. 1.

[16] *O'Reilly* above n. 3, p. 1134.

[17] Ibid.; see *Wandsworth London Borough Council* v. *Winder* [1985] AC 461 (HL). One would have to include also the possibility of a relator action: see Grubb, 'Two Steps Towards a Unified Administrative Law Procedure' [1983], *PL* 190, 192, 200.

[18] [1982] 3 All ER 1135.

the litigant wishes to overturn is not one alleged to infringe any existing right but a decision which, being adverse to him, prevents him establishing a necessary condition precedent to the statutory private-law right which he seeks to enforce'.[19] In reaching this decision their Lordships disapproved statements in an earlier case in the Court of Appeal which indicated that the duty in question could be enforced by action or by application for judicial review,[20] on thc basis that the court has no power to substitute its own decision for the (public-law) decision of the authority so as to establish the authority's (private-law) liability.

While this decision is no doubt correct in relation to its particular facts, it poses a difficulty. What is not clear from the above reasoning is why a distinction should be drawn in the manner indicated between public-law and private-law duties. On the analysis adopted in Chapter 1 the authority's duty to provide should be classified as a public, not a private, duty. It should not be regarded as a private duty merely because its breach gives rise to the possibility of a private-law type remedy, nor because only one person is affected; after all, the duty in question would certainly have been enforceable by the purely public-law remedy of mandamus and its breach compensable[21] under section 31(4) in the same application. If the reasoning in *O'Reilly* is to be carried to its logical conclusion, it should apply to all duties of public authorities which are capable of enforcement by mandamus, regardless of the remedy sought, because the question of breach of duty remains the same whether a public-law or private-law remedy is being sought, and should be determinable under Order 53 procedure; any other result runs counter to the whole purpose of section 31. The way in which the House of Lords in *Cocks* distinguished public and private duties would indicate, however, that where only the performance of a duty to provide is in question and the remedy is an injunction and/or damages, Order 53 is not appropriate;[22] however,

[19] Ibid. 1139.

[20] *De Falco* v. *Crawley Borough Council* [1980] QB 460; and see *Lambert* v. *Ealing London Borough Council* [1982] 2 All ER 397; *Thornton* v. *Kirklees Metropolitan Borough Council* [1983] 3 WLR 1; *An Bord Bainne Cooperative* v. *Milk Marketing Board* [1984] 2 CMLR 584; Grubb, above n. 17.

[21] Assuming, that is, that an action would lie. It is argued below Ch. 7.2(vii) that it would, but the question is by no means free of doubt.

[22] The logic of the public-duty/private-duty distinction is undoubtedly strained by cases involving an action for damages; see Cane, 'Public Law and Private Law: Some Thoughts Prompted by *Page Motors Ltd.* v. *Epsom and Ewell Borough Council*' [1984] *PL* 202. In that case, failure to enforce a possession order against some gypsies occupying the Council's land caused a nuisance to the plaintiff, who sued for damages; it was held that this was a 'private-law' case: (1982) 80 LGR 337.

Cocks does not decide this much, and, it is suggested, is confined in its effect to bringing within the ambit of Order 53, cases where a duty to provide depends on a duty to determine a question of entitlement; this is a common situation, and the result is desirable.[23]

(v) *Choice of remedy*

The reform effected by section 31(4) is of particular importance for public duties. Where, as in *Cocks*, the alleged breach of duty gives rise to a claim for damages, the claim can be coupled with an application for mandamus or a declaration to enforce the duty under Order 53. The basis of such a claim is of course in no way affected by section 31(4), and, as indicated above, not all actions for damages for breach of a public duty can be brought under Order 53. None the less the reform is a useful one, and could become even more useful if the action for breach of statutory duty is developed as a basis of public liability.[24]

A further development is that under section 31(5) mandamus is no longer necessary as an adjunct to *certiorari* where the applicant seeks a decision in his favour; instead the Court can merely remit the matter for reconsideration; and in fact this has been the practice under Order 53. However, mandamus will still be used in relation to review of decisions where there is an actual, as opposed to a constructive, refusal of jurisdiction.[25]

As regards the choice between mandamus and a declaration it is not yet clear on what principles either of these remedies will be preferred to the other, given the courts' discretion under section 31(2). In *Chief Constable of North Wales Police* v. *Evans*[26] the House of Lords preferred a declaration of the applicant's rights following his dismissal in breach of natural justice to mandamus to reinstate him, on the grounds that the latter remedy would usurp the Chief Constable's

[23] See also *Cicutti* v. *Suffolk County Council* [1980] 3 All ER 681; *Shah* v. *Barnet London Borough Council* [1983] 1 All ER 226 (HL); *R.* v. *Eastleigh Borough Council, ex p. Betts* [1983] 2 AC 613 (HL). One wonders, however, whether a logical development of *O'Reilly* and *Cocks* might not result in negligence or nuisance cases involving public authorities being drawn within O. 53; see Cane, above n. 22; *Davy* v. *Spelthorne Borough Council* [1983] 3 WLR 742; Cane, 'Public Law and Private Law Again: *Davy* v. *Spelthorne B.C.*' [1984], *PL* 16.

[24] See below Ch. 7.2.

[25] See e.g. *Insurance Officer* v. *Hamment* [1984] 1 WLR 857. In *Cocks*, p. 1128, Lord Bridge indicated that *certiorari* and mandamus are the proper remedies in relation to review of decisions. However, where the tribunal or authority decides that it has no jurisdiction, this decision may properly be challenged by *certiorari* and remitted under s. 31(5).

[26] [1982] 3 All ER 141 (HL).

functions; on the other hand in *Shah* v. *Barnet London Borough Council*[27] the House of Lords preferred *certiorari* and mandamus to a declaration where the authority erred in law in determining that the applicant was not entitled to a student grant. Both these decisions can be explained on the basis that the court cannot interfere in such a way as to compel the making of a particular decision, however timidly that principle may seem to have been applied. In general mandamus can be regarded as the appropriate remedy for enforcing a public duty under section 31, unless there are good reasons, such as those discussed above in Chapter 4, for granting a declaration in lieu.

5.3 ADMINISTRATIVE DECISIONS (JUDICIAL REVIEW) ACT 1977 (CWLTH)

(i) *Scope of the Act*

The Administrative Decisions (Judicial Review) Act 1977 (Cwlth) ['Judicial Review Act'] was brought into force in 1980, its purpose being 'to establish a single simple form of proceeding in the Federal Court of Australia for judicial review of Commonwealth administrative actions as an alternative to the present cumbersome and technical procedures for review by way of prerogative writ, or the present actions for a declaration or injunction'.[1]

As the title of the Act indicates, it is aimed at administrative decisions, which are defined by section 3(1), subject to certain exceptions, as decisions 'of an administrative character made, proposed to be made, or required to be made, as the case may be (whether in the exercise of a discretion or not) under an enactment'. It is not surprising that many of the cases decided under the Act have concerned the question whether the decision sought to be reviewed falls within this definition.[2] For our present purposes it is interesting to note that under section 3(2) the making of a decision includes (*inter alia*)

[27] Above n. 23.

[1] This was the purpose stated by the Cwlth A.-G. in his Second Reading Speech on the bill: *Parl. Deb.*, 30th Parl. 2nd Sess., p. 1394 (28 Apr. 1977). For commentary on the Act, see Flick, *Federal Administrative Law* (1983), pp. 171 et seq.; Griffiths, 'Legislative Reform of Judicial Review of Commonwealth Administrative Action' [1978], 9 *FLR* 42; Enright, *Judicial Review of Administrative Action* (1985), ch. 7; Aronson and Franklin, *Review of Administrative Action* (1987), ch. 11.

[2] See below for discussion of these cases, and Enright, above n. 1, ch. 7, which contains an exhaustive survey.

(a) making, suspending, revoking or *refusing* to make an order, award or determination; (b) giving, suspending, revoking or *refusing* to give a certificate, direction, approval, consent or permission; (c) issuing, suspending, revoking or *refusing* to issue a licence, authority or other instrument; . . . (f) retaining, or *refusing to deliver up, an article*; or (g) doing or *refusing* to do any other act or thing [emphases added]

and a reference to a failure to make a decision is to be construed accordingly.

Section 5 provides that a person who is aggrieved by a decision to which the Act applies may apply to the Federal Court for an order of review in respect of the decision on any one or more of various grounds,[3] which correspond more or less exactly with the grounds on which judicial review can be obtained at common law, except that 'error of law, whether or not the error appears on the record of the decision' is a ground which appears not to exist in Australia at common law, except where the error of law goes to jurisdiction or appears on the face of the record.[4]

Further, under section 6, where a person has engaged, is engaging, or proposes to engage in conduct for the purpose of making a decision to which the Act applies, a person who is aggrieved by the conduct may apply to the Federal Court for an order of review in respect of the conduct on any one or more of various grounds.[5]

Most important for our present purposes is section 7,[6] which provides as follows:

7(1) Where—
(a) a person has a duty to make a decision to which this Act applies;
(b) there is no law that prescribes a period within which the person is

[3] There are 17 grounds in all; 8 of the grounds are normal ones which apply at common law to the review of decisions, and 9 are designed to establish that 'the making of the decision was an improper exercise of the power conferred by the enactment in pursuance of which it was purported to be made' (s. 5(1)(e)), i.e. the normal 'abuse of discretion' grounds at common law. For comment on those grounds, see Enright, above n. 1, chs. 13–23.

[4] See above Ch. 2.2.

[5] *Mutatis mutandis* these grounds, again 17 in number, correspond exactly to those in s. 5, and more or less exactly to the grounds of review at common law.

[6] For cases in which this provision was applied, see *Thornton* v. *Repatriation Commission* (1981) 35 ALR 485; *Re O'Reilly, ex p. Australena Investments Ltd.* (1983) ALJR 36 (HCA). It is not always clear whether this provision or s. 3(2) applies: see Enright, above n. 1, pp. 163–8; *Fowell* v. *Ioannou* (1982) 45 ALR 491. The best view of s. 7, as suggested by Enright, is probably that it applies only to what he terms a 'passive failure to decide'; in other words it would apply an actual rather than a constructive refusal of jurisdiction (see above Ch. 2.2), but only where the decision-maker fails to exercise jurisdiction as opposed to deciding that he has none.

required to make that decision; and
(c) the person has failed to make that decision, a person who is aggrieved by the failure of the first-mentioned person to make the decision may apply to the Court for an order of review in respect of the failure to make the decision on the ground that there has been an unreasonable delay in making the decision.
(2) Where—
(a) a person has a duty to make a decision to which this Act applies;
(b) a law prescribes a period within which the person is required to make that decision; and
(c) the person failed to make that decision before the expiration of that period, a person who is aggrieved by the failure of the first-mentioned person to make the decision within that period may apply to the Court for an order of review in respect of the failure to make the decision within that period on the ground that the first-mentioned person has a duty to make the decision notwithstanding the expiration of that period.

Section 13 creates an important new general duty to give reasons for decisions which fall under the Act.[7]

The powers of the Federal Court are set out in section 16, under which the Court may in its discretion make all or any of the following orders:

1. an order quashing or setting aside the decision, or a part of the decision, with effect from the date of the order or from such earlier or later date as the Court specifies;
2. an order referring the matter to which the decision relates to the person who made the decision for further consideration, subject to such directions as the Court thinks fit;
3. an order declaring the rights of the parties in respect of any matter to which the decision relates;
4. an order directing any of the parties to do, or refrain from doing, any act or thing the doing, or the refraining from the doing, of which the Court considers necessary to do justice between the parties.

It will be seen that these orders correspond closely to *certiorari*, mandamus, declaration, and injunction at common law.

Where the application is in respect of conduct falling within section 6, the Court may, under section 16(2), make either or both of two orders corresponding to 3. and 4. above. Where the application is in respect of a failure to make a decision within the required period (in other

[7] See *Australian National University* v. *Burns* (1982) 43 ALR 25; Enright; above n. 1, ch. 11.

words it falls within section 7), the Court may make all or any of the orders corresponding to 3. and 4. above, and an order simply directing the making of the decision. The Federal Court has regarded itself as having a discretion to refuse any of these remedies in the same manner as at common law, even though section 16 cannot be read as expressly conferring such a discretion.[8]

The wide scope of the Act is obvious from the provisions referred to above. In effect, any decision of a minister, body, or officer[9] can be reviewed under the Act.[10] Under section 9 state courts have no jurisdiction to review any decision (or conduct, or failure to make a decision) to which the Act applies, or which is excluded from review by the Act.[11] However, under section 10 the rights conferred by sections 5, 6, and 7 are in addition to, and not in derogation of, any other rights to seek review that the applicant has, whether by the Federal Court, another court, or another tribunal, authority, or person.[12] The position under the Act is therefore quite different from the position under the Supreme Court Act, which, as is discussed above in Ch. 5.2, closes off other avenues of judicial review.

(ii) *Effect of the Act in public-duty cases*

What then is the effect of the Judicial Review Act on remedies to enforce public duties?

The first point here is that, like the Supreme Court Act (England and Wales), the Judicial Review Act gives a discretion to the court

[8] *Vangedal-Nielsen* v. *Smith* (*Commissioner of Patents*) (1980) 33 ALR 144; *Doyle* v. *Chief of General Staff* (1982) 42 ALR 283; *Visy Board Pty. Ltd.* v. *Attorney-General* (*Cwlth*) (1983) 51 ALR 705; *Lamb* v. *Moss* (1983) 49 ALR 533.

[9] While the Act is not expressly confined to review of decisions of Commonwealth ministers and officials, according to Powell J. it is so confined: see *Trimboli* v. *Onley* (1981) 37 ALR 38, 45. However, this view seems to overlook the possibility of delegation of federal powers to state officials: cf. Enright, above n. 1, p. 133. For discussion of the jurisdiction of the Federal Court and State courts, see below n. 11.

[10] There are some exceptions. Decisions of the Governor-General cannot be reviewed, and there are express exceptions under sch. 1: see s. 3(1); Enright, above n. 1, ch. 7.

[11] Again there are some exceptions; see s. 9(3–4). On jurisdiction, see generally *Nomad Industries Pty. Ltd.* v. *Commissioner of Taxation* (*Cwlth*) (1983) 51 ALR 94; *Lamb* v. *Moss*, above n. 8; Griffiths, above n. 1; Enright, above n. 1, ch. 6.

[12] *Domaine Finance Pty. Ltd.* v. *Federal Commissioner of Taxation* (1985) 61 ALR 375. However, the Federal Court can refuse to grant an application for review on the ground that the applicant has sought review otherwise than under the Act s. 10(2)(b); and the court in any proceeding otherwise than under the Act may refuse to grant an application on the ground that an application has been made under the Act: s. 10(2)(a). See Constitution (Cwlth), s. 75(v) (High Court jurisdiction); Judiciary Act 1903 (Cwlth), s. 39B (Federal Court jurisdiction).

to grant a remedy in the nature of mandamus (with or without *certiorari*), or a declaration or an injunction. The Federal Court is not subject to any restrictions in this regard, because no indication is given by section 16 to what it should have regard in deciding which remedy to grant. On the other hand section 16 is subject to one serious limitation to which the Supreme Court Act (England and Wales), section 31,[13] is not subject: the Federal Court cannot under section 16 award damages either alone or in conjunction with a mandamus-type order, or a declaratory order, in a case where an authority breaches its duty in such a way as to give rise to an action for damages. This, it is suggested, is an important defect in the Act from the point of view of public-duty enforcement, because it means that the advantages conferred by the Act are not available to a litigant who seeks damages for breach of a public duty; such a litigant would have to use the 'old' procedure, which is to say that he would have to sue for damages at common law; the only way in which he can obtain redress in the same proceeding which has the effects both of compensating his loss and compelling performance of the duty is to bring an action for statutory mandamus[14] and damages. While the procedure under the Supreme Court Act is by no means free from difficulty in cases of this kind,[15] it is clear that the Judicial Review Act is quite useless in such cases.

Another immediately obvious point of comparison between the Supreme Court Act and the Judicial Review Act is that the latter is aimed at 'decisions of an administrative character', whereas the former is aimed at administrative action generally. Does this mean that the Judicial Review Act is directed towards duties to determine, but not towards other duties?

The Act refers throughout to the concept of a decision, not to administrative action, and as is argued in this book, not all public duties are duties to decide things. None the less there are many reasons for supposing that the Act is not as restrictive in this regard as might at first appear. A refusal to do something can easily be construed as a decision not to do it, provided a broad view of the concept of a decision is adopted, and the wording of section 3(2) referred to above gives support to a very wide view of the meaning of 'decision'; in particular the words 'retaining, or refusing to deliver up, an article' and 'refusing to do any other act or thing' would appear to cover everything which is normally covered by the writ of

[13] See above Ch. 5.2.
[14] See above Ch. 4.5.
[15] See the discussion of the *Cocks* case, above Ch. 5.2(iv).

mandamus. In addition the Federal Court has taken just such a wide view of the term 'decision' as would be necessary to avoid the difficulty of matching it with the traditional ambit of mandamus. Fox ACJ said in *Evans* v. *Friemann*:

The making of a decision by a person is a mental process, which may be communicated orally or in writing, or be apparent from action taken. The making of a decision might precede, by a very short, or by a long period, communication or manifestation . . . In ordinary usage, the special feature of a decision is its conclusiveness or finality for the time being, and this is to be contrasted with the thought or consideration which precedes it. On the other hand a decision is not the same as a conclusion; the former normally has an objective, while the latter is more commonly associated with the end result of a process of thinking without the formation of an intention concerning future conduct. It would not be possible, even if the attempt were wise, to substitute a judicial exegesis for the word the legislature has used.[16]

However, in *Lamb* v. *Moss*[17] the Full Court of the Federal Court was unwilling to put even this 'conclusiveness' limitation on the word 'decision':

In our opinion, there is no limitation, implied or otherwise, which restricts the class of decision which may be reviewed to decisions which finally determine rights or obligations or which may be said to have an ultimate and operative effect. Such a conclusion is, in our opinion, in accordance with the plain legislative intention revealed by the words in the Act.[18]

The Court added, however,[19] that its discretion to refuse relief[20] would serve to exercise control over the circumstances in which and the stage at which judicial review will be embarked upon, and that any uncertainty in the scope of the word 'decision' could lead to the growth of a grey area between the jurisdictions of the Federal Court and state courts.

The position taken by the Federal Court is, it is suggested, wide enough to encompass, for example, the decision of the Metropolitan

[16] (1981) 35 ALR 428, 431. This dictum was cited with approval in *Rice Growers Co-operative Mills Ltd.* v. *Bannerman and Trade Practices Commission* (1981) 38 ALR 535, 543; *Legal Aid Commission of WA* v. *Edwards* (1982) 42 ALR 154; *Baker* v. *Campbell* (1982) 44 ALR 431, 435; *Fowell* v. *Ioannou* (1982) 45 ALR 491; *Nomad Industries*, above n. 11. See Enright, above n. 1, pp. 135 et seq.; *Mayer* v. *Minister for Immigration and Ethnic Affairs* (1984) 55 ALR 587, 592.

[17] (1983) 49 ALR 533.

[18] Ibid. 556.

[19] Ibid. 557.

[20] See above n. 8.

Police Commissioner in the *Blackburn* case[21] not to enforce the gaming laws in London, or the decision of the Inland Revenue Commissioners in the *National Federation* case[22] not to collect taxes owed by the Fleet Street casuals; to review a failure to enforce the law or collect taxes is equivalent to reviewing the decisions not to do those things. In *Visy Board Pty. Ltd.* v. *Attorney-General (Cwlth)*[23] the applicant sought review of a decision of the Attorney-General not to institute criminal proceedings against the applicants' business rivals under trade practice legislation. The application was dismissed because the decision was properly made, but it is interesting to note that the case could have been considered in just the same way on an application for mandamus at common law for the enforcement of a duty to enforce the law because, as with the latter kind of application, the court considered the factors to be taken into account in deciding whether to prosecute.

Since the Act is confined to 'decisions of an administrative character under an enactment' it is necessary to inquire whether the terms 'administrative' and 'enactment' affect the scope of public duties reviewable under the Act.

The term 'administrative' brings to mind two familiar difficulties in this area: the public-law/private-law distinction, and the distinction between executive, legislative, and judicial acts. Both these difficulties have featured in the case law on the Act.

The public-law/private-law distinction has not given rise to the difficulties which have occurred in England under the Supreme Court Act; nevertheless it is still necessary for the case to fall within the area of public law (though the term has been studiously avoided)[24] before the jurisdiction of the Federal Court can be invoked. Thus decisions to dismiss a university professor[25] or to transfer a police constable[26] have been held to arise under contracts of employment and are not

21 *R.* v. *Metropolitan Police Commissioner, ex p. Blackburn* [1968] 2 QB 118, for discussion of which see above Ch. 2.4.

22 [1982] AC 617 (HL), for discussion of which see above Ch. 2.4.

23 (1983) 51 ALR 705.

24 See, however, *Monash University* v. *Berg* [1984] VR 383, where the Victorian Full Court expressly approved *O'Reilly* v. *Mackman* [1982] 3 WLR 1096 (HL); their decision was, however, made under the Administrative Law Act 1978 (Vic.) (see below Ch. 5.4).

25 *Australian National University* v. *Burns*, above n. 7.

26 *Sellars* v. *Woods* (1982) 45 ALR 113; and see *Hamblin* v. *Duffy* (1981) 34 ALR 333; *Fowell* v. *Ioannou*, above n. 6; *Hawker Pacific Pty. Ltd.* v. *Freeland* (1983) 52 ALR 185; *Australian Film Commission* v. *Mabey* (1985) 59 ALR 25; contrast, however, *Chittick* v. *Acland* (1984) 53 ALR 143.

therefore reviewable under the Act.[27] A refusal to give reasons for an unsuccessful tender[28] has been held to fall outside the Act; but a decision to reject a tender based on an express statutory power to enter into a contract has been held to fall within the Act.[29] These decisions are, it is suggested, equivalent to the English decisions to the effect that such questions are not questions of public law,[30] and display a similar unwillingness to give weight to the public nature of such contracts,[31] emphasizing instead the formal basis of the decision, a result probably inevitable in view of the fact that a decision must be made 'under an enactment' to fall within the Act.

The term 'administrative' has been further interpreted to include decisions of magistrates hearing and deciding committal proceedings, in the course of which they are obliged to act judicially,[32] and does not therefore refer to 'purely' administrative decisions according to the traditional administrative-law classifications—classifications which have, in any event, been placed under severe pressure in recent years. Decisions of a legislative character obviously do not fall within the Act.[33]

Tied in with both of these problems is the problem of whether the decision in question was made 'under an enactment'.[34] Under section 3(1)(c) an enactment includes 'an instrument (including rules, regulations or by-laws) made under [an enactment]'. Decisions made under prerogative powers are presumably excluded from the scope of the Act, a fact which might cause some difficulty. A decision of a Royal Commissioner narrowly escaped such a description;[35] in another

[27] Cf. *Monash University* v. *Berg*, above n. 24, in which an arbitration under a building contract was held not to be reviewable under the Administrative Law Act 1978 (Vic.).

[28] *ABE Copiers Pty. Ltd.* v. *Secretary of Department of Administrative Services* (1985) 7 FCR 94.

[29] *ACT Health Authority* v. *Berkeley Cleaning Group Pty. Ltd.* (1985) 7 ALR 752 (*Burns*, above n. 25, distinguished).

[30] See above Ch. 5.2.

[31] See above Ch. 1.2(iv).

[32] *Moss* v. *Brown* (1983) 47 ALR 217; *Lamb* v. *Moss*, above n. 17; see also *Legal Aid Commission of WA* v. *Edwards*, above n. 16; *Baker* v. *Campbell*, ibid.; *Appliance Holdings Ltd.* v. *Lawson* [1984] 1 NSWLR 246; *Brewer* v. *Castles* (1984) 52 ALR 571; *Registrar of Motor Vehicles* v. *Dainer* (1985) 57 ALR 759.

[33] *Minister for Industry and Commerce* v. *Tooheys Ltd.* (1982) 42 ALR 260.

[34] See *Ross* v. *Costigan* (1982) 41 ALR 319; *Barbaro* v. *McPhee* (1982) 42 ALR 147; *Glasson* v. *Parkes Rural Distributions Pty. Ltd.* (1984) 55 ALR 179 (HCA); *Molomby* v. *Whitehead* (1985) 63 ALR 282 (access to documents; cf. above Ch. 2.6).

[35] *Lloyd* v. *Costigan* (1983) 48 ALR 241; cf. *Clyne* v. *Attorney-General (Cwlth)* (1984) 55 ALR 92, in which it was sought, unsuccessfully, to translate the prerogative power of *nolle prosequi* into a statutory duty.

case the deletion of a motel from a tourist accommodation guide was held not to be a decision under an enactment because there was no statutory authority, and none was needed, for the deletion.[36] In *Minister for Immigration and Ethnic Affairs* v. *Mayer*[37] a decision that a person was a refugee was held by the High Court to have been made 'under an enactment' rather than under prerogative powers, the court stressing the importance of the *source* of the power to make the decision. In view of the difficulties raised by the 'decision of an administrative character under an enactment' formula, there is a real danger that some public duties which ought to be reviewable under the Judicial Review Act will fall outside even the wide net of review which the formula can be regarded as casting; for example the decision in *Ex p. Lain*[38] would not have been possible if it had been an application under the Judicial Review Act, because the scheme in question in that case was not made under an enactment. The ascription of decisions not falling under the Act to the field of 'prerogative' is of doubtful justification and stems directly from the formula adopted by the Act. In this area it should be the subject-matter of the decision rather than its statutory basis which determines its reviewability.[39]

There is one further difficulty in reviewing public duties under the Act which may not be easily overcome, and this is the powers of the Federal Court with regard to the actual enforcement of public duties. In this respect much depends on the interpretation of section 16. It is clear that so long as the duty is a duty to make a decision the Court has full powers to force the making of the decision according to law. Where, however, the duty is essentially a duty of some other kind, it might be argued that the Court's powers are to be seen only as powers to compel the making of a decision, not powers to compel action. This is a proposition worth considering, because if it is true, then the Act would clearly have to be amended to enable the Court to compel action in a suitable case, as it would be able to at common law.

While it might be thought that the Act, being aimed at decisions, can only afford redress in terms of compelling the making of a decision according to law, the wording of section 16 gives scope for a wider

[36] *MacDonald* v. *Hamence* (1984) 53 ALR 136.

[37] *Minister for Immigration and Ethnic Affairs* v. *Mayer* (1985) 61 ALR 1009 (applying *Glasson*, above n. 34).

[38] [1967] 2 QB 864.

[39] See above Ch. 1.2(i). In cases of doubt it may be necessary to commence proceedings under both the Judicial Review Act and the 'old procedure' (i.e. an action or mandamus under Judiciary Act 1903 (Cwlth), s. 39B); see *Thurgood* v. *Director of Australian Legal Aid Office* (1984) 56 ALR 232.

interpretation of the Court's powers. Section 16(1)(b) allows the Court to refer 'the matter to which the decision relates to the person who made the decision for further consideration, *subject to such directions* as the Court thinks fit [emphasis added]'. Furthermore, under section 16(1)(c), the Court may declare 'the rights of the parties in respect of any matter to which the decision relates'; and, perhaps even more significantly, under section 16(1)(d) the Court may direct the doing of any act or thing the doing of which 'it considers necessary to do justice between the parties'. The cases involving public duties reviewed under the Act so far suggest that there will be little difficulty. In *Vickers* v. *Minister for Business and Consumer Affairs*[40] the applicant was refused the return of money which had been seized, unlawfully as it turned out, on suspicion that it was the proceeds of sale of narcotics, the relevant charges against him having been dismissed; Morling J. seemed to be prepared if necessary to order repayment under section 16(1)(d) rather than simply remit the decision for further consideration. Furthermore, in *Re O'Reilly, ex p. Bayford Wholesale Pty. Ltd.*,[41] an application for remitter of a mandamus application in the High Court to the Federal Court, Dawson J., referring to section 16(1)(d), said:

It cannot, I think, be contested that the jurisdiction of the Federal Court, under the provisions of the [Judicial Review Act] to compel the performance of a statutory duty . . . is of the same kind as that which might be exercised by this Court upon an application for a writ of mandamus.[42]

In an appropriate case it would seem therefore that the Court can either declare the rights of the parties in such a way as to compel the authority concerned to act in a particular way, or direct the way in which the remitted decision should be made, or even, if necessary, by appealing to the broad concept of 'justice between the parties', order the performance of particular acts, as is suggested by the above-mentioned cases. It is suggested that there is no public duty which cannot be effectually enforced in one of these ways, provided it is reviewable at all under the Act. If this is right, then the remedial limitations of the Act with regard to the enforcement of public duties are apparent rather than real. It may, however, be that the problems of standing and damages, as well as the problem of the scope of the Act, may yet render it not entirely as efficacious as it ought to have been in public-duty cases.

[40] (1982) 43 ALR 389.
[41] (1983) 57 ALJR 675.
[42] Ibid. 677.

5.4 ADMINISTRATIVE LAW ACT 1978 [VIC.]

The Administrative Law Act 1978 (Vic.) makes provision for judicial review in Victoria, but in less comprehensive a manner than the Judicial Review Act at the federal level. Like that Act it is aimed at decisions; section 3 says:

Any person affected by a decision of a tribunal may make application . . . to the Supreme Court or a Judge thereof for an order calling on the tribunal . . . to show cause why the [decision] should not be reviewed.

A 'decision' is defined by section 2 as

a decision operating in law to determine a question affecting the rights of any person or to grant, deny, terminate, suspend or alter a privilege or licence and includes a refusal or failure to perform a duty or to exercise a power to make such a decision.[1]

A 'tribunal' is defined by section 2 as

a person or body of persons (not being a court of law or a tribunal constituted or presided over by a Judge of the Supreme Court) who, in arriving at the decision in question, is or are by law required, whether by express direction or not, to act in a judicial manner to the extent of observing one or more of the rules of natural justice.[2]

The powers of the Court are dealt with by section 7:

Upon the return of the order of review, the Court or Judge may discharge the order or may exercise all or any of the jurisdiction or powers and grant all or any of the remedies which upon the material adduced and upon the grounds stated in the order might be exercised or granted in proceedings for or upon the return of any prerogative writ . . . or in an action for a declaration of invalidity in respect of the decision or for an injunction to restrain the implementation thereof and may extend the period limited by statute for the making of the decision but shall not exercise any other jurisdiction or power or grant any other remedy.

Under section 4(2) relief may be refused if the Judge is 'satisfied that no matter of substantial importance is involved or that in all the circumstances such refusal will impose no substantial injustice upon the applicant'.

[1] See *AB* v. *Lewis* [1980] VR 151; *Nicol* v. *Attorney-General* (*Vic.*) [1982] VR 353. The word 'duty' clearly refers to the later words 'to make a decision'.

[2] See *Nicol* v. *Attorney-General* (*Vic.*), above n. 1; *FAI Insurances Co. Ltd.* v. *Winneke* (1982) 56 ALJR 388 (HCA); *Trevor Boiler Engineering Co. Pty. Ltd.* v. *Morley* [1983] 1 VR 716; *Currie* v. *Road Traffic Authority* [1986] VR 401.

Section 8, like the Judicial Review Act, section 13, imposes a general duty on tribunals to furnish reasons for their decisions.

The jurisdiction of the Supreme Court under the Administrative Law Act appears from section 11 not to be exclusive, but the Full Court of the Supreme Court in *Monash University* v. *Berg*[3] has decided (applying *O'Reilly*) that only decisions of tribunals under public law, as opposed to contract, fall within the Act.

From the point of view of public-duty enforcement the Administrative Law Act does not appear to be such a beneficial reform as the Judicial Review Act. Since an injunction is only available under section 7 to restrain the implementation of the decision, and no other remedy is available than those specified under section 7, it follows that a mandatory injunction is not available; and since a declaration is only available in respect of invalidity of the decision, a declaration that an authority (even a tribunal within section 2) has failed to perform a duty is also unavailable. Like the Judicial Review Act the Administrative Law Act makes no provision for the award of damages. The inevitable conclusion seems to be that enforcement of a public duty, with the exception of mandamus to compel a tribunal to decide in accordance with law,[4] is impossible under the Administrative Law Act, which leaves a litigant seeking to enforce a public duty in Victoria to the mercy of the 'old procedure', which in this case is improved by the fact that under rules of court a declaration and statutory mandamus are available in an action for damages.[5]

5.5 PUBLIC LAW, PUBLIC DUTIES, AND STATUTORY REVIEW

The various remedies for enforcing public duties all have their advantages and disadvantages compared with mandamus. It is intolerable, however, that an applicant's prospect of enforcing a public duty should vary with his choice of remedy, particularly where even his standing is affected by this choice.[1] It follows from this that the new statutory review procedures offer a desirable solution to the problems of choice and cumulation of these remedies. It has been seen that the English solution seems superior to the Australian in that it retains the old mandamus jurisdiction beyond the realm of decisions

[3] [1984] VR 383; see also *Dominik* v. *Eutrope* [1984] VR 636; *Vowell* v. *Steele* [1985] VR 133; contrast *Robbins* v. *Harness Racing Board* [1984] VR 641.

[4] See *Keller* v. *Drainage Tribunal* [1980] VR 449.

[5] RSC (Vic.) O. 3, r. 1A and O. 53.

[1] See below Ch. 6.

and allows also a claim for damages. The first of these advantages may be insignificant if the Judicial Review Act is given a sufficiently wide interpretation; the second could pose some difficulties if the action for breach of a statutory public duty is developed in the manner suggested below in Chapter 7. On the other hand the Australian solution seems to be superior in avoiding the procedural finality of the *O'Reilly* rule. The public-duty/private-duty distinction is certainly rendered less critical if the benchmark for statutory review, be it 'public law' or 'a decision of an administrative character', is a prerequisite which permits rather than obliges the applicant to seek statutory review.

Whatever solution is adopted, and whether or not the *O'Reilly* rule is adopted, there still remains the problem of finding a suitable benchmark to define the scope of the new procedures. A very large number of the English cases have involved the problem of defining public law, and no fewer Australian cases have involved the problem of defining a decision of an administrative character. The proliferation of these cases seems to be a high price to pay for statutory review; it would perhaps have been preferable to allow the granting of any of the whole range of remedies within the context of an ordinary action, as suggested by Wade[2] and adopted, partially, in Victoria.[3] Given, however, the fact of statutory review and the necessity of defining in some way its scope, what approach ought to be taken in public-duty cases?

In the case of the duty to determine, it would seem necessary to retain the traditional mandamus jurisdiction as an implicit feature of the new procedures, but subject to two propositions which may not necessarily be regarded as consistent with that jurisdiction: first, that the legal or instrumental basis of the duty (prerogative, franchise, charter, and contract as well as statute) is irrelevant to ascertaining the public nature of the duty;[4] and secondly, that a duty to determine should be regarded as a public duty if administrative-law reasoning is normally applicable to it.[5] This second proposition is radical, because there are many decisions which answer this description which are not regarded as falling within the scope of statutory review. The advantage of adopting this view is that it avoids the enormous difficulties of defining the precise scope of 'state' decisions which arise from the many situations where it is not clear whether it is the state

[2] Wade, 'Procedure and Prerogative in Public Law' (1985), 101 *LQR* 180.

[3] RSC, O. 3, r. 1A; O. 53.

[4] See above Ch. 1.2.

[5] See above Ch. 2.2; cf. Beatson, ' "Public" and "Private" in English Administrative Law' (1987) 103 *LQR* 34, 53 et seq.

which is acting, but where the decision taken affects the public or a section of the public acutely. There is no point in defining a public duty to determine in accordance with a theory of the state, because the purpose of statutory review is either procedural only, or to enable public-law reasoning to be applied by judges who are specialists in applying it. Thus the decision-making duties of universities, trade unions, 'fringe organizations', and employers of employees entitled to a hearing before dismissal, would all come within the scope of statutory review. To adopt this position is merely in effect to carry the principles in *Ex p. Lain*[6] to their logical conclusion. The advantages are simplicity and broad access to the new procedures. The only disadvantage is that some of the duties reviewed are not state duties. However, any attempt in modern society to define state duties is doomed either to failure or circumvention. The important question is whether justice requires the application of administrative-law reasoning because of the public importance of the duty. The argument here is that the content of 'public law' should *reflect* the reasoning process adopted, not determine it.

In so far as is necessary the same reasoning can be applied to the other public duties, but poses much less difficulty. A duty to provide should be regarded as public except where it arises from a purely contractual obligation or a statutory obligation shared with members of the public.[7] Certain contractual obligations are not, however, purely private: a contractual obligation of a public authority which arises from a contract which is forced on a member of the public, for example in an emergency, should be regarded as a public duty;[8] and

[6] [1967] 2 QB 864; above Ch. 1.2. It would appear from *R.* v. *Panel on Take-overs and Mergers, ex p. Datafin PLC* [1987] 1 All ER 564 that the *Lain* principles *are* being carried to their logical conclusion. See also *Swain* v. *Law Society* [1983] 1 AC 598 (HL).

[7] The recent decision in *R.* v. *Secretary of State for Home Office, ex p. Dew* [1987] 1 WLR 881 would appear to contradict the statement in the text. Mandamus was sought to compel prison officials to provide medical treatment for a prisoner who sustained a bullet wound during his arrest. The allegation of delay or failure to treat was held not to be a decision of a public body in public law, but one properly to be raised in negligence, so the proceedings were struck out. The applicant did not claim damages. To similar effect is *Ettridge* v. *Morrell* (1987) 85 LGR 100, in which the Court of Appeal held that a duty to provide a schoolroom for an election candidate's meeting was enforceable in an action since no public-law decision was involved. These decisions have the effect of excluding public duties, as opposed to decisions, from O. 53 procedure and are therefore highly questionable. On the other hand *G* v. *Hounslow London Borough Council* (1987) 86 LGR 186, appears to force actions for damages involving impugned decisions into the public-law mould. Confusion reigns.

[8] See *R.* v. *Glamorgan County Council* [1899] 2 QB 536.

the same is true of contracts of employment of public servants,[9] and the contractual obligations of universities and trade unions towards their students and members.[10]

A duty to enforce the law arises normally under a statute conferring power to prosecute for a criminal offence, and this will be a public duty. However, there may be other modes of law enforcement which give rise to public duties. For example a power to impose a sanction of an administrative or purely civil character can entail a public duty to enforce the law if it is intended to ensure obedience to a statute. Such a duty should be regarded as a public duty even if imposed on a fringe organization rather than the state apparatus, because the reasoning applied will not differ in such a case from the ordinary police duty, revenue duty, or enforcement-agency duty.

A duty to disclose documents should presumably only be regarded as public where there is some constitutional significance involved. Preferably such duties should be defined by a comprehensive statute, as indicated in Chapter 2.

There remains the problem of duties arising in tort law, such as are apparent in the English housing cases of the *Cocks* variety.[11] This problem would not perhaps arise if prerogative relief or the equivalent were available in an ordinary action.

In so far as these duties involve duties to determine, for example whether to provide housing, there can be no doubt that they are proper subjects for statutory review. Quite apart from this kind of duty, it is quite possible that a duty simply to provide, or to enforce the law, could give rise to an action for breach of statutory duty, or arise under the law of negligence.[12] In these cases the question to be determined can vary from an acutely difficult administrative-law question such as whether an act occurs at the policy or operational level, to the quantum of damages or a question of limitation, which might be called purely tort questions, or could be purely evidential questions. If this kind of liability is going to increase, and if the competence of administrative-law judges is an important factor, it may become necessary to determine such cases within the framework of an application for judicial review, because the tort question becomes a question of judicial review. Since any of these questions can arise in the same action, it should be possible for a court to transfer the case, where necessary, to the appropriate forum. In practice this should mean that the case will be tried by a civil court except where

[9] See above Ch. 5.2, 5.3.
[10] See above Ch. 1.2(iv).
[11] [1982] 3 All ER 1135 (HL); and see above Ch. 5.2(iv), below Ch. 7.2(vii).
[12] See *Anns* v. *Merton London Borough Council* [1978] AC 728 (HL); below Ch. 7.

the court feels that the case involves, crucially, a question of administrative law better determined under statutory review procedure by judges versed in administrative law. Since this would be exceptional, the burden should be placed on the judge, perhaps after application by the public authority, not on the plaintiff. In England this could at present only be done where the judge hearing the case is able himself to grant leave and treat the case as an Order 53 application.[13] In Australia one would have to use the 'old procedure' because damages are not available under the Judicial Review Act or the Administrative Law Act.[14]

Even with the approach suggested here, there will of course continue to be cases in which the public nature of the duty is in doubt. In general these problems too should be resolved by the judges acting under discretionary powers, not left to the plaintiff to guess which is the correct court in which to enforce the duty.

Finally, one might well wonder what is the continuing usefulness of the law discussed in Chapters 3 and 4 in a situation where most public-duty cases will arise under the new procedures.

As far as Chapter 3 is concerned, the traditional principles of mandamus law should be regarded as incorporated *holus bolus* into the new procedural regime, which in neither England nor Australia tells us anything we did not know already about mandamus, except that it is possible to remove some of its awkward procedural difficulties. As far as the substantive law of mandamus is concerned, the problems discussed in Chapter 3 still remain, and the authorities are still highly relevant.

The effect of the new procedures on the remedies discussed in Chapter 4 is, most importantly, to make most of them, together with mandamus, available in the same application, and, as will be seen in Chapter 6, to put standing on a common footing. This does not mean that the law discussed in Chapter 4 will disappear; for example, the problem of when to order interlocutory relief to enforce a public duty, when to refuse a remedy in the court's discretion, or whether to interfere with the exercise of default powers, will remain in spite of the reforms. What it does mean is that the advantages of the various remedies will allow them to flourish under the beneficent shade of statutory review, while their disadvantages will eventually become irrelevant.

The net result, one hopes, will be that the remedies for enforcing public duties will be considerably simplified in procedure and reduced

[13] See Woolf, 'Public Law—Private Law: Why the Divide?' [1986], *PL* 220.

[14] See above Ch. 5.3, 5.4.

in number. The accumulated wisdom of the public duty cases, such as it is, is worth preserving in the era of statutory review. At the same time statutory review can be a means of unravelling much of their unwisdom.

6
Standing and Public Duties

6.1 INTRODUCTION

This chapter deals with standing for the various remedies which are available to enforce public duties: it attempts to answer the question, who can enforce a public duty? This is one of the most controversial aspects of the enforcement of public duties, and a few introductory remarks are necessary.

The main remedies used to enforce public duties are mandamus, declaration, injunction, and the action for damages. This chapter deals only with the first three of these remedies, because in tort law standing is merely a matter of whether liability can be established, so that the question of standing and the question of merits are fused. To a certain extent the same may be true, or may become true, of administrative law remedies,[1] but essentially, and traditionally, with these remedies the question of standing is a separate and preliminary matter which relates to the applicant's right to raise an argument rather than to the merits of the argument.

The rules of standing for mandamus evolved differently from the rules of standing for *certiorari* and prohibition, and standing for these public-law remedies evolved differently again from the rules of standing for declaration and injunction, which evolved as private law remedies and therefore had standing rules appropriate to private-law. One of the most telling criticisms of modern administrative law has been that the rules of standing depend on the remedy sought rather than on clear principles applicable to all public-law cases.

This discussion of standing is structured to deal with the development of standing rules at common law separately from their treatment under statutory judicial review. Whilst an understanding of the former is necessary for an understanding of the latter, it would be erroneous to maintain that the latter have replaced the former, because the common law in all probability informs and influences the content and interpretation of the statutory review provisions, and there are still

[1] See *Inland Revenue Commissioners* v. *National Federation of Self-Employed and Small Businesses Ltd.* [1982] AC 617 (HL); Craig, *Administrative Law* (1983), pp. 438–43; Cane, 'Standing, Legality and the Limits of Public Law' [1981], *PL* 322, 332–7; below Ch. 6.4.

cases which fall to be dealt with under the common-law principles; it may well be that the common-law principles will in turn be affected by cases decided under the statutory review procedures.[2]

Since the statutory rules have not yet received definitive interpretation, the basic principles of standing are open for radical reconsideration, and it will therefore be necessary to discuss the function and consequences of the rules of standing. This discussion raises issues which are wider than the issues raised by the enforcement of public duties, but there are in fact special considerations with regard to the enforcement of public duties which ought to influence the outcome of the general discussion of standing which is taking place during the current period of uncertainty.

The ensuing discussion of the common-law principles should not therefore be taken as totally descriptive of the law as it stands in the 1980s, but is descriptive of the law as it stood in England, Australia, and elsewhere prior to the advent of statutory judicial review.

6.2 STANDING FOR MANDAMUS AT COMMON LAW

(i) *Introduction: an overview*

The history of standing in relation to mandamus has been presented as one of gradual liberalization.[1] In fact it is not at all clear that the rules have been made substantially more liberal, at least up to 1982,[2] because many of the modern decisions have been as illiberal as some of the old ones. The decision of the House of Lords in the *National Federation*[3] case has undoubtedly prepared the way for a more liberal approach, but as will be seen,[4] the position is little clearer than it was prior to that case, and it is not clear how the case will be interpreted, or even whether it will be followed in jurisdictions other than England. Liberalization apart, the development of the rules of standing for mandamus has been concealed behind a mass of largely unhelpful terminology and hampered by a deplorable confusion of

[2] Since O. 53 procedure in England applies to all public-law cases, the prediction in the text could only be fulfilled in Australia or other jurisdictions which retain the common-law principles; see above Ch. 5.

[1] See e.g. Wade, *Administrative Law*, 5th edn. (1983), pp. 640–1; Thio, *Locus Standi and Judicial Review* (1971), Ch. 6.

[2] The date of the decision in *Inland Revenue Commissioners* v. *National Federation of Self-Employed and Small Businesses Ltd.* [1982] AC 617 (HL).

[3] Above n. 2.

[4] See below Ch. 6.3.

utterance and inconsistency of purpose and result. While it would be optimistic to say that the position is now much clearer even than it was two hundred years ago, it is likely that the general trend is away from using mandamus purely 'for the benefit of the subject' towards using it also 'for the advancement of justice',[5] so that the standing rules are not exclusively based on individual rights, but support the emergence of mandamus as an important part of modern administrative law, in which the securing of legality is paramount.

With the exception of the notion that mandamus would not lie for the benefit of a stranger,[6] no clear doctrine emerged in England until the late nineteenth century. The most influential decision was that in *R.* v. *Lewisham Union Guardians*.[7] In that case a sanitary authority, responsible for preventing outbreaks of diseases such as smallpox, applied for mandamus to compel the guardians of the poor for the district to enforce vaccination legislation. They failed on the ground that they had no specific legal right to performance of the duty: they were, as it were, non-Hohfeldian plaintiffs, having no right as the correlative of the duty in question. This unfortunate decision formed the basis of standing for mandamus in England, until it was overruled by the House of Lords in the *National Federation* case in 1982, although in two contemporaneous cases with similar facts mandamus had been obtained by the Local Government Board.[8] In a number of later cases the concept of a specific legal right was given a broad interpretation based on an interest in the performance of the duty greater than that of the public generally, variously described as 'a special interest'[9] or 'a sufficient legal interest'[10] or the like.[11] The development has been seen to be a liberal one, but in fact both formulas have persisted until the present day and scrutiny of the cases

5 Both phrases are taken from the famous dictum (see above Ch. 3.1) of Lord Mansfield in *R.* v. *Barker* (1762) 97 ER 823, 824–5, indicating two quite distinct, though overlapping, purposes of mandamus.

6 Tapping, *Mandamus* (1848), pp. 27–8, and cases cited therein. This passage was relied on in the *Lewisham Guardians* case, below n. 7, and also in *R.* v. *Manchester Corporation*, below n. 9.

7 [1897] 1 QB 498.

8 *R.* v. *Keighley Union Guardians* [1876] 40 JP 70; *R.* v. *Leicester Union Guardians* [1899] 2 QB 632. The Local Government Board can be regarded as a superintending authority, which the sanitary authority in the *Lewisham Guardians* case was not.

9 *R.* v. *Manchester Corporation* [1911] 1 KB 560.

10 *R.* v. *London Union Assessment Committee* [1907] 2 KB 764.

11 For Australian terminology, see Stein, ed., *Locus Standi* (1979), p. 84. In *R.* v. *St Pancras Vestry* (1890) 24 QBD 371, 378 the term 'actual and personal interest' was used. In Malaysia the Specific Relief Act, Cap. 137, s. 44, confines standing for a mandatory order to cases where a property, franchise, or personal right would be injured.

reveals that the 'sufficient interest' formula can be applied as restrictively as the 'legal right' formula; there is indeed no clear distinction between the two, and the tendency in recent cases has been to apply either or both of them as though they were interchangeable.[12] The approach adopted here will be to look at the type of situation rather than the formula used. Broadly two categories of situation arise: first, situations in which the issue concerns the effect on the applicant of administrative proceedings to which he was not a party; and secondly, situations in which the duty affects a large group of people and the issue is whether a member of that group can enforce the duty. These two categories correspond closely to the duty to determine (in the first type of situation) and the other duties (in the second type of situation), that is the duty to provide, the duty to enforce the law, and constitutional duties.[13] The possibility of a 'citizen's action' will also be discussed.

(ii) *Administrative or judicial proceedings: the duty to determine*

In the case of a duty to determine, the improper exercise of the duty may affect persons other than the party immediately affected by the determination and even persons who are not in fact parties to the proceedings at all. The problem of standing here is to decide which of these persons should be allowed to challenge the determination by mandamus to hear and determine according to law.

In two well-known English cases of the *Lewisham Guardians* period, *R.* v. *Bowman*[14] and *R.* v. *Cotham*[15] mandamus issued at the instance of persons who had objected to the granting of liquor licences and alleged that the justices had taken into account irrelevant considerations. In *R.* v. *Bowman* it was said that the objectors had a statutory right to be heard, which had been infringed; in *R.* v. *Cotham* standing was not even discussed. In *R.* v. *London Union Assessment Committee*,[16] however, a rating authority was denied standing to challenge by mandamus the exemption of certain properties from the rating list by the parish overseers because it had no particular ground for claiming due performance of the duty. No doubt the apparent

[12] e.g. *R.* v. *Customs and Excise Commissioners, ex p. Cook* [1970] 1 WLR 450; *R.* v. *Hereford Corporation, ex p. Harrower* [1970] 1 WLR 1424; *Environmental Defence Society Inc.* v. *Agricultural Chemicals Board* [1973] 2 NZLR 758.

[13] See above Ch. 1.5.

[14] [1898] 1 QB 663.

[15] [1898] 1 QB 802. *Bowman* and *Cotham* have been followed by the High Court of Australia in *Sinclair* v. *Mining Warden at Maryborough* (1975) 5 ALR 513; see also *R.* v. *Liquor Control Commission, ex p. Dickens (S. E.) Pty. Ltd.* [1983] VR 303.

[16] [1907] 2 KB 764.

inconsistency of these decisions reflects the distinction between judicial and administrative determinations, which has since disappeared, and the greater willingness of the courts to award mandamus in liquor licensing cases,[17] as well as the fact that the objectors in *R.* v. *Bowman* and *R.* v. *Cotham* had actually participated in the proceedings impugned.

The uncertainty of the English law has been considered in a number of modern Australian cases which unfortunately do not clarify the width of the standing rules to challenge determinations by mandamus. The most awkward is the Victorian case *R.* v. *Whiteway, ex p. Stephenson*.[18] S was an objector to the granting of a liquor licence to X. The licensing court had adjourned the proceedings because a bill was before the legislature which, if passed, would have had the effect of removing certain procedural requirements, X's failure to observe which, formed the basis of S's objection. It appeared however that S's real intention was to have all applications dealt with in order to facilitate the hearing and granting of a similar application of his own. On his application for mandamus to hear and determine X's application it was held that he had 'no sufficient legal special right' in the matter and that his position was the same as that of any other ratepayer or resident. However, in *Ex p. NSW Rutile Mining Co., re Burns*[19] the New South Wales Court of Appeal gave standing to landowners to compel a mining warden to hear and determine according to law a successful application by R Co. for mineral claims over their land, because they were entitled to receive notice of the application and were impeded by the determination from making any effective bargain for the sale of the land with anyone other than R Co.

The Western Australian Full Court took a liberal view of standing, which can be contrasted with *Ex p. Stephenson*, in *R.* v. *Harlock, ex p. Stanford and Atkinson Pty. Ltd.*[20] S Ltd. applied to a mining warden for forfeiture of twenty-six mineral claims belonging to X Ltd. on grounds of alleged failure to comply with labour conditions. The application being dismissed, S Ltd. applied for mandamus to hear and determine. As in *Ex p. Stephenson*, the applicant had an ulterior motive: forfeiture would have given S Ltd. the opportunity of applying for the mineral

[17] This results from the unavailability of *certiorari*: see Wade, above n. 1, p. 639. See however *Ettingshausen* v. *Lal* [1981] 1 NSWLR 503 (standing for mandamus in liquor licence case refused to a body with an historical interest).

[18] [1961] VR 168; for the Australian position generally, see Stein, above n. 11, ch. 4.

[19] [1967] 1 NSWLR 545.

[20] [1974] WAR 101; see also *Ex p. Rice, re Hawkins* (1956) 74 WN (NSW) 7.

claims for itself or its associates, and in fact S Ltd. had earlier successfully opposed an application by X Ltd. for exemption from the labour conditions in question. Although the judgments differ somewhat, all three judges regarded the decisive factor as being the fact that S Ltd.'s complaint against X Ltd. was properly made under the statute, so that S Ltd. were regarded as one of a class of persons entitled by law to ask for the exercise of the warden's jurisdiction and therefore had a 'sufficient personal interest' in the matter. It was also regarded by two of the judges as highly relevant that S Ltd. would, if successful, obtain a chance of reversing the warden's decision and a chance to apply for the claims itself:

> A too nice inquiry into what the interest of the prosecutor is in the performance of the command, or what chance he has of real advantage, would be to take this Court into realms not contemplated by the law which creates the jurisdiction and itself creates the standing of the prosecutor to ask for its performance.[21]

This decision can be contrasted with *Ex p. Northern Rivers Rutile Pty. Ltd., re Claye*[22] in the New South Wales Court of Appeal, in which precisely such a 'too nice inquiry' was embarked upon. The facts were similar, except that the company seeking forfeiture had previously had an application for a lease rejected, so that its interest was 'flimsy and unreal'. The Western Australian decision is surely, however, much preferable. It is not for the court hearing the application for mandamus to judge the prospects of the applicant's success in proceedings of an entirely different nature, but to consider whether the applicant should be allowed to challenge the decision impugned.

One other Australian case deserves mention because it shows that even a party to proceedings may not always have standing. In *Ex p. Mullen, re Wigley*[23] in the New South Wales Court of Appeal, M laid an information against N alleging that she was guilty of incitement not to register for national service. N was convicted but refused to pay the fine. M then applied to the clerk of petty sessions for a warrant for N's committal, but the warrant was refused. When M applied for mandamus to hear and determine the application it became apparent that the whole proceedings were brought in bad faith because N's counsel was instructed not to oppose the order. Accordingly M was held to have 'no legal, pecuniary or special interest' in the matter 'so as to place himself in a superior position to

[21] [1974] WAR 113–14 per Wickham J.
[22] (1968) 72 SR (NSW) 165.
[23] [1970] 2 NSWR 297; see above Ch. 3.4(iv) for further discussion of this case.

that of a common informer'.[24] It is hard to see, however, how this decision can be supported on the basis of standing. Having lawfully brought a private prosecution, M was, one would have thought, entitled to ensure that the punishment was enforced; in any event the court could have used its discretion to refuse an order without distorting the rules of standing.

As a result of these decisions it is not possible to assert that participation, or the right to participate, in proceedings of a judicial or administrative character, carries with it an automatic right to enforce a decision according to law by mandamus. There must in addition be some right or interest considered worthy of protection. To use the terminology of Stein, there is a quantitative as well as a qualitative aspect of the interest to be satisfied before standing is given.[25]

(iii) *Membership of a group of persons: duties other than the duty to determine*

Membership of a group of persons similarly affected has not in general been sufficient to achieve standing. On the other hand there are no clear principles establishing the nature of the right or interest which will distinguish a member from the rest of the group for the purpose of standing.

A pre-existing right has been held to form a basis for standing, and it is clear that an equitable right as well as a legal right will suffice,[26] for the equitable right of a mortgagee of unregistered land was held in *R.* v. *Registrar of Titles, ex p. Moss*[27] to be sufficient for standing to compel registration of the land.

The suffering of special damage over and above that suffered by other members of the group will generally give standing, and such damage need not, apparently, be such as to support an action in tort, though the principle can be regarded as analogous to the rule that standing to sue in public nuisance depends on the suffering of special damage. The best example of this 'special damage' principle is *State (Modern Homes Ltd.)* v. *Dublin Corporation*[28] in the Supreme Court of Ireland. In this case the authority failed to give effect with all convenient speed, as enjoined by statute, to its decision to make a planning scheme, and as a result the applicants' business as developers was adversely affected; it was held that they had standing to compel

[24] Above n. 23, p. 300.
[25] See Stein, above n. 11, p. 85; *Ex p. Cook*, above n. 12.
[26] See Thio, above n. 1, pp. 114–15.
[27] [1928] VLR 411.
[28] [1953] IR 202.

performance of this duty. In *R.* v. *Manchester Corporation*[29] the authority was charged by statute with the duty of enacting by-laws prescribing distances within which trams in their area might lawfully follow one another. By-laws were enacted, but enforcement in the city centre was left to the discretion of the police. The police took no action, and there was a resulting increase in the number of collisions. An insurance company, which had been responsible for the insertion of the relevant clause in the statute and had suffered an increase in the number of claims made against it, was granted standing to apply for mandamus to compel the authority to amend the by-law to comply with the statute. Two of the judges stressed that the company was in a different position from the public generally; the third, dissenting, judge, relying on the *Lewisham Guardians* decision, held that the authority owed no duty to the company with regard to the by-law. Similarly a newspaper was given standing to compel due exercise of a power to restrict reporting of a criminal case because it ran the risk of penalties for infringing a restriction order even if it was invalid.[30]

However, even this minimal liberality in according standing has not always been forthcoming. In the interesting case of *R.* v. *Hereford Corporation, ex p. Harrower*[31] the applicants, in their capacity as ratepayers, were accorded standing to compel the authority to observe its standing orders with regard to contracts, though their real interest was that they were electrical contractors aggrieved by the unlawful awarding of contracts to their competitors. Some older cases were even more restrictive: as was mentioned above, a rating authority was denied standing to compel proper preparation of a rating list by an assessment committee,[32] and a union of shop workers to compel enforcement of shop-hours legislation by a local authority.[33]

On the other hand, membership of a group of ratepayers or electors has traditionally been a sure basis for standing, because ratepayers have a financial, and electors a constitutional, interest in the due performance of duties under local-government legislation. Ratepayers have standing, after *Ex p. Harrower*, to compel observance of standing orders,[34] and also to compel due preparation of a valuation list.[35]

[29] [1911] 1 KB 560.

[30] *R.* v. *Russell, ex p. Beaverbrook Newspapers Ltd.* [1969] 1 QB 342; cf. *R.* v. *Horsham Justices, ex p. Farquharson* [1982] QB 762.

[31] [1970] 1 WLR 1424; cf. *Hawker Pacific Pty. Ltd.* v. *Freeland* (1983) 52 ALR 185, 189.

[32] *R.* v. *London Union Assessment Committee*, above n. 16.

[33] *R.* v. *Rathmines Urban District Council* [1928] IR 260 (SCI).

[34] See also *R. (McKee)* v. *Belfast Corporation* [1954] NI 122.

[35] *R.* v. *Paddington Valuation Officer, ex p. Peachey Property Co. Ltd.* [1966] 1 QB 380.

The liberality towards electors is well illustrated by *R.* v. *Licensing Commission, ex p. McAnalley*,[36] in which the Queensland Full Court gave standing to a person who had signed a petition for a local liquor option poll to challenge by mandamus the removal of certain signatures from the petition, even though his own signature was not impugned; and standing is granted automatically where the franchise itself is denied, on the basis that the duty to hold an election is owed to each member of the electorate.[37]

It is interesting to observe that the courts have sometimes found it difficult to face reality in relation to business competitors or competitors for a scarce resource: the ambivalence of their attitude here is easily seen in the mining cases mentioned above and in *Ex p. Harrower*. In a Canadian case an intending purchaser of public lands had no standing to compel the vendor authority to seek specific performance or rescission of a contract for the sale of land to a competitor who failed to comply with a contractual term;[38] in an English case, developers were unable to enforce against a local authority a duty to provide a water supply because the duty, being for the benefit of the whole district, could not be enforced so as to benefit the developers at the expense of the general body of ratepayers.[39] As will be seen, a similar approach has been taken with extra-statutory concessions, so that a taxpayer has no standing to compel strict enforcement of revenue law against other taxpayers[40] or the criminal law against business competitors.[41] In *R.* v. *Commissioners of Customs and Excise, ex p. Cook*[42] two bookmakers sought mandamus to compel the Commissioners to collect an excise duty from their competitors by two half-yearly instalments as demanded by statute rather than by one month's duty and eleven post-dated cheques; Lord Parker CJ held that the bookmakers had no specific right, nor any interest over and above that of the community as a whole, and refused standing, even though he expressed alarm that the word of the Minister, who

[36] [1972] Qld R 522.

[37] *Re Mckay and Minister of Municipal Affairs* (1973) 35 DLR (3d) 627; see also *Tonkin* v. *Brand* [1962] WAR 2; *Re Peters* (1978) 22 N & PEIR 54.

[38] *Hughes* v. *Henderson* (1963) 42 DLR (2d) 743.

[39] *Hey & Croft Ltd.* v. *Lexden and Winstree Rural District Council* (1972) 70 LGR 531 (declaration).

[40] See the *National Federation* case, above n. 2.

[41] *Hayes* v. *Montreal Nord* [1944] Que. SC 415, in which a business operator, compelled to abandon his business because of the threat of prosecution for contravention of by-law, had no standing to compel enforcement against others in the same position; contrast the similar case of *R.* v. *Braintree District Council, ex p. Willingham* (1982) 81 LGR 70, where standing was not even raised.

[42] [1970] 1 WLR 450.

authorized the concession, was outweighing the law of the land. On the other hand, a ratepayer has standing to challenge favourable treatment of another ratepayer.[43]

(iv) *Citizen actions and public duties*

The law as revealed in this discussion clearly does not allow a citizen's action, or *actio popularis*, in which a citizen *qua* citizen seeks to enforce a duty on behalf of the public at large. Yet again, examination of the cases reveals a curiously ambivalent attitude. With the increased activism of the judges in recent years in opening up the administration to scrutiny some citizen actions have been brought, and with some limited success.[44]

Most notable are a group of cases brought by Mr Blackburn against London authorities. His case against the Metropolitan Police Commissioner to enforce the gaming laws was mentioned in Chapter 2.[45] The question of standing was raised in this case, and all three judges in the Court of Appeal doubted whether Blackburn had sufficient interest to claim standing,[46] but the point was not decided only because in the event mandamus was unnecessary, the Commissioner having changed his policy with regard to enforcement of the gaming laws. Yet Edmund Davies LJ was able to stress the importance of the case, which was useful

> (a) in highlighting the very real anxiety which many responsible citizens manifestly entertain as to the adequacy of the steps taken hitherto to exterminate a shocking and growing cancer in the body politic; and (b) in clarifying the duty of the police in relation to law enforcement generally.[47]

In Blackburn's case against the Commissioner to enforce the obscenity laws, also mentioned in Chapter 2,[48] standing was not raised on

[43] *Arsenal Football Club Ltd.* v. *Ende* [1979] AC (HL).

[44] For discussion, see Harlow, 'Public Interest Law in England: the State of the Art', ch. 4 of Dhavan and Cooper, eds., *Public Interest Law* (1986).

[45] *R.* v. *Metropolitan Police Commissioner, ex p. Blackburn* [1968] 2 QB 118.

[46] Lord Denning MR appeared to require the applicant to be 'adversely affected' (p. 137). This formulation is unique and can be regarded as a solipsism. The discussion of standing in this case does however support a liberal approach.

[47] [1968] 2 QB 149. In *R.* v. *Metropolitan Police Commissioner, ex p. Blackburn* [1973] QB 149 both Lord Denning MR (at p. 254) and Phillimore LJ (at p. 258) expressed their gratitude to Mr Blackburn for bringing the matter of enforcement of the obscenity laws before the courts. The importance of the *Blackburn* cases in securing proper enforcement of the law is convincingly argued by Lord Denning in *The Discipline of Law* (1979), pp. 118–27.

[48] *R.* v. *Metropolitan Police Commissioner, ex p. Blackburn* (*No. 3*) [1973] 1 QB 241; and see *R.* v. *Metropolitan Police Commissioner, ex p. Blackburn, The Times*, 7 Mar. 1980.

behalf of the Commissioner, and was not therefore discussed at all; but in a later case brought by Blackburn against the Greater London Council for prohibition to prevent the application of a wrong test by the authority in performance of their duties as film censors,[49] he was given standing as a citizen and a father, and his wife, who joined in the application, as a ratepayer.[50]

The result of this confusion seems to be that, in England at least, a citizen's action has *in practice* been allowed to develop, almost by conspiracy, because of the beneficial results which accrue from such actions. Yet among all the high-flown language of the judges there is hardly anything which provides a sure legal basis for a citizen's action to enforce a public duty. In fact the statements of the judges in the first of the cases mentioned above tend to discourage rather than encourage a citizen's action, because it seems quite clear that, whatever their lordships' sympathies, they would, on authority, have been obliged to refuse an order on the ground of lack of standing even if Blackburn had otherwise established a case for an order. It is no comfort to a citizen to reflect that the hazard and cost involved in bringing such an action is offset by a mere possibility of an order being granted even if there is a good case on the merits, and a mere possibility of a favourable order for costs if he fails, whatever the court's sympathies, and a mere possibility of the authority concerned conceding the standing question.

Of course there is no difference in principle between a person of doubtful standing and an aggregation of such persons.[51] As a result, even a society or an interest group with a genuine concern may not achieve standing for mandamus. The classic instance is the New Zealand case of *Environmental Defence Society Inc.* v. *Agricultural Chemicals Board*,[52] in which the applicants, a society whose objects were research

[49] *R.* v. *Greater London Council, ex p. Blackburn* [1976] 1 WLR 550; cf. *Attorney-General, ex rel. McWhirter* v. *Independent Broadcasting Authority* [1973] QB 629; below Ch. 6.3; *R.* v. *Independent Broadcasting Authority, ex p. Whitehouse, The Times*, 14 Apr. 1984.

[50] Bridge LJ was able to accord standing only to Mrs Blackburn as a ratepayer (p. 567).

[51] *National Federation* case, above n. 2, pp. 633, 646–7, 662–3.

[52] [1973] 2 NZLR 758; cf. *Re Canadians for the Abolition of the Seal Hunt and Minister of Fisheries and the Environment* (1980) 111 DLR (3d) 333 (animal rights group refused standing to compel enforcement of seal hunting regulations); contrast *Re District of North Vancouver and National Harbours Board* [1978] 89 DLR (3d) 704 (residents accorded standing to compel enforcement of houseboat mooring regulations). The first of these cases illustrates very well the injustice which can be caused by standing rules: the only persons (as opposed to animals) directly affected by the regulations were the hunters, in whose interests it was that the regulations should not be enforced; by denying standing to the abolitionists the judge effectively denied a hearing to the only parties (seals being unable, perhaps regrettably, to invoke the court's jurisdiction)

into environmental law and the preservation and study of natural resources, applied for mandamus to compel the Board to exercise its powers to review the registration of a chemical, exposure to which was alleged to have caused deformities in human foetuses. Although it was held that the society had no case on the merits, it was also held that they had no standing to bring the case at all, since the duty in question was not owed to them.

It will be argued below that the law is seriously defective on this point of a citizen's standing for mandamus, and the effect of recent cases will be discussed.

6.3 STANDING AND PUBLIC DUTIES: THE PRIVATE-LAW REMEDIES

The question, who can enforce a public duty by way of application for an injunction or a declaration, is one of some interest, but of less practical significance than might be imagined. This is because (1) since public duties are subject to the regime of mandamus, this remedy will be available in all cases where an injunction or a declaration might be sought;[1] (2) the standing rules developed in relation to injunctions and declarations are very restrictive, though they have been relaxed in recent cases; and (3) the modern reforms of administrative-law remedies discussed above in Chapter 5 have reduced very considerably the area where the rules of standing for prerogative orders, as opposed to injunctions and declarations, are different. The proper role of the injunction and the declaration in relation to public duties is therefore likely to become that of supplementing the prerogative orders within the regime of statutory judicial review; to this the problem of standing for injunction and declaration at common law is largely irrelevant.

For these reasons there has not been a great deal of case law involving public duties which is relevant to this section, and the continued relevance of the cases is probably marginal. None the less there are some areas where these remedies are of some importance, notably those areas where the new statutory procedures are inapplicable, and the standing rules for injunction and declaration still exert some influence, if only residually or by side wind.

able to argue the point in question; note that the hunters would undoubtedly have had standing to challenge the constitutionality of the regulations.

[1] Except, that is, cases against the Crown, where mandamus is not available but a declaration is; see above Chs. 3.2(i), 4.3(iv).

The injunction and the declaration are generally discussed together[2] in the context of standing because they are often sought together and because the standing rules are broadly similar. As Thio points out, there is no case in which a plaintiff has applied for both remedies and the question of standing has been treated differently in relation to each remedy;[3] in fact the courts have now expressly stated that the position is the same for both remedies.[4]

Since both the injunction and the declaration developed as private-law remedies they were generally available only where a private right has been or might be infringed. The courts have shown reluctance to modify the basic principles so as to take account of the specifically public-law uses of the two remedies.

These basic principles were laid down in *Boyce* v. *Paddington Borough Council*,[5] a notorious case which has shown an extraordinary degree of persistence, being applied to declarations by the House of Lords in 1942,[6] approved by the House of Lords in *Gouriet* v. *Union of Post Office Workers*[7] in 1978, and by the High Court of Australia in *Australian Conservation Foundation Inc.* v. *Commonwealth*[8] in 1979, and even used as the basis of the action for breach of statutory duty by the House of Lords in *Lonrho Ltd.* v. *Shell Petroleum Co. Ltd.* (*No. 2*)[9] in 1981.

Boyce's case, as it happens, is one of the relatively few cases on standing for an injunction which involves a public duty. The Council were under a statutory duty to keep a churchyard as an open space. They erected a hoarding in it to prevent the plaintiff, an adjoining owner, from acquiring a prescriptive right to light. The plaintiff sued for an injunction to restrain the erection of the hoarding. The plaintiff's action failed on the facts, but Buckley J. laid down two situations in which an injunction would issue: (1) where interference with a public right also involves interference with a private right, and (2) where

[2] For discussion see De Smith, *Judicial Review of Administrative Action*, 4th edn. by Evans (1980), pp. 450–61; Whitmore and Aronson, *Review of Administrative Action* (1978), pp. 330 et seq.; Thio, *Locus Standi and Judicial Review* (1971), ch. 7; Stein, ed., *Locus Standi* (1979), ch. 2; Craig, *Administrative Law* (1983); Kyrou, 'Locus Standi of Private Individuals Seeking Declaration or Injunction at Common Law' (1982), 13 *Melb. ULR* 453.

[3] Thio, above n. 2, pp. 131–3.

[4] *Gouriet* v. *Union of Post Office Workers* [1978] AC 435; *Australian Conservation Foundation Inc.* v. *Commonwealth* (1980) 54 ALJR 176, 180.

[5] [1903] 1 Chanc. 109 (revd. [1903] 2 Chanc. 556).

[6] *London Passenger Transport Board* v. *Moscrop* [1942] AC 332.

[7] Above n. 4.

[8] Ibid.

[9] [1982] AC 173. *Boyce* still applies to declarations even after the O. 53 reforms: see *Barrs* v. *Bethell* [1982] 1 All ER 106.

the plaintiff suffers 'special damage peculiar to himself'. In the Court of Appeal the Attorney-General was joined and the plaintiff succeeded.

These principles are of course the criteria for the actionability of a public nuisance, and indeed are derived from that tort.[10] It is perhaps not surprising that they have been construed in a private-law sense, so that generally the plaintiff must show that he has an actual or prospective cause of action of some kind, or at least a right which is protected by private law.[11] However, it is unclear whether this is always the case, and whether the concept of 'special damage' indicates damage over and above the rest of the public, or damage which gives rise to an action in tort; the confusion is not assisted by the fact that special damage may result in an action for public nuisance or breach of statutory duty, which would render the special-damage principle identical to the private-right principle, but the case law and the judgments in *Gouriet*'s case, which is the leading modern authority on standing in injunction and declaration cases generally,[12] do not seem to suggest that the courts have any intention of confining the *Boyce* principles to situations where there is a cause of action. None the less the cases do indicate a very narrow view which suggests a distinction between the protection of private and public rights. In fact *Gouriet* was a decision which was primarily concerned to restrict the right of an individual, as opposed to the Attorney-General, to secure observance of the criminal law by injunction or declaration precisely because no private right is involved.[13] The *Australian Conservation Foundation* case, approving *Gouriet*, and subsequent Australian cases[14] involve a slight relaxation of the *Boyce* rules in that the requirement of special damage means that the plaintiff must have some special interest in the matter over and above that enjoyed by the public generally, a reformulation

[10] See Craig, above n. 1, p. 427.

[11] De Smith, above n. 1, pp. 450–2, esp. n. 46.

[12] [1978] AC 435.

[13] Ibid. 481, 494, 500, 511–12, 518–19, 521.

[14] *Ingram* v. *Commonwealth* (1980) 54 ALJR 395 (HCA); *Onus* v. *Alcoa of Australia Ltd.* (1981) 149 CLR 27 (HCA); *Coe* v. *Gordon* [1983] 1 NSWLR 419; contrast the more liberal Canadian position set out in *Thorson* v. *Attorney-General of Canada* (*No. 2*) (1974) 43 DLR (3d) 1 (SCC). The point of the reformulation of the second of the *Boyce* principles, according to Brennan J., in *Onus* (p. 647) is 'the better to express the principle which now governs the standing of a private person to sue to enforce the performance of a public duty'. However, the superiority of this approach lies in the fact that it is more liberal rather than in its clarity; while no 'mere emotional or intellectual interest' (Gibbs CJ in *Australian Conservation Foundation*, above n. 4, at p. 181) will suffice, the requirement of a special interest involves 'a curial assessment of the importance of the concern which a person has with a particular subject matter and of the closeness of that person's relationship to that subject matter' (Stephen J. in *Onus* at p. 42, and see also Gibbs CJ at p. 42 to the same effect).

which approximates to the old common-law standing rule for mandamus, and is obviously inconsistent with the equation of standing with actionability. On this question the judgments in *Gouriet* are quite unclear, but this may not be important now that more liberal rules apply to Order 53 cases.[15]

The cases which involve public duties are especially hard to reconcile with the *Boyce* rules unless those rules are construed much more liberally than they have been in cases involving excess of powers.

Consistent with the narrow approach in *Boyce* are cases such as *Glossop* v. *Heston and Isleworth Local Board*,[16] decided before *Boyce*, in which an injunction was refused to a plaintiff suing in respect of failure to fulfil a duty under the Public Health Act 1875 (England and Wales) because mandamus was the proper remedy. This sharp division between public and private duties seems to ignore the fact that a breach of a public duty can also give rise to a private cause of action. However, these cases can also generally be explained on another basis.[17]

Another post-*Boyce* strand of authority is, however, quite inconsistent with the *Boyce* principles. These cases all concern duties arising under education legislation. In *Gateshead Union Guardians* v. *Durham County Council*[18] in 1918, local guardians *in loco parentis* to poor children were granted a declaration and were held entitled to an injunction to enforce the education authority's duty to provide the children with education free of charge, because the guardians as parents had a statutory right in that regard. It could be argued that this is in effect a case of special damage, but clearly no private right of any kind is involved, and in later English cases educational duties have been enforced by injunction in circumstances where there is even more clearly no special damage in the *Boyce* sense; in *Meade* v. *Haringey London Borough Council*[19] the Court of Appeal was prepared to grant

[15] See Craig, above n. 1, pp. 430–1; below Ch. 6.4.

[16] (1879) 12 Chanc. D. 102; *Attorney-General* v. *Dorking Union Guardians* (1881) 20 Chanc. D. 595; *Attorney-General* v. *Clerkenwell Vestry* [1891] 3 Chanc. 527; *Davies* v. *Gas, Light, and Coke Co.* [1909] 1 Chanc. 708; *Clark* v. *Epsom Rural District Council* [1929] 1 Chanc. 287. In *Holland* v. *Dickson* (1888) 37 Chanc. D. 669 and *Jeanneret* v. *Hixson* (1890) 11 LR (NSW) Eq. 1 mandamus was refused because a mandatory injunction was available; cf. *Wentworth* v. *Woolahra Municipal Council* (1982) 149 CLR 672 (HCA). Consistent with this sharp division is the rule that the Attorney-General can only sue in respect of a breach of a public not a private duty: *Attorney-General* v. *Poole Corporation* [1938] Chanc. 23.

[17] See above Ch. 3.4(ii), and below Ch. 7.2(v).

[18] [1918] 1 Chanc. 146, 167.

[19] [1979] 1 WLR 637, 647, per Lord Denning MR, and 649 per Eveleigh LJ; see also *Wood* v. *Ealing London Borough Council* [1966] 3 All ER 514; *Bradbury* v. *Enfield London Borough Council* [1967] 1 WLR 1311; *Lee* v. *Enfield London Borough Council* (1967)

an injunction and a declaration at the instance of parents who alleged that the education authority was in breach of its duty to provide education, even though 37,000 children were affected, on the ground that they were 'particularly damnified'. Thus the concept of special damage is capable of extending beyond the confines of private law in England as well as Australia.

The high water mark is undoubtedly *Attorney-General, ex rel. McWhirter* v. *Independent Broadcasting Authority*,[20] in which Lord Denning MR, in language reminiscent of the *Blackburn* cases discussed above[21] and repeated in the Court of Appeal in *Gouriet*, expressed himself prepared to grant an injunction to a member of the public, a television licence holder, to restrain the broadcasting of a film in breach of the authority's duty to satisfy itself that programmes were not offensive. *McWhirter* was, however, overruled in *Gouriet*. There are also cases that indicate that failure to perform a duty to deliver mail entitles a member of the public to an injunction or a declaration.[22]

Thus the authorities are not clear with regard to the question whether an action for breach of statutory duty, nuisance, or negligence must lie before a mandatory injunction can be granted, but the weight of modern cases favours a standing rule which is not based on actionability.

The interesting feature of these public-duty cases is that *Boyce* is hardly ever even referred to in the judgments, let alone applied. It is suggested therefore that there is authority which would exempt the enforcement of public duties from the *Boyce* rules on the basis that an injunction or a declaration can be obtained by a member of the public to enforce a public duty where it is clear that the statute intends to confer a right on him, even though this right may be incapable of creating an action for damages for breach of statutory duty or any other cause of action in tort. It is suggested that this approach would enable the injunction and the declaration to be put on a par with mandamus for the purposes of standing, a result which

66 LGR 195 (one plaintiff given standing as a parent and ratepayer, another as a school governor); *Legg* v. *Inner London Education Authority* [1972] 1 WLR; *Winward* v. *Cheshire County Council* (1978) 77 LGR 172. *Watt* v. *Kesteven County Council* [1955] 1 QB 408, 425, 429 suggests that a cause of action is required, and an injunction was refused in that case to compel an education authority to provide tuition fees; however, the *Gateshead* case was not doubted.

[20] [1973] QB 629; overruled by the House of Lords in *Gouriet*.

[21] See Ch. 6.2(iv).

[22] *Bradley* v. *Commonwealth* (1973) 128 CLR 557 (HCA); *Fairfax (John) Ltd.* v. *Australian Postal Commission* [1977] 2 NSWLR 124; *Fairfax (John) Ltd.* v. *Australian Telecommunications Commission* [1977] 2 NSWLR 400.

appears virtually to have been achieved in some cases prior to *Gouriet*,[23] but which is not pre-empted by that decision. The Australian cases have shown how the second limb of the *Boyce* principles can be used flexibly.

In cases where no individual has standing under the *Boyce* rules for an injunction or a declaration, *Gouriet*'s case decides that only the Attorney-General can sue, either *propria motu* or at the relation of a member of the public.[24] In practice the Attorney-General does not proceed against government departments and only quite rarely against other public authorities,[25] and his decision not to proceed is not reviewable in the courts.[26]

The standing rules for enforcing public duties by injunction and declaration reveal a tentative approach which hovers between the restrictive *Boyce* position and the citizen action. The cases are inconsistent with the view that these remedies should only protect rights secured by private law, but fall far short of establishing standing rules which are appropriate to public law. The factors which influence the courts are not very apparent in the cases; no explanation is given why the position should be different for injunction or declaration from mandamus, or why the *Boyce* rules should be considered even remotely relevant in these cases. No convincing analysis of the Attorney-General's role is given. Recent cases do not clarify the law or how it will be applied. The whole area requires light, fresh air, and open spaces.

[23] See *Morton* v. *Eltham Borough* [1961] NZLR 1; *Tonkin* v. *Brand* [1962] WAR 2; *Woodcock* v. *South Western Electricity Board* [1975] 1 WLR 983; *Gravesham Borough Council* v. *British Railways Board* [1978] Chanc. 379; *Booth & Co. (International) Ltd.* v. *National Enterprise Board* [1978] 3 All ER 624; *Wilson* v. *Independent Broadcasting Authority* 1979 SLT 279; *Finlay* v. *Minister of Finance* (1983) 146 DLR (3d) 704, all of which are cases which are arguably inconsistent with the *Gouriet* approach; the *Morton*, *Woodcock*, and *Booth* cases can be explained on the basis that an action would lie for breach of statutory duty, if an unusually broad view of that tort is taken (i.e. one which might not have been taken if damages had been in question: see below Ch. 7.2), but the others cannot be explained on any basis except a broad view of the *Boyce* 'special damage' requirement; the *McWhirter* case cannot be explained even on that basis, but was overruled in *Gouriet*. Another interesting case is *Barber* v. *Manchester Regional Hospital Board* [1958] 1 WLR 181, in which a declaration was granted that a Minister had breached his statutory duty by refusing to hear an appeal, but no right of action was in question.

[24] [1978] AC 494, 500, 511–12, 518–19, 521; see e.g. *Attorney-General* v. *Cockermouth Local Board* (1874) LR 18 Eq. 172; *Attorney-General* v. *Birmingham Drainage Board* [1910] 1 Chanc. 48; *Attorney-General* v. *Lewes Corporation* [1911] 2 Chanc. 495.

[25] See above Ch. 4.3(v).

[26] [1978] AC 482, 488–91; cf. *Barton* v. *R.* (1980) 147 CLR 75 (HCA); *Clyne* v. *Attorney-General (Cwlth)* (1984) 55 ALR 92.

6.4 STANDING UNDER THE SUPREME COURT ACT 1981 (ENGLAND AND WALES)

Applications for judicial review in England are now governed exclusively by the Supreme Court Act 1981 (England and Wales). Section 31(3)[1] governs the standing requirements, which are based on a 'sufficient interest in the matter to which the application relates'.

This provision has been dealt with in the case of *Inland Revenue Commissioners* v. *National Federation of Self-Employed and Small Businesses Ltd.*,[2] an important decision of the House of Lords which is relevant to almost all aspects of standing for mandamus and other remedies for the enforcement of public duties. The facts of that case are set out and discussed above in Chapter 2. The problem discussed in this section is how far the new statutory rule has modified standing for mandamus at common law in the light of the House of Lords' decision.

The first point which needs to be made clear is that the decision was actually made in relation to Rules of the Supreme Court, Order 53, rule 3(5), which is identical to section 31(3); the section was enacted later because of doubts concerning the competence of the rule committee to enact rule 3(5).[3] It is therefore possible that their Lordships were constrained by having to interpret rule 3(5) so as to make no alteration in the substantive law, though this consideration appears to have had little if any practical effect on the reasoning adopted; it could, however, be used to circumvent any unduly restrictive aspects of that reasoning in future.

On the face of it, section 31(3) is directed only to the question of leave. However, their Lordships seem to have assumed[4] that the test of 'sufficient interest' is applicable *inter partes* as well as on the ex-parte application for leave, a position which is entirely satisfactory because

[1] For full text of s. 31, see App.

[2] [1982] AC 617. For interesting discussions of this case see Harlow and Rawlings, *Law and Administration* (1984), pp. 300–7; Cane, 'Standing, Legality and the Limits of Public Law' [1981], *PL* 322; Griffiths, 'Mickey Mouse and Standing in Administrative Law' [1982], *CLJ* 6; Wade, *Administrative Law*, 5th edn. (1983), pp. 587–91.

[3] '[I]t does not—indeed, cannot—either extend or diminish the substantive law' per Lord Scarman [1982] AC 647. Both Lord Scarman (ibid.) and Lord Diplock (p. 638) regarded standing as a matter of procedure. Lord Wilberforce considered (p. 631) that r. 3(5) and the impending legislation (s. 31(3)), merely preserved the common-law rules.

[4] The assumption is apparent throughout the judgments. The House of Lords considered that the Divisional Court had been right to grant leave to apply, but should have rejected the application on the question of standing.

the adoption of different criteria of standing at the two stages would be patently absurd.

It is important to bear in mind also that, as a consequence of section 31(2), the test of 'sufficient interest' is applicable not just to the prerogative orders, but also to declarations and injunctions. In the words of Lord Diplock:

> if, before the new Order 53 come into force, the court would have had jurisdiction to grant to the applicant any of the prerogative orders it may now grant him a declaration or injunction instead, notwithstanding that the applicant would have no locus standi to claim the declaration or injunction under private law in a civil action against the respondent to the application, because he could not show that any legal right of his own was threatened or infringed.[5]

It is suggested that this simple reform sweeps away a great deal of the unnecessary complication involved in applications for remedies to enforce public duties. Whatever test is adopted for standing, that test should not vary with the remedy sought. It is surely right that in such an application the advantages of all the relevant remedies should be available, and not their disadvantages.[6] It may, however, be the case that in practice a different result will be reached in mandamus cases, because of the nature of the subject-matter involved, at least where the duty involved is not a duty to determine.[7]

[5] [1982] AC 639. In the case under discussion the applicants sought a declaration that the Commissioners had acted unlawfully, and mandamus to assess and collect taxes from the casual workers. *Gouriet* v. *Union of Post Office Workers* [1978] AC 435 was explained as being a case in private law, which would make it of little significance now that all public-law cases must be brought under O. 53; see above Chs. 5.2, 6.3. For public-duty cases where standing was granted, on this altered basis, for a declaration, see *R.* v. *Secretary of State for the Environment, ex p. Ward* [1984] 2 All ER 556; *Steeples* v. *Derbyshire County Council* [1984] 3 All ER 468.

[6] See above Ch. 5.2.

[7] Lord Wilberforce said at [1982] AC 631: 'It would seem obvious enough that the interest of a person seeking to compel an authority to carry out a duty is different from that of a person complaining that a judicial or administrative body has, to his detriment, exceeded its powers. Whether one calls for a stricter rule than the other may be a linguistic point: they are certainly different and we should be unwise in our enthusiasm for liberation from procedural fetters to discard reasoned authorities which illustrate this.' (cf. Lord Scarman at p. 648). This dictum has been explained by Cane, above n. 1, in three ways: (1) as referring to the nonfeasance/misfeasance distinction; (2) as a suggestion that detriment is not required for mandamus; and (3) as a proposition that 'sufficient interest' may exist only where the duty is owed to the applicant. Cane rejects all three approaches, concluding that standing for mandamus is the same as for *certiorari* and prohibition. However, Lord Wilberforce's dictum merely recognizes the fact that *certiorari* and prohibition apply only to determinations, whereas mandamus applies to a much wider variety of situations; for this reason his Lordship points out that 'the fact that the same words are used

Turning now to the test of 'sufficient interest' itself, one might ask whether it represents merely a codification of the old rules, or a new global formula. It was argued in the lower courts that Order 53 had left the basis of standing for each of the individual remedies unaltered. Their Lordships side-stepped this issue by simply reinterpreting the common law.[8] Lord Diplock put forward a striking justification for this approach:

> The rules as to 'standing' for the purpose of applying for prerogative orders, like most of English public law, are not to be found in any statute. They were made by judges, by judges they can be changed; and so they have been over the years to meet the need to preserve the integrity of the rule of law despite changes in the social structure, methods of government and the extent to which the activities of private citizens are controlled by governmental authorities, that have been taking place continuously, sometimes slowly, sometimes swiftly, since the rules were originally propounded. . . . It would, in my view, be a grave lacuna in our system of public law if a pressure group, like the federation, or even a single public-spirited taxpayer, were prevented by outdated technical rules of locus standi from bringing the matter to the attention of the Court to vindicate the rule of law and get the unlawful conduct stopped.[9]

Their Lordships had therefore no compunction in overruling the 'deplorable'[10] decision in *R.* v. *Lewisham Union Guardians*,[11] and endorsing the liberal views of Lord Denning MR in *R.* v. *Greater London Council, ex p. Blackburn*.[12]

Although their Lordships were *ad idem* in ratifying the generally felt need for less restrictive standing rules, they were not agreed what test should be adopted of a 'sufficient interest', and their entire reasoning was influenced, it is suggested unduly so, by the manner in which the case came before them. The lower courts had assumed that the issue of standing was a preliminary issue distinct from the merits of the application, and for this reason the illegality of the Commissioners' arrangement with the Fleet Street casuals was conceded. This concession was withdrawn before the House of Lords,

to cover all the forms of remedy allowed by the rule does not mean that the test is the same in all cases'. Lord Diplock on the other hand (p. 640) emphasizes the similarities between the remedies. If Lord Wilberforce's point had been given more attention by the House of Lords, we would perhaps be much nearer a definition of standing for mandamus. Whether the wider scope of mandamus demands a *narrower* standing rule is perhaps doubtful.

[8] See esp. Lord Diplock at [1982] AC 637, and Lord Scarman, ibid. 648.

[9] Ibid. 639–40, 644.

[10] Ibid. 653, per Lord Scarman, who also described it as 'heresy'.

[11] [1897] 1 QB 498; see above Ch. 6.2.

[12] [1976] 1 WLR 550.

which held that the arrangement was not illegal. Their Lordships followed the lead of Lord Wilberforce in finding an essential linkage between standing and illegality, such that standing ceases to be necessarily a preliminary issue:

> There may be simple cases in which it can be seen at the earliest stage that the person applying for judicial review has no interest at all, or no sufficient interest to support the application: then it would be quite correct at the threshold to refuse him leave to apply. The right to do so is an important safeguard against the courts being flooded and public bodies harassed by irresponsible applications. But in other cases this will not be so. In these it will be necessary to consider the powers or the duties in law of those against whom the relief is asked, the position of the applicant in relation to those powers or duties, and to the breach of those said to have been committed. In other words, the question of sufficient interest can not, in such cases, be considered in the abstract, or as an isolated point; it must be taken together with the legal and factual context. The rule requires sufficient interest in the matter to which the application relates. This, in the present case, necessarily involves the whole question of the duties of the Inland Revenue and the breaches or failure of those duties of which the respondents complain.[13]

In the result the judgments formed a curious patchwork: Lord Wilberforce held that the applicants had no sufficient interest because the respondents owed no duty towards them;[14] Lord Diplock held that the applicants had shown no unlawful conduct;[15] Lords Fraser and Roskill that they had no sufficient interest;[16] and Lord Scarman that the respondents had performed their statutory duty, therefore the applicants had no sufficient interest.[17] In so deciding, at least three of their Lordships appeared to suggest that the interest of an applicant depends on the extent or gravity of the illegality.[18]

It is not clear why standing and legality should be linked in this way, and indeed the precise nature of the link is itself unclear. In

[13] [1982] AC 630. For discussion of the distinction between standing and merits, see Cane, above n. 1; Stein, ed., *Locus Standi* (1979), ch. 1. The latest pronouncements of the High Court of Australia on this matter seem to reflect the traditional view: see *Onus* v. *Alcoa of Australia Ltd.* (1981) 149 CLR 27, 38.

[14] Above n. 13, pp. 633, 635–6.

[15] Ibid. 644.

[16] Ibid. 647 and 663–4, respectively.

[17] Ibid. 654–5.

[18] Lord Wilberforce referred to 'a case of sufficient gravity' (ibid. 633); Lord Diplock to 'flagrant and serious breaches of the law . . . which are continuing unchecked' (p. 641); Lord Fraser to 'exceptionally grave or widespread illegality' (p. 647); Lord Roskill to 'some grossly improper pressure or motive' (p. 662). It may be that their Lordships did not intend to link standing with the extent of illegality except in the sense that the duty in issue in the case was a very limited one; if so a clear indication of the extent of the duty would surely have been helpful.

considering this difficult question it is interesting to observe that their Lordships did not consider that standing was only a question relating to the granting of leave. Indeed Lord Diplock regarded the matter of standing as

> a 'threshold' question in the sense that the court must direct its mind to it and form a prima facie view about it on the material that is available at the first stage. The prima facie view so formed, if favourable to the applicant, may alter on further consideration in the light of further evidence that may be before the court at the second stage, the hearing of the application for judicial review itself.[19]

As Cane[20] has pointed out, it makes sense to consider *all* aspects of the case at the leave stage, not just standing. It is none the less a mistake to conclude that standing is not a logically prior issue to the merits of the case, unless perhaps one wishes to take a novel view of the function of the standing rules, i.e. that they exist not to decide who should be allowed to argue the case, whatever its merits, but to decide what interests the law should protect in reviewing administrative decisions. Even on that view one might ask how the breach of duty alleged is in fact relevant to determining which interests should be protected.[21] This is pursued below, but whatever the merits of the linkage of standing and legality, the principles adumbrated by the House of Lords will be much harder to apply even than the old rules of standing, confused, illogical, and restrictive as they were. In the particular form in which their Lordships put those principles,[22] the problem is particularly acute, and indeed the approach is quite inconsistent with the reasoning generally adopted in reviewing the performance of duties other than the duty to determine: as we have seen,[23] the problem of 'technical' breaches of duty has been dealt with not by denying standing to the person claiming to be affected,

[19] [1982] AC 642. It would appear from the judgments of Lord Diplock (p. 638), Lord Fraser (p. 646), and Lord Scarman (pp. 648, 653) that standing is a matter of discretion. Lord Wilberforce (p. 631) was adamant that it is not. In practice it probably makes little difference, but in principle Lord Wilberforce's view is preferable, because a person's right of access to the courts is in question; see, however, Yardley, 'Prohibition and Mandamus and the Problem of Locus Standi' (1957) 73 *LQR* 534; *R.* v. *Monopolies and Mergers Commission, ex p. Argyll Group* [1986] 1 WLR 763, 773 per Donaldson MR.

[20] Above n. 1.

[21] It can plausibly be argued that although the factual context is relevant, the relevant facts are those relating to the applicant's interest, not the breach of duty by the respondent. Such facts are rarely in dispute, and therefore standing can almost always be treated as a preliminary issue.

[22] Above n. 18.

[23] Above Chs. 2.3, 2.4.

but by treating the duty as one of open texture, so that, provided the authority has acted reasonably, it will be said to have complied with its duty. Furthermore, it is in any case hard to see any logical connection at all between standing of an applicant and the extent, as opposed to the effect on him, of the illegality of which he complains.

In addition to the problem of standing and legality, their Lordships, having bent over backwards to set in motion the reform of the law of standing, seem to have been of different views as to the appropriate test of 'sufficient interest'. As Lord Fraser put it:

> All are agreed that a direct financial or legal interest is not now required, and that the requirement of a legal specific interest laid down in [*Lewisham Union Guardians*] . . . is no longer applicable. There is also general agreement that a mere busybody does not have a sufficient interest. The difficulty is, in between those extremes, to distinguish between the desire of the busybody to interfere in other people's affairs and the interest of the person affected by or having a reasonable concern with the matter to which the application relates.[24]

The number of tests proposed is bewildering: whether the statute 'gives any express or implied right to persons in the position of the applicant to complain of the alleged unlawful act or omission' (Lord Fraser),[25] whether the applicant is 'within the scope or ambit of the duty . . . a good working rule though not perhaps an exhaustive one' (Lord Wilberforce);[26] whether the applicant has a 'genuine grievance reasonably asserted' (Lord Scarman);[27] Lord Diplock did not commit himself, but Lord Roskill expressly disapproved Lord Denning MR's view in *R.* v. *Greater London Council, ex p. Blackburn*[28] because

> if applied to all applications for judicial review, [it] would extend the individual's right of application for that relief far beyond any acceptable limits, and would give a meaning so wide to a 'sufficient interest' . . . that they would in practice cease to be, as they were clearly intended to be, words of limitation upon that right of application.[29]

The effect of the decision is merely therefore to restrict standing to enforce the Revenue's duty of equal treatment[30] to situations, probably very rare, of widespread and flagrant illegality. Apart from the vague

[24] [1982] AC 646.

[25] Ibid. For the difficulty of construing legislative intent, see below Ch. 7.2.

[26] Above n. 25, pp. 631.

[27] Ibid. 654.

[28] Above n. 12.

[29] [1982] AC 661.

[30] See above Ch. 2.4. A taxpayer has since been given standing in *R.* v. *HM Treasury, ex p. Smedley* [1985] 1 All ER 589.

and controversial linkage of standing and legality, the formulas adopted by their Lordships tell us very little about the purpose or application of the standing rules. What is lacking is a discussion of the policies which are appropriate to be secured under the standing rules. The pronouncements in the House of Lords are a general encouragement to applicants but give no guidance how their assertion of a sufficient interest will be dealt with.

It is hard to say whether the liberal sentiments of the House of Lords will be translated into correspondingly liberal rules, but the indications are that they will. In *R.* v. *Secretary of State for Social Services, ex p. Greater London Council and Child Poverty Action Group*[31] the applicants argued that a supplementary benefit regulation required the Ministry to identify some 16,000 persons who were entitled to receive payments. The Ministry refused to act because the cost of identifying those entitled would outweigh by more than ten times the amounts to be paid. Woolf J., following the *National Federation* case, held that the CPAG but not the GLC had standing to enforce the duty because the former was designed to represent the interests of unidentified claimants, whereas the latter were not guardians of the public interest, had no special responsibility to claimants over and above that owed to all ratepayers, and were not financially affected. The decision is applauded on the first point, but surely the GLC should have been allowed standing, especially if they alone had applied for relief, on the basis that at least some of their ratepayers would be likely to be affected by the Ministry's inaction? Quite possibly no one would have been able to raise the matter if the CPAG had not acted, because the persons affected by the failure to perform the duty were not aware of the fact.

Outside the area of public duties the cases seem to indicate a willingness to adopt a broad interpretation of 'sufficient interest'.[32] It may be that in time the uncertainties arising out of the *National Federation* case will be resolved satisfactorily.

6.5 STANDING UNDER STATUTORY REVIEW PROCEDURES IN AUSTRALIA

Under the Administrative Decisions (Judicial Review) Act 1977 (Cwlth), section 5(1), standing for the purposes of review under the

[31] *The Times*, 16 Aug. 1984. On appeal the Court of Appeal held that the Ministry had not breached their duty, so that the question of the CPAG's standing did not arise; the court dismissed the GLC's appeal on the question of their standing: *The Times*, 8 Aug. 1985; cf. *R.* v. *Secretary of State for Social Services, ex p. Child Poverty Action Group, The Times*, 15 Feb. 1988.

[32] See Aldous and Alder, above n. 1, p. 96; *R.* v. *Independent Broadcasting Authority, ex p. Whitehouse, The Times*, 14 Apr. 1984 (television licensee); *R.* v. *Secretary of State for the Environment, ex p. Ward* [1984] 2 All ER 556.

Act is restricted to 'a person who is aggrieved by a decision to which this Act applies'; and under section 3(1) such a decision is defined as a decision of an administrative character made under an enactment.[1] Similarly, the Administrative Law Act 1978 (Vic.), section 3, restricts standing to a 'person affected by a decision of a tribunal'; however, the term 'person affected' is further defined by section 2:

> a person whether or not party to the proceeding [sc. before the tribunal], whose interest (being an interest that is greater than the interest of other members of the public) is or will or may be affected, directly or indirectly, to a substantial degree by a decision which has been made or is to be made or ought to have been made by the tribunal.

Unlike the English provision, these formulas do not of course displace standing for mandamus at common law.[2] They seem to have been treated, as in England,[3] as an excuse for a new start: none of the cases so far has referred to the cases decided under the common-law standing rules, save those in relation to the particular statutory provision being reviewed.[4] The indications so far are that a liberal approach will be taken to standing under statutory review procedures.[5]

(i) *Administrative Decisions (Judicial Review) Act 1977 (Cwlth)*

In *Tooheys Ltd.* v. *Minister for Business and Consumer Affairs*[6] the applicants bought some machinery from an American company, and had paid the importers the amount of customs duty which the latter had been charged. They sought an order of review under the 1977 Act of the Minister's refusal to make a determination under the customs legislation, which would have had the effect of rendering the

[1] See above Ch. 5.3 for discussion of the scope of this definition. The 'person aggrieved' formula naturally applies to conduct and to any failure to make a decision which falls under the Act: see s. 3(4).

[2] Ibid. In this context it is interesting to note that the standing rules for *statutory* mandamus in Australia (see above Ch. 4.5) require a 'personal interest': see *Bilbao* v. *Farquhar* [1974] 1 NSWLR 377, where this was interpreted not be equivalent to the 'legal specific right' test for mandamus at common law, and was applied to a purely familial interest; *Grzybowicz* v. *Smiljanic* [1980] 1 NSWLR 627.

[3] See above Ch. 6.3.

[4] See below n. 8; and Enright, *Judicial Review of Administrative Action* (1985), ch. 15.

[5] Griffiths, in 'Legislative Reform of Judicial Review of Commonwealth Administrative Action' (1978), *FLR* 42, 47–9, feared that the formula of 'person aggrieved' was too restrictive (see also *Report of Commonwealth Administrative Review Committee* (Parl. Paper No. 144, 1971); these fears have not been borne out, it seems.

[6] (1981) 36 ALR 64 (affirmed by the Full Court s.n. *Minister for Industry and Commerce* v. *Tooheys Ltd.* (1982) 42 ALR 260); cf. *Fordham* v. *Evans* (1987) 72 ALR 529.

goods exempt from duty, with the further result that the applicants would be able to claim a refund from the importers. The Minister objected to the competency of the proceedings on the ground, *inter alia*, that the applicants were not a person aggrieved within section 3(1). The matter of standing therefore arose as a preliminary matter and was decided by Ellicott J. as follows:

> If I thought it was necessary for the applicant to establish its right to a refund in the event of a determination being made by the Minister I would decline to determine this issue as a preliminary point. The question whether an applicant is a person aggrieved is one of mixed law and fact and in many cases would best be determined at a final hearing when all the facts are before the court and the court has the benefit of a full argument on the matter. This is so in this case.
>
> However, I have formed the view that it is unnecessary for the applicant to show that it has a right to a refund, in the circumstances mentioned, in order to establish locus standi to bring these proceedings. The words 'a person who is aggrieved' should not, in my view, be given a narrow construction. They should not, therefore, be confined to persons who can establish that they have a legal interest at stake in the making of the decision. It is unnecessary and undesirable to discuss the full import of the phrase. I am satisfied from the broad nature of the discretions which are subject to review and from the fact that the procedures are clearly intended in part to be a substitution for the more complex prerogative writ procedures that a narrow meaning was not intended. This does not mean that any member of the public can seek an order of review. I am satisfied, however, that it at least covers a person who can show a grievance which will be suffered as a result of the decision complained of beyond that which he or she has as an ordinary member of the public. In many cases that grievance will be shown because the decision directly affects his or her existing or future legal rights. In some cases, however, the effect may be less direct. It may affect him or her in the conduct of a business or may, as I think is the case here, affect his or her rights against third parties . . .[7]

This approach has been approved and applied in several subsequent cases,[8] and comes as a breath of fresh air after the awkward and

[7] Above n. 6, p. 79. The requirement of suffering beyond that of an ordinary member of the public was also demanded in *Hawker Pacific Pty. Ltd.* v. *Freeland* (1983) 52 ALR 185 (applicant given standing on narrow ground that it had been invited to tender for a contract awarded to another company).

[8] *Safadi* v. *Minister for Immigration and Ethnic Affairs* (1981) 38 ALR 399, 403; *Ricegrowers Co-operative Mills Ltd.* v. *Bannerman and Trade Practices Commission* (1981) 38 ALR 535, 540; *Vickers* v. *Minister for Business and Consumer Affairs* (1982) 43 ALR 389, 408; *Doyle* v. *Chief of General Staff* (1982) 42 ALR 283, 287; *Fowell* v. *Ioannou* (1982) 45 ALR 491, 504; *Ralkon Agricultural Co. Pty. Ltd.* v. *Aboriginal Development Commission* (1982) 43 ALR 535, 543; *Lamb* v. *Moss* (1983) 49 ALR 533, 554; *Kioa* v. *Minister for Immigration and Ethnic Affairs* (1984) 53 ALR 658, 665–7.

conflicting cases decided on common-law principles discussed above. It is interesting to note the similarity with the turn taken by English law during a comparable period: Ellicott J. did not go so far as to merge standing and legality, but he did make it clear that standing is not necessarily only a preliminary issue;[9] it is still true that a mere busybody or stranger cannot sue, and that standing will not be refused on narrow grounds.

In *Alcoa of Australia Ltd.* v. *Button*[10] Woodward J. appeared to fuse standing and merits completely by holding that the applicants were not entitled to a refund of duty and therefore had no standing to argue such entitlement.

(ii) *Administrative Law Act 1978* (*Vic.*)

The provisions of the Administrative Law Act are potentially more restrictive than those of the Judicial Review Act because there must be an interest in the decision and the interest must be affected to a substantial degree; these restrictions could be regarded as introducing a quantitative as well as a qualitative element. Presumably, however, a wide interpretation of the word 'interest', in line with the previous case law, as well as the words in parentheses in section 2, will be adopted.

In *Charlton* v. *Members of Teachers Tribunal*[11] the Tribunal considered a claim for a salary increase made by an association representing high-school principals. It upheld the claim, but the Minister of Education requested a stay of the operation of the decision until Parliament had considered a government proposal to disallow the regulation embodying the Tribunal's decision and a review of the decision. The Tribunal then made the regulation embodying its decision, refusing to review the decision, but provided by a further regulation that the first would not come into operation until 31 days after it was laid before Parliament. In the event Parliament disallowed the first regulation. The applicant, the President of the association, sought an order of review of the Tribunal's decision on the ground that it had not given the association a hearing on the application for a stay. On the question of standing, the respondents contended that the word 'substantial' in section 2 meant 'considerable, big or weighty',

[9] It may, however, be a preliminary issue in relation to applications under s. 11(1)(c) of the Act for an extension of time for bringing the application: see *Doyle* v. *Chief of General Staff*, above n. 8.

[10] (1984) 55 ALR 101.

[11] [1981] VR 831; see also *R.* v. *Liquor Control Commission, ex p. Dickens* (*S. E.*) *Pty. Ltd.* [1983] VR 303.

as opposed to 'significant, in the sense of being more than trifling' as suggested by the applicant. McGarvie J. appeared to favour the applicant's view, but held that on either view the applicant had standing not merely because of his official position, but because the Tribunal's decision affected his salary to the extent of $330. It is suggested that the applicant's view of section 2 is preferable because it minimizes the quantitative aspect of standing under the section, but that, bearing in mind that the issue was one of natural justice, the point should have been determined solely on a consideration of the applicant's position; if the effect on his salary is relevant then any high-school principal would have standing to challenge the decision even if the association considered it had had fair hearing.

It is interesting to note that in none of these Australian cases referred to was the type of order sought regarded as relevant. This suggests that with regard to decisions the law of standing is the same no matter what *type* of order is sought, and to that extent the differences between standing for mandamus and for *certiorari* have been abolished—a highly desirable result. However, the criteria of standing adopted in these cases are no guide to the solution of the problems of standing for the enforcement of duties other than the duty to determine, and the Australian courts will have to consider whether they wish to follow the House of Lords on this matter.

On the whole the modern Australian cases show the same trends as the modern English cases. There is agreement on liberalizing the standing rules, but no real indication where to draw the line.

6.6 STANDING AND PUBLIC DUTIES: A SOLUTION

Having considered the question of standing for the various remedies by which public duties are enforced, and bearing in mind that an opportunity is now afforded by the new statutory review procedures to formulate new standing rules for public duties, let us consider what rules would be appropriate. It will not be possible here to suggest a definitive solution to the problem of standing, which of course covers all administrative action, not just the performance of public duties, but it may be possible to suggest a suitable general approach to standing for public duties which could be incorporated within a more general test of standing.[1]

[1] For recent discussions concerning the recasting of the standing rules see Vining, *Legal Identity* (1978); Stein, ed., *Locus Standi* (1979), ch. 9; Cane, 'The Function of Standing Rules in Administrative Law' [1980], *PL* 303. See also *Law Commission Working Paper*, No. 40 (1970) (UK); *Australian Law Reform Committee Discussion Paper*,

There are two points which are obvious: first, the question of standing to enforce a public duty should not involve a different enquiry in relation to each remedy, but a common test should be applied; and secondly, a test based on private rights is insufficient for effective enforcement of public duties by individuals.

There is of course some doubt about what might be meant by a private right in this context. If standing rules are to be formulated so as to protect persons with an existing or prospective action in tort, then they are, in the context of public duties, extremely restrictive, especially when one considers that the action for breach of statutory duty has not been developed in public-law cases.[2] If one construes the notion of a private right more widely, or adds to it the notion of special damage, one is confronted with some acutely difficult questions. What rights or interests do we wish to recognize other than those already protected by the law of tort? Why should we allow only those directly affected in the relevant respects to argue that the duty has not been performed?

The case law makes it clear that there is no good reason why the standing rules should be confined in this way, and to do so would run counter to the general trend of liberalization (albeit cautious liberalization) which can be seen in recent English and Australian cases. A view of administrative law which sees it as designed merely to protect rights, unless the term is very widely construed, would seem to be outmoded and lacking in any cogent justification. There are so many statutes designed to confer benefits on the public or sections of it that such a view of administrative law would render huge areas of administrative action effectively unreviewable in the courts.

It is not in fact immediately obvious that standing rules actually perform any useful function at all.[3] Is there any good reason for supposing that there are cases where the courts should not even consider an argument that a public duty is not being performed because of the identity of the party raising the argument? A number of propositions are regularly put forward to support standing rules. Strangely enough, and regrettably, these propositions are hardly ever articulated by the judges, but tend to be assumed in the judgments and articulated only by commentators.

First, it is said that the courts should not entertain frivolous or vexatious applications. However, if such applications are applications

No. 4 (1977); New Zealand Public and Administrative Law Reform Committee, *Standing in Administrative Law* (1978); Taylor, 'Individual Standing and the Public Interest: Australian Developments' (1983), *2 Civ. JQ* 353.

[2] See below Ch. 7.2.

[3] See Cane, above n. 1.

of no merit, then they can be dealt with as such. A frivolous application can be brought by a person who has standing as well as by one who has not. The requirement of leave, where applicable, is the best way of filtering out such cases at an early stage. It must be said, however, that the very existence of frivolous applications in public-duty cases is exceedingly doubtful.

Secondly, it is argued that the ability to argue a question should be confined to those who are directly affected by the outcome of the argument because they are the most effective advocates. This is clearly contradicted in public-duty cases. Indeed it would appear that in cases discussed in this chapter where standing was not granted, or was in danger of not being granted, the merits were extremely well argued, and the public undoubtedly benefited from the ventilation or resolution of the issue.[4] It might be truer to say that the quality of the argument might benefit if it is put by someone who has only the public interest at heart and takes a great risk of having to pay costs with no tangible benefit to himself. In the case of interest or pressure groups, such as environmental groups, it may well be that the quality of the arguments and evidence brought is equal or even superior in some cases to those of the public authority itself.

Thirdly, it might be thought that standing is a useful way of controlling the justiciability of certain issues. It does seem as though many decisions on public-duty questions could be described in this way. However, standing should not be used for this purpose because issues of justiciability can and should be dealt with on their merits as a part of the substantive law, not hidden behind propositions which rest on the identity of the party applying for relief. If an issue is not justiciable, then it is not justiciable at the instance of anyone.

Fourthly, it is said that broad standing rules would force the courts to entertain excessively hypothetical questions. This proposition too is seen not to be valid in public-duty cases. The arguments raised in many cases discussed above where standing was denied were quite clearly moot, and their mootness is in any event quite distinct from the question of standing; a hypothetical question can be raised by a person who seeks to protect private rights prospectively as well as by one who comes to the court merely as a citizen to further his view of the public interest.

Fifthly, it is said that if those affected directly by administrative action see no cause to complain, then those not directly affected should not be allowed to complain. This argument seems unpromising

[4] See above Ch. 6.2(iii).

because the conclusion in no way necessarily follows from the premise. However, it requires some careful consideration.

The argument assumes first of all that there is always a distinction between those directly affected and those not directly affected. While this is true of most administrative action, it is certainly not true of most administrative action which occurs in the performance of public duties. For example, a duty to enforce the law may or may not affect some groups more than others, and may affect different groups in different ways. In many instances the public generally are affected, but no member of the public is affected more than any other. The very term 'affected' assumes that there are some interests which are worthy of protection and some which are not, so that we need to inquire what interests we wish to protect rather than assume that there is already a category of protected interests, and simply give effect to this view in the rules of standing. Assuming, however, that the basic premise is true, it still does not follow that those not directly affected should not be heard. It may be that those directly affected do not complain for reasons which do not appeal to others; or they may be unable or unwilling to risk litigation; or they may not be affected in precisely the same way as others; it is even possible that those directly affected benefit from the illegality.[5] To give some interests precedence over others in deciding who can argue that a public duty has not been performed involves a value judgment which the law should be loath to undertake; judgments of value should be confined to the merits of the case and be made after hearing the argument, not be used to deny a hearing to a person who thinks there is a grievance to be aired. When standing is denied, the court not only denies a hearing to a person or a section of the community on a matter which is likely to be of public importance, but it also allows the administration to behave in a manner which may be unlawful; the smaller the number of persons able to seek relief in the courts, the greater, in practice, is the immunity of the administration from attack when it is guilty of unlawful action or inaction.

Even more importantly, in many public-duty cases, especially those concerned with the duty to enforce the law, it may well be that if a citizen has no standing then no one has, and the dereliction of duty will go unchecked purely for this reason. It would indeed be a strange view of administrative law which allowed such a situation to obtain, and it seems clear that the courts will not allow it.[6]

[5] e.g. *Re Canadians for the Abolition of the Seal Hunt*, above, Ch. 6.2 n. 52.

[6] e.g. the *Child Poverty Action Group* case, above Ch. 6.4 n. 31.

It does not, however, follow, as Craig indicates,[7] that a citizen's action should be available in literally all cases. It may well be that there are cases where standing should not be granted to a person not directly affected by a failure to perform a public duty. This is seen most clearly in relation to purely procedural duties: one might well argue that X cannot complain that Y was not accorded a hearing or given statutory notice of a certain matter if Y himself does not complain, because the objection here is purely technical in the sense that there is no connection between X and the illegality he is seeking to establish, except perhaps an incidental one.[8] The same may be true in cases which fall outside the area of decision-making; would one wish to allow X to argue that the housing authority was in breach of its duty to provide Y with a house, when X himself derives no benefit from Y being given a house (assuming X does not in some sense represent persons in Y's position)? In this case the distinction is not one between those directly affected and those indirectly affected, but between those directly affected and those not affected at all: the public can claim to be affected by a failure to implement housing legislation generally, but cannot claim to be affected by a refusal to give Y a house. The acquiescence of the only person or persons affected must in these cases preclude others from disturbing the status quo. The considerations mentioned in the last paragraph are not decisive here because there is a party capable of enforcing the duty, but he chooses not to do so. Why should this status quo be disturbed, even for his benefit, if he is insufficiently aggrieved to complain?

Is it possible, however, to distinguish between these two kinds of case when one reverts to the analysis of public duties which has been used in this book?

A duty to determine need not be specially dealt with here because the persons affected will be the same whether the determination is challenged by mandamus or *certiorari* or any other remedy, and there are no specifically public-duty questions involved in standing to challenge administrative decisions. Where the complaint is simply of a total failure to determine, the only legitimate complainant is a party which has brought the matter up for determination. There is no evidence that cases of this latter kind have caused any standing problems.

A duty to provide may be such as to affect a small or a large group according to the circumstances. A duty to provide a service, that is,

[7] *Administrative Law* (1983), pp. 459–60.

[8] Craig uses the example of a person complaining of a failure to accord natural justice to another person.

a duty to provide something for the benefit of persons statutorily entitled, such as a house, only affects those entitled. A duty to provide a public good, that is, a duty to provide something to which no person in particular is entitled, but which is provided for the benefit of the public in general, affects all equally; an example would be a duty to provide a fire brigade.

A duty to enforce the law is similar to a duty to provide, but is properly regarded as a duty to provide a public good. It may be that particular regulatory laws are enacted for the benefit of particular groups, but at a fundamental level all regulatory laws are designed to benefit the public in general and all members of the public have an interest in such laws being enforced. Of course some people also have an interest in the law being enforced without discrimination, even though they do not benefit from the enforcement as such of the law in question. For example, I may benefit slightly if rating laws are enforced against my neighbour, but not if tax laws or gaming laws are enforced against him; the benefit or lack of it is not, however, crucial, because the interest is really a constitutional interest in equal protection being observed rather than an economic interest.

Categorization of public duties in this manner shows that there are indeed two different kinds of case: first there are cases where failure to provide a public good affects all members of the public equally, or at least not to a significantly different extent; and secondly there are cases where only one person or a group of persons is entitled to the provision of a public service. In the first kind of case, if one refuses to allow a citizen's action, there may be no one who can enforce the duty; in the second kind of case it seems right to restrict standing to those entitled. The pursuit of legality for its own sake is certainly laudable, but it seems odd to allow a challenge to a person who is in no way affected by a failure to perform a public duty when the person for whose benefit the duty exists is insufficiently motivated (as opposed to merely unable) to complain. The public-good/public-service distinction appears to provide an adequate and sensible means of distinguishing between the two kinds of case.

It is suggested that the standing formulas are capable of embracing both a citizen's action in the first kind of case and a restriction of standing to those entitled in the second.[9] This could be done, as

[9] See e.g., the remarks of Lord Wilberforce discussed above Ch. 6.4 n. 7. However, the thrust of his Lordship's dictum is to the effect that mandamus covers a wider field and should therefore perhaps be narrower in its scope; the position advocated here is that for this very reason a wider view of mandamus may be necessary. Under O. 53 procedure it is not of course necessary to advert to the type of remedy sought, but rather to the type of situation involved.

suggested by Craig[10] and Whitmore and Aronson,[11] by making standing a matter of right in the case of 'protected classes' (Craig) or 'equitable or statutory rights' (Whitmore and Aronson), and discretionary in other cases. Preferably, however, the distinction should be woven into the 'sufficient interest' or 'person aggrieved' formulas rather than depending, even only partially, on the court's discretion, which is already too overloaded for comfort.[12] The possibility of some bifurcation within the rules of standing appears to have been recognized by Lord Wilberforce in the *National Federation* case,[13] and seems to be workable in practice if the distinction suggested above is used. Of the two slightly different ways of drawing the distinction put forward by Craig, and Whitmore and Aronson, it would seem that Craig's is more appropriate for public-duty cases because it recognizes the distinction drawn above between duties to provide a public service and duties to provide a public good. The fusing of standing and merits in the manner suggested in the *National Federation* case, discussed above, would not appear to assist the enforcement of public duties; it is open to the objection that persons seeking to enforce public duties will be deterred from doing so because they do not know whether the illegality which they are to establish will be regarded as serious enough to warrant the court granting them standing.

Whatever position is adopted, the important task is to rid the law once and for all of the illogical and possibly dangerous notion that public duties can only be enforced by the Attorney-General. The office of Attorney-General does not have an illustrious history of enforcing public duties. Attorneys-General are unwilling to act against public bodies and the courts seem unwilling to review their reasons, however poor, for failing to act.[14] Individuals and pressure groups have shown that they can effectively fulfil the role of invoking the court's process to get the duty enforced, and it is impossible to argue that in the cases where they have done so anything but good has resulted from the ventilation of the issue. In complex and democratic societies the raising of issues affecting the public at large cannot be confined to governments and persons with private rights to protect. With the increase in dangers to the public arising from activities, public or private, which require control, and the extension of state

10 Above n. 7.

11 *Review of Administrative Action* (1978), p. 474

12 See above Ch. 3.4.

13 See above n. 9.

14 *Gouriet* v. *Union of Post Office Workers* [1977] 1 QB 729 (revd. [1978] AC 435 (HL)).

provision for individuals, a corresponding extension of standing appears to be both necessary and desirable.

7
Damages for Failure to Perform a Public Duty

7.1 INTRODUCTION

As was indicated in Chapter 1, there are many public duties which are created by the common law.[1] The most important type of common-law public duty is of course the duty imposed by liability in tort. Although tortious duties are regarded as duties in private law, increasingly public authorities find that their purely public functions such as maintenance of prisons or inspection of buildings can give rise to liability in tort. The effect of decisions imposing liability in situations of this kind is to impose a duty of care in providing or enforcing the law, thus creating an affirmative public obligation which can be enforced by an action for damages. Whether or not one defines tort law as a system which creates duties, there is no doubt that the practical effect of most torts is to do precisely that.

In the present inquiry it is not possible to include a thorough survey of all the public duties which can be created in this way, which would in itself require a lengthy treatise of the kind undertaken by Aronson and Whitmore in a recent book.[2]

The course adopted here will be to pursue those areas which are of greatest concern to the law of public duties, namely breach of statutory duty and negligence. Breach of a statutory public duty has not been treated at any great length in standard texts on tort or administrative law;[3] negligence of course has been so treated,[4] but the law is in a particularly interesting state of flux, and also the significance of the distinction between statutory powers and statutory duties is one of the crucial questions in this area. Other areas of relevance are the law of nuisance, the emergent tort of misfeasance in public office, and the rule in *Rylands* v. *Fletcher*.[5] The position with regard to nuisance is somewhat similar to the position in negligence,[6]

[1] See above Ch. 1.2.

[2] *Public Torts and Contracts* (1983).

[3] For the literature, see below Ch. 7.2 nn. 2–3.

[4] For the literature, see below Ch. 7.3 n. 6.

[5] [1861–73] All ER Rep. 1.

[6] See Craig, *Administrative Law* (1983), pp. 527–31; Wade, *Administrative Law*, 5th edn. (1983), pp. 650–5.

and the question of liability in nuisance for failure to perform a statutory duty is dealt with below in relation to the nonfeasance rule in breach of statutory duty. Although misfeasance in public office can, in spite of its name, include nonfeasance in public office, cases of such a kind are rare.[7] Liability of public authorities in *Rylands* v. *Fletcher* is also quite rare.[8]

7.2 BREACH OF A STATUTORY PUBLIC DUTY

(i) *Introduction*

An action for damages not only serves to compensate loss, but serves to deter the defendant and those similarly situated from the kinds of act or omission causing the loss. Thus in deciding whether a public authority is liable in damages for failure to perform a public duty the courts are in effect reviewing the performance of the duty and regulating the actions of the authority. At present the circumstances in which a public authority can be held liable for failure to perform a public duty are somewhat limited, but the likelihood is that instances of such liability will increase; it is therefore important to decide first, when it is imposed, and second, when it ought to be imposed.

It is strange that while the liability of public authorities in negligence in relation to the exercise of statutory powers has been developed to a considerable extent over the last ten years, the tort of breach of statutory duty is still to a large extent ignored by the courts and by academic writers in relation to the problem of public liability. For example, Harlow in *Compensation and Government Torts*[1] describes breach of statutory duty as a missing tort, but devotes only two pages to it and makes no suggestions as to how it might be developed as a form of public liability.

Unfortunately, in considering the tort of breach of statutory duty, we are straightaway plunged into an area of the law which is

[7] Except for cases of refusal of a licence, but here, as in other cases in this tort, the act must be done maliciously; see *David* v. *Abdul Cader* [1963] 1 WLR 835 (PC); *Campbell* v. *Ramsay* (1968) 70 SR (NSW) 327 (refusal of licence); *Dunlop* v. *Woolahra Municipal Council* [1982] AC 173 (PC) (malice required); *Bourgoin SA* v. *Minister of Agriculture* [1985] 3 All ER 585; Craig, above n. 6, pp. 548–50; Aronson and Whitmore, above n. 2, pp. 120–31.

[8] See Craig, above n. 6, pp. 531–3.

[1] (1982) (otherwise a stimulating discussion). Aronson and Whitmore's *Public Torts and Contracts* (1983) is similarly shy of this question.

notoriously unclear and controversial.[2] It will be impossible in this discussion to separate completely a number of problems of general significance in this tort from those which affect public duties in particular, but the discussion will be structured so far as possible to throw light on public duties, and indicate how the general principles have been applied to them.

The main problem in the tort of breach of statutory duty is that of defining the criteria which decide when the breach of a statutory provision is actionable. Apart from this problem of actionability, the other potential problems of what constitutes a breach of duty, what kind of damage is actionable, and the rules of remoteness, have not posed too much difficulty,[3] and will not be dealt with exhaustively here.

(ii) *The 'broad' approach and the 'narrow' approach*

It has never been the case that a breach of any statutory provision gave rise automatically to liability in damages. None the less the courts seem originally to have favoured an approach which allowed a plaintiff who had suffered special damage from a breach of a statutory duty to recover.[4] This notion has been evident even in public-duty cases. In *Ferguson* v. *Kinnoul* (*Earl*) it was said:

[2] For discussion of breach of statutory duty generally, see Fricke, 'The Juridical Nature of the Action upon the Statute' (1960), 76 *LQR* 240; Williams, 'The Effect of Penal Legislation in the Law of Tort' (1960), 23 *MLR* 233; Buckley, 'Liability in Tort for Breach of Statutory Duty' (1984), 100 *LQR* 204; Rogers, *Winfield and Jolowicz on Tort*, 12th edn. (1984), ch. 7; Trindade and Cane, *The Law of Torts in Australia* (1985), ch. 22; Stanton, *Breach of Statutory Duty in Tort* (1986).

[3] For breach of statutory public duties, see above Ch. 2; Stanton, above n. 2, pp. 73 et seq, 148 et seq. Buckley, above n. 2; Phegan, 'Breach of Statutory Duty as a Remedy Against Public Authorities' [1974-6], *UQLJ* 158; Finn, 'A Road not Taken: the Boyce Plaintiff and Lord Cairns' Act' (1983), 57 *ALJ* 493, 571; Robinson, *Public Authorities and Legal Liability* (1925), ch. 4; Garner and Jones, *Administrative Law*, 6th edn. (1985), pp. 224-6; Hotop, *Principles of Australian Administrative Law*, 6th edn. (1985), pp. 465-8; Wade, *Administrative Law*, 5th edn. (1983), pp. 665-8; De Smith, *Judicial Review of Administrative Action*, 4th edn. by Evans (1980), pp. 530-6; and for remoteness and other aspects of breach of statutory duty not covered here, see the literature cited above n. 2.

[4] *Schinotti* v. *Bumstead* (1796) 101 ER 750; *Cullen* v. *Morris* (1819) 117 ER 741; *Lyme Regis* (*Mayor*) v. *Henley* (1834) 5 ER 1097 (HL) ('peculiar damage beyond the rest of the King's subjects'); *Barry* v. *Arnaud* (1839) 113 ER 245; *Ferguson* v. *Kinnoul* (*Earl*) (1842) 8 ER 412 (HL); *Couch* v. *Steel* (1854) 118 ER 1193; *Pickering* v. *James* (1873) LR 8 CP 489. *Fulton* v. *Norton* [1908] AC 451 can also be included, though it is not clear on what basis the Privy Council was prepared to find liability in damages for failure by a Provincial Secretary to submit the plaintiff's Petition of Right to the Lieutenant-Governor. For a full discussion of the early public duty/special damage rule, see Finn, above n. 3.

when a person has an important public duty to perform, he is bound to perform that duty; and if he neglects or refuses so to do, and an individual in consequence sustains injury, that lays the foundation for an action to recover damages by way of compensation for the injury that he has sustained.[5]

This very broad statement was qualified only by the further proposition that no action would lie for an erroneous exercise of discretion.[6]

However, since about the 1870s the most marked tendency has been to confine quite narrowly the area of potential liability. This has been done in a number of ways, notably by the use of the concept of legislative intention as an umbrella principle which allowed the use of various other arguments to show that an action for damages was not available. While the concept of legislative intention seems unexceptionable as a starting-point, in cases of the kind in question any real indication of legislative intention is *ex hypothesi* missing, and therefore reference to it serves only to obscure the underlying policy reasons for the court's decision. For this reason legislative intention has been rejected by commentators[7] and been either the object of mere lip-service or else a cloak for distinct policy decisions in the courts.[8]

The policy factors which have been decisive have not been pursued with any consistency, and have not, except in certain areas,[9] been analysed with any candour by the judges. Lord Denning MR felt impelled to say in *Ex p. Island Records Ltd.*: 'The dividing line between the pro-cases and the contra-cases is so blurred and so ill-defined that you might as well toss a coin to decide it. I decline to indulge in such a game of chance.'[10]

Even though the performance of the common law in this area is far from impressive, it must be conceded that there is no consensus as to what criteria would be appropriate, and legislatures have not tried to impose any general solutions in the form of statutory presumptions or principles for the courts to follow; this kind of solution is probably in any event of little use because the real policy decision would either be hampered by too much precision or in effect left to the courts anyway.[11]

[5] Above n. 4, p. 423; see also *Ashby* v. *White* (1703) 92 ER 126, 136, per Holt CJ; *Rowning* v. *Goodchild* (773) 98 ER 425. Some of the cases cited here and above n. 4 required the duty to be a 'ministerial' one.

[6] (1842) 8 ER 427.

[7] See esp. Buckley, above n. 2.

[8] See e.g. the cases cited below n. 26.

[9] See the discussion of nonfeasance cases below.

[10] [1978] 1 Chanc. 122, 135.

[11] See Buckley, above n. 2, pp. 230–2; *Law Commission Report* No. 21 (1969), para. 38.

The practical result of the cases is that as a general rule statutes which are intended to promote health or safety, with the exception of regulations for the construction or use of motor vehicles, are actionable, but other statutes are not.[12] However, the courts have never in fact decided this, and there are examples which show that liability outside the area of safety legislation is perfectly possible.[13]

The seminal authority for the narrow approach is *Atkinson* v. *Newcastle and Gateshead Waterworks Co.*,[14] in which a statutory undertaker was held not liable for breach of a statutory provision which required it to maintain the water pressure in a fire-plug at a specified level, when the breach resulted in failure by firefighters to prevent damage to property. The statute provided criminal penalties for this failure, but was held to be in the nature of a private legislative bargain between the legislature and the undertaker which did not envisage the undertaker being burdened with large claims in damages. Although this case is quite specific as to the policy factors involved, its emphasis on legislative intention has been used as the basis of a general and restrictive approach to liability for breach of statutory duty.[15]

This restrictive approach involves two different arguments which occur either separately or in combination: first, the argument that the provision of another remedy forecloses any legislative intention to create an action for damages; and secondly, the argument that there can only be an action for damages when the statute is intended to benefit or protect a particular class of persons rather than the public in general. Each of these arguments has a long history (the second, however, was not evident until after *Atkinson*) and is clearly very relevant to the position of public authorities. In addition there is a third argument which has been used in relation to public authorities only, namely that they can be held liable for misfeasance but not nonfeasance. In order to assess the position it will be necessary to look at these three approaches in turn.

[12] See the illuminating discussion in Trindade and Cane, above n. 2.

[13] See discussion below; the best example is perhaps *Pickering* v. *James*, above n. 4 (damages awarded to defeated election candidate when officer failed to provide ballot papers); see also *Morton* v. *Eltham Borough*, below n. 31; *Ministry of Housing* v. *Sharp* [1970] 2 QB 223; cases cited below n. 22.

[14] (1877) 2 Ex. D. 441; cf. *Johnston* v. *Consumers' Gas Company of Toronto Ltd.* [1898] AC 447 (PC).

[15] The most notable examples are *Phillips* v. *Britannia Hygienic Laundry Co.* [1923] 2 KB 365; *Cutler* v. *Wandsworth Stadium Ltd.* [1949] AC 398 (HL); *McCall* v. *Abelesz* [1976] QB 585.

(iii) *The 'alternative remedy' approach*

The first argument, the 'alternative remedy' approach, was first set out by Lord Tenterden CJ in *Doe* v. *Bridges*:

where an Act creates an obligation, and enforces the performance in a specified manner, we take it to be a general rule that performance cannot be enforced in any other manner. If an obligation is created, but no mode of enforcing its performance is ordained, the common law may, in general, find a mode suited to the particular nature of the case.[16]

We have seen that the same argument, when used to exclude the availability of mandamus, is quite unconvincing and seems to be in the process of decline.[17] Does it stand up to scrutiny when applied to the action for damages for breach of statutory duty?

Lord Tenterden's dictum involves two propositions: (1) that the existence of a remedy in the statute excludes the availability of any other remedy, and (2) that if no remedy is prescribed the courts can imply the availability of a suitable remedy (i.e. for our purposes, mandamus or an action for damages).

The first proposition is now quite untenable, at least as a universal principle. It has been very regular practice for more than a century, in spite of *Atkinson*, for the courts to imply the existence of an action for damages even where the statute provides a penalty, for example criminal sanctions;[18] the existence of administrative remedies, it is true, has sometimes been held to exclude an action for damages, but it is far from clear that the reason for this is that the statutory remedies are sufficient.[19]

The second proposition is also far from clear. In a sense it is uncontroversial to say that the courts can imply a remedy *ex silentio*, and of course in the case of mandamus the implication is so strong that it does not generally require argument. However, we are very far from a position in which there is any kind of presumption that an action for damages will be implied, and to that extent the dictum is most unhelpful; the policy questions involved are much too intricate

[16] (1831) 109 ER 1001, 1006; the three rules formulated by Willes J. in *Wolverhampton New Waterworks Co.* v. *Hawkesford* (1859) 141 ER 486, 495 reflect a similar approach.

[17] See above Ch. 3.4(ii).

[18] The crucial case here was *Groves* v. *Wimborne* (*Lord*) [1898] 2 QB 402; cf. *Solomons* v. *Gertzenstein* (*R.*) *Ltd.* [1954] 2 QB 243.

[19] For case law see below nn. 22–4; above Ch. 3.4(ii). 'Sufficient' is perhaps an ambiguous term: sufficient for the plaintiff, or sufficient to secure performance of the duty generally? Since administrative remedies do not involve compensation they can never be sufficient except where the action is designed in effect to obtain review of the duty. For a contrary view, see Stanton, above n. 2, pp. 21, 73 et seq.

to be decided by resort to such crude principles. Neither is it helpful to argue, as does a recent commentator, that the absence of a remedy in an administrative statute generally means that no action for damages is available;[20] such an argument might well result in many public duties being mere pious aspirations, a position which, it is argued in this book, the courts have no desire to adopt, and should not adopt. Justice requires the development of a system of public liability, not a system of widespread public immunity.

In public-duty cases the most important issue is whether the existence of administrative default powers excludes the availability of an action for damages as it has sometimes been held to exclude mandamus. The principle in *Pasmore*'s case, discussed above[21] is the equivalent in public-duty cases of Lord Tenterden's dictum, and in cases arising under education legislation the English courts have applied the *Pasmore* principle in excluding an action for damages based on an allegation of failure to provide education because of the availability of default powers under the Education Act 1944, section 99.[22] There is, however, some authority for the view that an action would lie at the suit of an individual who can show that he is 'particularly damnified' by a failure to provide education,[23] and section 99 has been held to be an inadequate remedy in cases of personal injury arising from failure to comply with building regulations relating to place of education.[24] It may or may not be that there are good reasons for allowing an action for personal injuries but not for failure to provide education, but the 'alternative remedy' approach does not help to explain those reasons or predict the outcome of future cases. The very important issues which are at stake in cases such as those mentioned cannot be resolved simply by looking at whether

[20] Buckley, above n. 2, p. 217. Cases of this kind are actually very rare, but see *Gnapp (Eric) Ltd.* v. *Petroleum Board* [1949] 1 All ER 980; *Wool Shipping and Scouring Co. Ltd.* v. *Central Wool Committee* (1920) 28 CLR 51 (HCA) (both of these cases concerned wartime regulations); cf. *Booth & Co. (International) Ltd.* v. *National Enterprise Board* [1978] 3 All ER 624.

[21] Ch. 3.4(ii).

[22] *Watt* v. *Kesteven County Council* [1955] 1 QB 408; *Chapman* v. *Essex County Council* (1957) 55 LGR 28; *Wood* v. *Ealing London Borough Council* [1966] 3 All ER 514; *Cumings* v. *Birkenhead Corporation* [1972] Chanc. 12; and cf. *Southwark London Borough Council* v. *Williams* [1971] Chanc. 734; *Wyatt* v. *Hillingdon London Borough Council* (1978) 76 LGR 727.

[23] *Gateshead Union Guardians* v. *Durham County Council* [1918] 1 Chanc. 146, 167 per Scrutton LJ; *Meade* v. *Haringey London Borough Council* [1979] 1 WLR 637, 647; *Watt* v. *Kesteven County Council*, above n. 22; see, however, *Wood* v. *Ealing London Borough Council*, ibid. And see above Ch. 6.3.

[24] *Reffell* v. *Surrey County Council* [1964] 1 WLR 358; see also *Ching* v. *Surrey County Council* [1910] 1 KB 736; *Morris* v. *Carnarvon County Council* (1910) 79 LJKB 670.

there is an alternative remedy; the real question is whether policy requires that an action should or should not lie, and to this the existence or absence of an alternative remedy is essentially irrelevant.

(iv) *The 'class of persons' approach*

The second strand of argument which runs through the cases is based on the 'class of persons' approach. By this approach the case is decided according to whether the plaintiff is a member of the class of persons intended to be protected or benefited by the statute. This approach was used as a means of creating in *Groves* v. *Wimborne* (*Lord*)[25] the very important liability for breach of statutory duty under industrial legislation. However, it is a corollary of this approach that if the statute is intended for the welfare of the general public there is no relevant class of persons. Most notably in *Phillips* v. *Britannia Hygienic Laundry Ltd.*[26] This corollary was used, in the teeth of a strong dissenting judgment from Atkin LJ, to prevent the circumvention of the fault principle in road-accident cases.

These principles have been regularly employed in public-duty cases. There are, however, a number of difficulties inherent in this approach.

First, is it clear that all statutes can be analysed in this rather simplistic way? The courts have been led down this path into some strange reasoning. A good example is *Read* v. *Croydon Corporation.*[27] The first plaintiff, a ratepayer, suffered economic loss when the second plaintiff, his daughter, contracted typhoid as a result of the authority's breach of its duty 'to provide a supply of pure and wholesome water sufficient for the domestic use of all the inhabitants of the town'. The daughter recovered in negligence, and the father in breach of statutory duty, but it was held that the daughter had no action for breach of statutory duty because unlike the father she was not one of the class of persons, the ratepayers, who were entitled to demand performance of the duty, in spite of the reference to 'inhabitants' in the statute. There are statutes, such as the one considered in *Read*'s case, which

[25] [1898] 2 QB 402.

[26] [1923] 2 KB 832; and see *Badham* v. *Lambs Ltd.* [1945] 2 All ER 295; *Commerford* v. *Board of School Commissioners of Halifax* [1950] 2 DLR 207; *Tan Chye Choo* v. *Chong Kew Moi* [1970] 1 WLR 147 (PC).

[27] [1938] 4 All ER 631; cf. *Sephton* v. *Lancashire River Board* [1962] 1 WLR 623; *Owen* v. *Kojonup Shire Council* [1965] WAR 3. For the possibility of illogically restrictive decisions, see e.g. *Evenden* v. *Manning Shire Council* (1929) 30 SR (NSW) 52 (widow of a drowned man unable to recover on the basis of failure to observe a river safety regulation because the duty was owed to the public generally); contrast, however, *Dawson & Co.* v. *Bingley Urban District Council* [1911] 2 KB 149, and *Maceachern* v. *Pukekohe Borough* [1965] NZLR 330.

can be said to protect or benefit the public and particular classes of persons in different ways.[28]

Secondly, one may ask why a person who has suffered damage should be deprived of a remedy merely because he is a member of the public, especially when he is the only member of the public who has suffered damage, or, for that matter, why a person who is fortunate enough to enjoy membership of a class of persons should automatically get a remedy. Should Read be able to sue in breach of statutory duty for economic loss if the whole town receives no water, merely because he is a ratepayer? Should his daughter be unable to sue in breach of statutory duty for contracting typhoid merely because she is not a ratepayer?

Thirdly, how does one define a class of persons as opposed to the general public? Are public-utility consumers a class of persons or are they the general public, and should their rights of action depend on this distinction?

The 'class of persons' approach is particularly crucial in public-duty cases, because public duties tend to fall into the category of duties which are intended to benefit the general public rather than a class of persons. It is not surprising that most of the cases have involved public utilities.

In *Clegg, Parkinson, & Co.* v. *Earby Gas Co.* an action against a public utility for damages for failure to supply gas was refused because the duty was 'an obligation created by statute to do something for the benefit of the public generally or of such a large body of persons that they can only be dealt with practically, en masse'.[29]

This kind of reasoning has been applied in number of cases,[30] but if applied to all public-duty cases would make the possibility of an action for damages for failure to perform a statutory duty virtually an academic one. Fortunately it has not been applied in such a way as consistently to deny a remedy, even in cases involving public utilities. In the New Zealand case of *Morton* v. *Eltham Borough*[31] the authority refused to comply with its duty to provide a gas supply because it was proving uneconomic to do so; it was held that an action would lie against the authority, and *Clegg* was distinguished on the ground that the statute did not, as in *Clegg*, provide a penalty for failure to comply. *Read*'s case also shows how a large class of

[28] See *Solomons* v. *Gertzenstein (R.) Ltd.*, above n. 18, p. 264, per Romer LJ.

[29] [1896] 1 QB 592, 594–5, per Wills J.

[30] See *Stevens* v. *Aldershot Gas, Water, and District Lighting Co.* (1932) 102 LJKB 12, and the cases cited above n. 15.

[31] [1961] NZLR 1.

persons can still be a class, even though one might say that it can only be dealt with 'practically en masse'.

Buckley observes[32] that the 'class of persons' approach serves two objectives: (1) to reinforce the remoteness principle established in *Gorris* v. *Scott*,[33] and (2) to protect statutory undertakers from wide liability. However, it is not clear that either of these points is very persuasive.

The rule in *Gorris* v. *Scott* denies an action in cases where the statute is in principle actionable but the harm which occurs is one other than that envisaged by the statute; in *Gorris* itself, for example, the statute provided that animals on board ship should be kept in pens, and the provision existed for sanitary reasons; it was held that the plaintiff had no action when his animals were washed overboard. The 'class of persons' approach overlaps with this rule to some extent, and imposes a further restriction on liability rather in the way that the duty concept in negligence overlaps with the remoteness rules based on foreseeability. None the less the argument rather assumes that the *Gorris* rule is desirable, which may be open to question, and Buckley seeks to justify the approach further by saying that it is consistent with the foreseeability principle; this last point is untenable because there are cases (*Gorris* is one) where the harm is foreseeable but not within the kind of harm envisaged by the statute; a better way of securing even development of tort law would be to apply the foreseeability principle instead of the *Gorris* rule to breach of statutory-duty cases.

The more important point about public liability is fundamental to the problem discussed in this chapter, and is dealt with below. If, however, immunity is sought, the 'class of persons' approach is not necessarily an effective or just way of achieving it. The basic difficulty with this approach is that it is at once too broad and too narrow: we may wish to compensate a person who has suffered special damage even though he has no particular status or entitlement under the statute, but is simply a member of the public; on the other hand we may wish to avoid unlimited liability to a large class of persons who have all suffered some damage, particularly if that damage is economic. If the notion of a class of persons has any role to play in public-duty cases—and it seems unlikely for the reasons indicated above to be a suitable candidate if a global test is sought—it will clearly have to applied thoughtfully.

[32] Above n. 2, pp. 210–14.

[33] (1874) LR 9 Ex. 125; cf. *Knapp* v. *Railway Executive* [1949] 2 All ER 508 (railway crossing regulations held not to be intended to protect a train driver—a case surely wrong on its facts).

(v) *The 'nonfeasance' approach*

At the root of the question of liability in damages for failure to perform a public duty is an argument that can best be described as the 'nonfeasance' argument. This argument regards public authorities performing public duties as equivalent to rescuers, inasmuch as a public duty is imposed on an authority for the purpose of ameliorating conditions which, but for its intervention, would be dangerous to persons or property. At common law there is no liability for a pure omission, in the sense that in the absence of a pre-existing duty relationship with the plaintiff, a defendant cannot be held liable for failing to confer what amounts to a benefit on the plaintiff; thus he is under no obligation to attempt to improve a dangerous circumstance, nor, having decided to make the attempt, to do everything reasonably practicable to remove the danger; his only duty is not to worsen the plaintiff's situation. According to the 'nonfeasance' approach the position of a public authority is similar to that of a rescuer, so that it cannot be held liable for 'nonfeasance', but only for 'misfeasance' or 'malfeasance'.[34]

The most common application of the nonfeasance principle has been in cases based on an allegation of non-repair of a highway, and in these cases it is still extant except where it has been abrogated by statute.[35] The history of the rule is interesting and sheds some light on the policy factors which affect liability for failure to perform public duties.

The rule was first articulated in 1788 in *Russell* v. *Men of Devon*, in which Lord Kenyon CJ held that the ratepayers could not be held liable for nonfeasance because it would be 'productive of an infinity of actions'.[36]

In fact even with a very consistent application of the rule over some 200 years there has been, if not an infinity, at least a very large number of cases involving liability of highway authorities; on the whole the courts have been inclined to call whatever fault the highway authority is guilty of, nonfeasance rather than misfeasance, and the result has been a very considerable degree of immunity in spite of the persistence of attempts to establish liability. The most crucial decision

[34] See *Winfield*, above n. 2, pp. 80–7; Trindade and Cane, ibid., pp. 305–7; Fleming, *The Law of Torts*, 6th edn. (1983), pp. 402–6. For discussion of the nonfeasance rule in public law, see Sawer, 'Nonfeasance Revisited' (1955), 18 *MLR* 541, and id., 'Nonfeasance under Fire' (1966), 2 *NZULR* 115.

[35] See below n. 62.

[36] (1788) 100 ER 359, 362; see, however, *Hartnall* v. *Ryde Commissioners* (1863) 112 ER 494.

is that in *Cowley* v. *Newmarket Local Board*,[37] in which the House of Lords refused to apply the broad 'special damage' approach to a breach of a statutory duty to keep a highway in repair. However, since the early part of this century there has been a tendency to restrict the operation of the rule in various ways, and this, as is noted by Sawer,[38] has prevented the nonfeasance rule from being developed into a rule of general application. These restrictions are first, that the rule will not apply where the danger arises from the construction of the highway[39] or the exercise of powers or duties other than those relating to repair and maintenance of highways, for example those derived from legislation concerning traffic regulation or drainage works;[40] and secondly, that the rule will not apply to artificial structures placed in the highway.[41] At the same time the courts have resisted attempts to make highway authorities liable as occupiers for failing to deal with dangers on the highway placed there by a stranger or a predecessor highway authority.[42] They have also been unwilling to impose liability on successor authorities for the misfeasance of their predecessors.[43]

The policy reasons behind the nonfeasance rule are not hard to divine. As can be seen from *Russell*'s case, it has always been a priority to protect highway authorities from vexatious litigation; another factor

[37] [1892] AC 345. See also *Municipality of Pictou* v. *Geldert* [1893] AC 524 (PC); *Sanitary Commissioners of Gibraltar* v. *Orfila* (1890) 15 App. Cas. 400 (PC); *Saunders* v. *Holborn District Board of Works* [1895] 1 QB 64.

[38] Above n. 34 (1955), p. 555. In *Dawson & Co.* v. *Bingley Urban District Council*, above n. 27, the English Court of Appeal seems to have thought that the nonfeasance rule could be applied to a duty to provide accurately marked fire-plugs; misfeasance was found on the facts, but why was the test applied at all? Fortunately this is a rare instance of the nonfeasance rule straying outside its territory; even in the case of a bridge the rule has been rejected: *Guilfoyle* v. *Port of London Authority* [1932] 1 KB 336.

[39] *Woolahra Municipal Council* v. *Moody* (1913) 16 CLR 353 (HCA); *Willoughby Municipal Council* v. *Halstead* (1916) 22 CLR 352 (HCA). The same applies to drains: *Benalla Shire Council* v. *Cherry* (1911) 12 CLR 642 (HCA).

[40] See *Skilton* v. *Epsom Urban District Council* [1937] 1 KB 112; *Polkinghorn* v. *Lambeth Borough Council* [1938] 1 All ER 339.

[41] *Bathurst Borough* v. *Macpherson* (1879) App. Cas. 256 (PC); *Donaldson* v. *Sydney Municipal Council* (1924) 24 SR (NSW) 408; *South Australian Railways Commissioner* v. *Barnes* (1927) 40 CLR 179 (HCA); *Buckle* v. *Bayswater Road Board* (1937) 57 CLR 259 (HCA); *Skilton*, above n. 40; *Connolly* v. *Minister of Transport* (1965) 63 LGR 372; *Webb* v. *South Australia* (1982) 56 ALJR 912 (HCA).

[42] See *Gorringe* v. *Transport Commission* (1950) 80 CLR 357 (HCA); *Bretherton* v. *Hornsby Shire Council* [1963] SR (NSW) 334.

[43] *Maguire* v. *Liverpool Corporation* [1905] 1 KB 767; *Nash* v. *Rochford Rural District Council* [1917] 1 KB 384; *Baxter* v. *Stockton-on-Tees Corporation* [1959] 1 QB 441; *Florence* v. *Marrickville Municipal Council* [1960] SR (NSW) 562.

which has probably become even more important is that public funds designated for the development of the highway system should not be eaten away by large claims for damages. This was perhaps especially important during the second half of the nineteenth century;[44] it must also be remembered that in England the imposition of liability would have meant a heavy burden on the ratepayers, who at some periods were few in number, and may even have necessitated a special rate in some instances, with the possibility of imposing liability on persons who could not be regarded as even indirectly responsible for the damage. Policy considerations of this kind do not of course necessarily have the same relevance in jurisdictions such as Australia where the history of highway authorities has been quite different. None the less the nonfeasance rule has, at least in its essentials, survived in these jurisdictions, probably because of what could be seen as the unnecessary inhibition which liability for nonfeasance would impose on development.[45] One might have expected that, whatever the utility of the policy considerations, they would point to complete immunity rather than liability for misfeasance, but the misfeasance/nonfeasance line provided a convenient and logically attractive method of restricting liability.

The nonfeasance rule was also applied to drainage authorities, in cases where they were sued for property damage arising from flooding, for reasons very similar to those which motivated the courts in highway cases.[46] It is no coincidence that the problematical and influential *Pasmore* case[47] was one which involved drainage duties. The tone was set by James LJ in *Glossop* v. *Heston and Isleworth Local Board*:

If the neglect to perform a public duty for the whole of the district is to

[44] See Sawer, above n. 34 (1966), p. 115.

[45] The nonfeasance rule was held to be applicable in Australia in *Sydney Municipal Council* v. *Bourke* [1895] AC 433 (PC); *Birch* v. *Australian Mutual Provident Society* (1906) 4 CLR 324 (HCA); and *Shield* v. *Huon Municipality* (1916) 21 CLR 109. The relevance of the history of the nonfeasance rule to Australia has been discussed in *Nielsen* v. *Brisbane Tramways Co. Ltd.* (1912) 14 CLR 354, 363 (HCA); *Buckle* v. *Bayswater Road Board*, above n. 41; *Hayes* v. *Brisbane City Council* (1979) 5 QL 269, 277–9. Although one might say that the history has no direct relevance outside England, it is no doubt the developmental reasons which have caused the remarkable persistence of the rule. In Canada the Supreme Court argued convincingly as early as 1911 that the rule did not apply (*Vancouver City* v. *McPhalen* (1911) 45 SCR 194); however, it has been the legislatures, not the judges, who have been keener to abolish the rule (see below n. 62).

[46] With the addition perhaps of the further reason that the Minister had supervisory powers under the Public Health Act 1875 (England and Wales); see above Ch. 3.4(ii).

[47] *Pasmore* v. *Oswaldtwistle Urban District Council* [1898] AC 387 (HL).

enable anybody and everybody to bring a distinct action . . . because he has not had the advantages he otherwise would be entitled to have if the Act had been properly put into execution, it appears to me the country would be buying its immunity from nuisances at a very dear rate indeed by the substitution of a far more formidable nuisance in litigation and expense that would be occasioned by opening such a door to litigious persons, or to persons who might be anxious to make profit and costs out of this Act. . . . It seems to me that if this action could be sustained, it would be a very serious matter indeed for every ratepayer in England . . . [and] I do not see why [such an action] could not in a similar manner be maintained by every owner of land . . . who could allege that if there had been a proper system of sewage, his property would be very much improved . . . and why he would not be entitled to bring it on the very day the [local drainage] board was constituted and the duties cast upon it.[48]

It may be that in 1879 policy required decisions of this kind,[49] especially as the duties referred to by James LJ were often new ones imposed by the Public Health Act 1875 (England and Wales). However, the potential liability involved is greatly exaggerated, because James LJ seems to have forgotten that non-repair or failure to act does not in itself mean that the authority is liable in negligence or nuisance. The nonfeasance must be negligent according to the usual principles, or else the authority must be liable for failing to rectify a nuisance which arises on its land. It is inconceivable that liability could be imposed in the kind of situation he envisaged.[50]

The drainage cases are considerably complicated by factors not present in the highway cases: the statutes in question are usually ones creating new duties, and often the authority is an occupier of land on which the nuisance arises. In addition, questions of liability can arise out of the overloading of an adequate system due to development for which the authority may or may not be responsible; while the same might be said to be true of highways, there is not the same element of reliance on the system by the party suffering the loss as there is in highway cases.

[48] (1879) 12 Chanc. D. 102, 109, 114–15.

[49] See *Hammond* v. *St Pancras Vestry* (1874) 43 LJCP 157; *Attorney-General* v. *Dorking Union Guardians* (1881) 20 Chanc. D. 595; *Robinson* v. *Workington Corporation* [1897] 1 QB 619. In both these latter cases the alternative remedy rule in *Pasmore*, above n. 47, was also relied on. A similar view seems to have prevailed in Australia: see *Essendon Corporation* v. *McSweeney* (1914) 17 CLR 524 (HCA); *Madell* v. *Metropolitan Water Board* (1935) 36 SR (NSW) 68. In *Hesketh* v. *Birmingham Corporation* [1924] 1 KB 260 the rule was also applied to statutory powers.

[50] This is so whether the case is based on negligent nonfeasance (because duties of this kind are not absolute: see above Ch. 2.3), or nuisance (where liability for nonfeasance is based on negligence: *Sedleigh-Denfield* v. *O'Callaghan* [1940] 3 All ER 349 (HL)).

Whatever the differences, the nonfeasance rule flourished during the period of development by the drainage boards in England,[51] but has never been so entrenched as the highway nonfeasance rule, and *Glossop* and similar cases have tended to be distinguished on various grounds.[52] In the most recent cases the courts seem to be disowning the nonfeasance rule entirely.[53] This has not resulted in a flood of litigation, but whether it is the law or sanitary engineering which has prevented the flood is perhaps open to doubt.

It is pertinent at this point to ask what is meant by nonfeasance and misfeasance. It is obvious that the case of a danger of which the authority was unaware and which it therefore made no attempt to remove is one of nonfeasance,[54] and it is equally obvious that if in carrying out its duties the authority actively creates a danger the case is one of misfeasance.[55] Between these two cases there are other kinds of case which are not obviously to be described by either term, cases which have been aptly called cases of 'inadequate amelioration', i.e. cases in which the authority has made some attempt to rectify matters, but in such a way as to leave the situation still dangerous.[56]

The application of the nonfeasance rule to cases in this 'no man's land' between nonfeasance and misfeasance has resulted in some

[51] This coincides with the period covered by the Court of Appeal decisions cited above nn. 48–9.

[52] In *Jones* v. *Llanrwst Urban District Council* [1911] 1 Chanc. 393, and *Haigh* v. *Dendraeth Rural District Council* [1945] 2 All ER *Glossop* was distinguished on the ground that the system was not an inherited one as it was in *Glossop*. In *Attorney-General* v. *St Ives Rural District Council* [1960] 1 QB 312 (affd. [1961] 1 QB 336) the statute was construed as being intended to benefit a class of persons; cf. *R.* v. *Marshland, Smeeth, and Fen District Commissioners* [1920] 1 KB 155. In *Sephton* v. *Lancashire River Board*, above n. 27, there was no other remedy available. In *Rippingdale Farms Ltd.* v. *Black Sluice Internal Drainage Board* [1963] 1 WLR 1347 the duty was described as a positive and continuing one.

[53] See esp. *Pride of Derby and Derbyshire Angling Association* v. *British Celanese Ltd.* [1953] Chanc. 149, 188 per Denning LJ; applied in *Smeaton* v. *Ilford Corporation* [1954] Chanc. 450.

[54] *Thompson* v. *Brighton Corporation* [1894] 1 QB 332.

[55] See *Blackmore* v. *Mile End Old Town Vestry* (1882) 9 QBD 451; *Sydney Municipal Council* v. *Bourke*, above n. 45; *McClelland* v. *Manchester Corporation* [1912] 1 KB 118; *Woolahra Municipal Council* v. *Moody*, above n. 39; *Shoreditch Corporation* v. *Bull* (1904) 90 LT 210 (HL); *Short* v. *Hammersmith Corporation* (1910) 104 LT 70; *Newsome* v. *Darton Urban District Council* [1938] 3 All ER 93; *Simon* v. *Islington Borough Council* [1943] 1 All ER 41; *Grafton City Council* v. *Riley Dodds (Australia) Ltd.* [1956] SR (NSW) 53; *McDonogh* v. *Commonwealth* (1985) 73 ALR 148.

[56] See *Campisi* v. *Water Conservation and Irrigation Commission* (1936) 36 SR (NSW) 631; *Wilson* v. *Kingston-upon-Thames Corporation* [1949] 1 All ER 679; *Burton* v. *West Suffolk County Council* [1960] 2 QB 72; *Hocking* v. *Attorney-General* [1963] NZLR 513, *Culcairn Shire Council* v. *Kirk* [1964–5] NSWR 909.

strange decisions which cast doubt on the relevance of the distinction. The leading case in this area is *Burton* v. *West Suffolk County Council*.[57] A dip in the highway was frequently flooded, and when flooded was marked with red flags during the daytime and red lights at night. On one occasion on which a flood had subsided, these warnings were removed, but water left in the dip froze, causing an accident. These facts, held the Court of Appeal, amounted to nonfeasance, and the authority were not liable. *Hocking* v. *Attorney-General*[58] also concerned a part of the highway which was susceptible to flooding; the authority twice put in a temporary culvert which was inadequate to deal with all foreseeable flooding; an unusually heavy flood occurred which was so great as to create a chasm in the road which was the cause of an accident. The New Zealand Court of Appeal held that the facts amounted to misfeasance; having decided to intervene, the authority were under a duty of care to guard against foreseeable dangers.

One might have expected that the removal of the warnings in *Burton* would be described as misfeasance, and the half-hearted repair in *Hocking* would be described as nonfeasance.[59] However, justice surely requires at the very least that if an authority is aware of a danger and does not take reasonable steps to warn road users if not remove the danger, it should be liable; the description of the situation as one of misfeasance or nonfeasance does not seem to help the attainment even of this minimal protection.[60] Liability should presumably also occur where the acts or omissions of the authority constitute a trap or inducement to the road user who reasonably relies on the authority having made the road safe; this proposition seems not, however, to be assimilated within the nonfeasance doctrine.[61]

Dissatisfaction with the nonfeasance rule as applied to highways

[57] Above n. 56. One might have difficulty in saying why it is misfeasance to fail to prevent a stranger from excavating the highway (*Hitchins* v. *Port Melbourne* (*Mayor*) (1888) 14 VLR 748), or why it is nonfeasance to leave tar on the road which melts in hot weather (*Holloway* v. *Birmingham Corporation* (1905) 69 JP 358). The duty to enforce the law against highway obstructors is not apparently considered as an aspect of the nonfeasance rule, but no right of action lies, according to *Lynch* v. *Mudgee Shire Council* (1981) 46 LGRA 204. (Is this sound, if the plaintiff suffers special damage? If so, the only remedy is mandamus (see above Ch. 2.4).)

[58] Above n. 56.

[59] The argument that since the authority was not obliged to act at all, they should not be liable for stopping short of what they could do, seems shaky after *Anns* v. *Merton London Borough Council* [1977] 2 All ER 492 (HL): see below Ch. 7.3; this does not of course mean that the case of inadequate amelioration is not a case of nonfeasance, but rather that the immunity in nonfeasance cases is unjustifiable.

[60] *Moul* v. *Croydon Corporation* (1918) 88 LJKB 505.

[61] Sawer, above n. 34 (1966), pp. 122–3; *Hocking* above n. 56.

has led to its abolition in England and in parts of Canada.[62] The scheme established under the English Act provides for a special statutory defence for the highway authority, which appears to have the effect that the burden of proof is shifted to the authority, a matter which has been the subject of adverse comment.[63]

A much simpler position than that established by the English statute is to make highway authorities liable for nonfeasance as well as misfeasance (in other words for negligence simple), and leave it to the courts to decide in each instance whether the authority was in fact negligent. This would avoid the unfortunate effect of putting the burden of proof on the authority, but would not necessarily result in unreasonable liability. In fact the available evidence suggests that the courts will take into account the resources of the authority and the possibility of inspection or prevention or warning of the danger, and there is no reason why the ordinary principles of negligence should not accommodate both the vast mileages of the Australian outback and the crowded streets of London.[64] The same is true of drainage authorities, and indeed it seems that the courts are heading for something like the negligence approach in drainage cases. Perhaps the same can be done for highways; if so the nonfeasance rule will become a relic of the past rather than a comfortable anomaly.

[62] See Highways (Miscellaneous Provisions) Act 1961 (England and Wales), s. 1 (now Highways Act) 1980 (England and Wales), s. 58(2); Sawer, above n. 34 (1955), pp. 548–9.

[63] Notably by Dworkin, (1962) 25 MLR 336, and Sawer, above n. 34 (1966), pp. 117–20.

[64] English decisions, particularly of the Court of Appeal, since 1961 do not support the fears of wide liability. In fact the simple negligence test, even hampered by the strange position with regard to the burden of proof, has proved quite successful: see *Griffiths* v. *Liverpool Corporation* [1966] 2 All ER 1015; *Meggs* v. *Liverpool Corporation* [1968] 1 All ER 1139; *Burnside* v. *Emerson* [1968] 3 All ER 741; *Bird* v. *Tower Hamlets London Borough Council* (1969) 67 LGR 682; *Whiting* v. *Hillingdon London Borough Council* (1970) 68 LGR 437; *Rider* v. *Rider* [1973] 1 All ER 294; *Haydon* v. *Kent County Council* [1978] QB 343; *Pitman* v. *Southern Electricity Board* [1978] 3 All ER 901; *Tarrant* v. *Rowlands* [1979] RTR 144. *Haydon*'s case (in which *Anns* was distinguished) in particular shows that it is far from the intention of the courts that highway authorities should act as guarantors of the safety of every way under all conditions. Sawer's comment (above n. 34 (1966), p. 129) that the 1961 Act would be 'too big a jump into liability for Australian and New Zealand authorities' discloses the same error as that of James LJ (above n. 48 and text). There have been tentative moves towards the abolition of the nonfeasance rule in Western Australia and South Australia: see Trindade and Cane, above n. 2, p. 504.

(vi) *The 'negligence* per se' *approach*

Yet another approach to breach of statutory duty cases is the 'negligence *per se*' doctrine,[65] according to which the statute merely fixes the standard of care for duties which already exist at common law. While there is something to be said for this doctrine in terms of the even development of tort law, it is either inapplicable or unhelpful in public-duty cases, for the simple reason that in these cases there is no substratum of common law to fall back on. It makes no sense to say, for example, that there can be no liability for failure to provide housing or education because the authority is under no *common-law* duty to do so. There are cases, which are discussed below,[66] when failure to perform a public duty can give rise to liability in negligence, but the manner of the development of the law in this area makes it virtually impossible for the theory of liability developed there to be used as a basis for liability in cases of the kind under discussion, and in fact the courts in England and Australia at least have shown no inclination to pursue the doctrinal path represented by the negligence *per se* approach.[67] It has also been criticized for putting breach of statutory duty into a conceptual strait-jacket.[68]

(vii) *Breach of statutory duty: a synthesis*

Having considered four somewhat unpromising, or at least inconclusive, bases for liability for failure to perform a public duty it is now opportune to look at current approaches to the problem and see what conclusions can be reached.

The crucial question can be stated in this way: granted that, as a general rule, the courts will be inclined to find liability mainly in cases where the statutory provision is intended to provide for health or safety, to what extent does the law allow for public liability where the provision is not of this kind, or where the loss suffered is economic, or at any rate not personal injury or property damage?[69]

The latest decision of the House of Lords which discusses the principles applicable in breach of statutory-duty cases, *Lonrho Ltd.* v. *Shell Petroleum Co. Ltd.* (*No. 2*), attempts to consolidate some of the arguments discussed above. Lord Diplock, in refusing an action based

[65] For discussion of this approach, see Thayer, 'Public Wrong and Private Action' (1914), 27 *HLR* 317; Williams, above n. 2; Phegan, above n. 3; Buckley, above n. 2; *R.* v. *Saskatchewan Wheat Pool* (1983) 143 DLR (3d) 9 (SCC).

[66] See below Ch. 7.3.

[67] See Buckley, above n. 2.

[68] Ibid.; and see *Monk* v. *Warbey* [1935] 1 KB 75 (HL).

[69] See below, nn. 78–80 and text.

on breaches of provisions imposing sanctions on the illegal regime in Southern Rhodesia, reiterated Lord Tenterden's dictum, but nominated the following exceptions: (1) cases where the obligation is imposed for the benefit of a particular class of individuals, and (2) cases where a public right is in question and a member of the public suffers particular, direct, and substantial damage other than and different from that which was common to the rest of the public.[70]

This attempt to reconcile the various principles seems to produce a result which is radically different from that produced by their operation singly. The exceptions are so wide as to obliterate entirely, for all practical purposes, the 'rule' contained in Lord Tenterden's dictum. This may well be desirable, because, as is noted above, this dictum bears very little relation to the present reality both of precedent and of policy. The result of adopting Lord Diplock's approach would obviously be that the possibility of an action for breach of statutory duty is greatly increased, but of course much depends on what is meant by the terms 'class of persons' and 'public right'. In relation to public duties this approach has the immediate advantage that the 'class of persons' argument can be circumvented by the 'public right/special damage' exception. Since membership of a class of persons does not require the additional factor of special damage, one might wonder whether liability is being fixed too widely.

What kind of practical outcome, we may ask, does policy require? In general it would seem that there should be no presumption that a public duty should not be actionable, and indeed the courts have never taken such a position. If one approaches the problem from the point of view of how best to enforce the duty, the result will be that an action for damages will only be necessary when there is no other remedy; it has been shown that this approach is unsatisfactory because different remedies suit different purposes. A more sensible starting-point would be to ask whether there are any reasons for refusing an action, and this is in fact the approach taken in cases involving common-law duties.[71] The only policy reasons for refusing an action in public-duty cases are: (1) the fear of unlimited liability which would prevent the authority from fulfilling the purposes of the

[70] [1982] AC 173, 185. Finn, above n. 3, discussing the *Wentworth* case, below n. 72, would prefer to develop the potential contained in Lord Cairns' Act rather than recast the action for breach of statutory duty. After *Lonrho* and *Wentworth* it seems unlikely that this course will be taken, whatever its attractions, and in England it is probably not possible. Since the action for breach of statutory duty clearly does need recasting, this course would seem in any event a preferable one.

[71] See *Home Office* v. *Dorset Yacht Co. Ltd.* [1970] 2 All ER 294 (HL); *Anns*, above n. 59.

statute or would otherwise be an unreasonably heavy burden on public funds; (2) the fear that persons who were not intended to benefit from the statute may be able to claim damages at the expense of those whom it was intended to benefit; (3) that a public authority should not be held liable in respect of a lawful exercise of discretion; and (4) that large amounts in damages may be charged on public funds for what are essentially technical breaches, liability in breach of statutory duty being strict. It does not appear that there is any particular type of statutory provision which, because of its general content, should never be the subject of liability. The statutes imposing public duties which one might think would come nearest to being held to be of this type are education statutes, but even here the balance of authority favours the 'special damage' approach.[72]

The first two points are taken care of by the 'public right/special damage' exception put forward in *Lonrho*, provided that the notion of a class of persons is kept within strict limits, and provided that the courts do not take an 'absolute duty' approach to the question of breach of duty;[73] this would account also for the fourth policy argument mentioned above. The use of the term 'public right' can presumably be taken to mean that the statute is intended to protect or benefit the public in some particular respect, and that an action for damages will only be allowed when it would further that intention; it would not, for example, further the intention of the statute in *Lonrho* itself to allow an action for damages in favour of one whose losses

[72] See above nn. 22–3. In *Becker* v. *Home Office* [1972] 2 QB 407 it was said *obiter* that a breach of prison rules, which are purely regulatory, could not result in an action, as this would undermine prison discipline. It is hard to take this proposition very seriously; if the court had found a breach of duty in that case (the prisoner claimed that the prison authorities had unlawfully withheld a cheque belonging to her), there would probably have been no effective remedy. Other breaches of prison regulations could result in personal injury; surely these at least must be actionable? In fact the courts have never held that purely regulatory or administrative statutes are not actionable; indeed in *Arthur Barnett Ltd.* v. *Dunedin Metropolitan Fire Board* [1964] NZLR 305 it was even held that the Board's duties to train firemen efficiently and to be fully informed of what was necessary to fight a fire in particular premises in accordance with prescribed standards were actionable; contrast *Bennett & Wood Ltd.* v. *Orange City Council* (1967) 67 SR (NSW) 426; and see *Wentworth* v. *Woolahra Municipal Council* (1982) 149 CLR 672 (HCA). Statutes involving economic regulation might be regarded as outside the realm of breach of statutory duty, but liability is not precluded even here: see *Booth* v. *National Enterprise Board*, above n. 20; *Garden Cottage Foods Ltd.* v. *Minister of Agriculture* [1984] AC 130 (HL); *Bourgoin SA* v. *Minister of Agriculture* [1985] 3 All ER 585.

[73] See above Ch. 2.3; *Ministry of Housing* v. *Sharp*, above n. 13, in which there is a divergence of view as to whether the duty was strict or only a duty to take care. *Haydon* (above n. 64), *Wyatt* (above n. 22), and *Meade* (above n. 23) also make this point plain.

flowed from the restriction of the activities intended to be discouraged by the statute. To take a harder case, to allow an action in favour of a person who registers a land charge, as in *Ministry of Housing* v. *Sharp*,[74] does further the intention of the statute, even though it is principally the purchaser who is intended to benefit.

The problem involved in the third point is highlighted by some recent cases arising under the Housing (Homeless Persons) Act 1977 (England and Wales). In *Thornton* v. *Kirklees Metropolitan Borough Council*[75] the plaintiff claimed damages for distress and inconvenience allegedly caused by the failure of the authority to provide him with temporary accommodation in accordance with the Act, under section 3(4) of which the authority were obliged to provide temporary accommodation for a person whom they had reason to believe was homeless, pending inquiries and a final decision as to his entitlement. The action was commenced as an ordinary action for damages and not by way of application for judicial review, but was struck out on the ground that the claim disclosed no cause of action. The Court of Appeal held that an action could lie under the statute, and that the claim should not be struck out.

The main problem involved in this case is that the public duty became executory only when the authority reached a decision that the plaintiff was entitled to provision, which is a very common situation in modern statute law. The result is that the cause of action, if any, is dependent on the purely public-law question of whether the authority were entitled in law to reach a decision that the plaintiff was not entitled; he could at best sue for damages only if it was shown that the decision was impermissible and that a decision in his favour was imperative, which, given the essentially discretionary nature of the function involved (one is reminded here of the *Ferguson* case mentioned above), rendered the plaintiff's task very difficult, though not impossible; if, for example, he were able to show that the decision was invalid because the authority had applied the wrong test of homelessness, and that if they had applied the right test they must have come to a decision in his favour, then the duty would be in effect executory, and, if the statute is capable at all of founding a civil action, he would be able to sue.

Buckley, in order to solve the problem created by cases of this kind, invokes the *Anns* v. *Merton London Borough Council*[76] distinction between

[74] Above n. 13.

[75] [1979] 3 WLR 1; see also *De Falco* v. *Crawley Borough Council* [1980] QB 460; *Lambert* v. *Ealing London Borough Council* [1982] 2 All ER 394; *Cocks* v. *Thanet District Council* [1982] 3 All ER 1135 (HL).

[76] Above n. 59.

policy and operation and comments as follows: 'if hopeless confusion is to be avoided, it is essential that the action for breach of statutory duty is kept carefully distinct from the law relating to damages for loss caused by unlawful administrative action. . . .'[77]

He goes on to argue that an action for breach of statutory duty is more likely to lie the more specific the nature of the statutory obligation.

The answer provided by Buckley does not seem quite satisfactory. The policy/operation distinction has in fact no relevance to the situation in *Thornton*, because the decision is not a discretionary one in the *Anns* sense (whether to act, how many resources to employ, and so on), but a decision as to an entitlement; once the decision is made in favour of the applicant, or else the facts are such that the authority cannot validly decide against him, then the duty must be performed to the best of the authority's ability, and can be enforced by mandamus. In this situation there seems to be no good reason why a *Thornton*-type plaintiff should not recover his losses arising from the failure to provide accommodation, if he can prove the matters referred to above. Negligence is not in fact in issue in such a case, but liability for breach of statutory duty; it is not a question of whether the authority has exercised reasonable care, but whether they have caused damage by unlawfully refusing to perform their statutory duty. If there is no liability in such a case, any attempt to construct a system of public liability must indeed be futile.

Specificity, moreover, is not likely to assist us any further. While it is true that most provisions which result in liability are quite specific, it is equally true that some are very unspecific; the explanation is that specific provisions tend often to be ones which are designed to prevent personal injury, and it is that fact and not their specificity which has determined the result. Most of the provisions which we are concerned with are decidedly unspecific, but that in itself is no good reason to deny a remedy. Is it a good answer to a child who sues for failure of the education authority to provide her with education that the statutory provision is unspecific even though clearly breached? If there is any case at all for denying liability, a much better answer must be given.

What perhaps lies at the root of arguments against liability of public authorities for breach of statutory duty is that the kind of loss involved is usually, at least in difficult cases, pure economic loss, a kind of loss which in the tort of negligence has only recently come to

[77] Above n. 2, p. 220.

be compensable, and which the courts are being careful to keep within reasonable limits.[78]

As far as breach of statutory duty is concerned, there has never been any restriction that economic loss must flow from personal injury or damage to the plaintiff's property. In *Read*[79] and *Morton*,[80] for example, the loss was purely economic; in *Ministry of Housing* v. *Sharp*[81] a local authority was held liable to the Ministry, which had the benefit of a development charge, when it breached its duty to issue an accurate certificate to a purchaser showing all charges affecting the land, with the result that the Ministry lost the benefit of its charge; although this was one of the first economic-loss cases in negligence, it was not questioned that economic loss is recoverable in breach of statutory duty.

It is necessary, however, to ensure that the liability does not become unduly wide, and in the case of economic or non-physical losses there is a danger that it will be. It is surprising to see, for example, that two members of the Court of Appeal in *Meade* v. *Haringey London Borough Council*[82] were prepared to find that an action for breach of statutory duty would lie for failure to provide education in the borough during the six weeks of a caretakers' strike. It is hard to imagine that the authority could really be held liable to each of 37,000 children who suffered in their education. Here, if anywhere, is a case of infringement of a purely public right where no individual can be said to have suffered special damage over and above the rest of the public.

It is suggested that if Lord Diplock's formulation in *Lonrho* represents in practical terms a return to where we started the discussion, namely the 'special damage' approach of the mid-nineteenth-century public-duty cases[83]—and it could be so regarded and developed—it will serve better than any of the other formulations to allow development of public liability without imposing liability to an unreasonable extent. As is mentioned above, this is only the case if

[78] See Trindade and Cane, above n. 2, pp. 296 et seq.; *Minister for Environmental Planning* v. *San Sebastian Pty. Ltd.* [1983] 2 NSWLR 221; *The Aliakmon* [1986] 2 WLR 902 (HL); *The Mineral Transporter* [1986] AC 1 (PC).

[79] Above n. 27.

[80] Above n. 31.

[81] Above n. 13. In *Gaultier* v. *South Norfolk Rural Municipality* [1972] 1 WWR 258 economic loss was recovered in respect of inability to use roads which the authority had failed to repair.

[82] Above n. 23; and for fuller discussion see Ch. 2.3.

[83] See above n. 4 and text. The 'special damage' approach is also evident in public-nuisance cases: see *Boyce* v. *Paddington Borough Council* [1903] 1 Chanc. 109 (relied on in *Lonrho* itself).

a narrow view is taken of what constitutes a 'class of persons'. There may be instances where loss is inflicted on all members of the class, but it would be unreasonable in view of the numbers involved to impose liability: it would be unfortunate if the children of Haringey in *Meade*, or for that matter the consumers of Eltham in *Morton*, were regarded as a class. The result would be a multiplicity of suits for a single breach of duty to recover economic losses which, if they had occurred in a negligence case, would clearly not be recoverable. It is interesting to note that both the *Meade* and *Morton* cases concerned actionability in the context of the availability of an injunction, so that the prospect of unlimited liability was not directly relevant, although it might conceivably have become so. However, there is no reason why this approach should not be taken in cases of the *Thornton* variety, where each member of the class of persons has to demonstrate his entitlement to be regarded as such; it is suggested that the 'class of persons' approach be confined to such instances, although they could equally and more simply be regarded as instances of the 'special damage' approach. One can define these instances as ones which fall within the notion of a duty to provide a public service as opposed to a public good, cases, that is to say, in which a member of the public must show that he is within the class of the public intended to be benefited or protected, except where the class comprises substantially the whole of the public. One does not for example have to demonstrate one's entitlement to education or an electricity supply; these are essentially goods to which all are entitled.

To summarize, the advantages of the 'special damage' approach are as follows: (1) the undesirable effects of the 'alternative remedy' approach are avoided (it would be best if the courts now expressly disavowed this approach instead of restricting it); (2) the nonfeasance cases can be subjected to the 'special damage' formula by simple legislative amendment, where necessary, to remove the hard core of the nonfeasance doctrine; (3) the 'class of persons' approach can be circumvented in cases in which it would cause injustice; (4) the way is clear to create a general form of liability for failure to perform a public duty without imposing unreasonably extensive liability.

It is not entirely surprising that the courts have adopted for about a hundred years a restrictive approach to cases of breach of statutory duty involving purely public duties, when one considers the immunities which have prevailed not only in relation to public authorities but in relation to non-physical damage generally. However, the general trend is towards opening up new areas of public liability, and the tort of breach of statutory duty is clearly of great importance as one such area. It should not be ignored merely because the courts have generally refused in the past to allow it to develop as a form of public liability. The broad approach manifest in the nineteenth-century cases provides useful guidance. It is to be hoped that the courts will return to the relative freedom which they represent.

7.3 NEGLIGENCE, STATUTORY POWERS, AND PUBLIC DUTIES

(i) *Introduction*

Since the tort of breach of statutory duty deals with liability in respect of failure to perform a statutory duty, one might think that is the end of the question of damages for failure to perform a public duty. However, as is indicated in Chapter 1, public duties exist by virtue of the common law as well as by statute, and many of these take the form of a duty of care established by the law of negligence. Because of the availability of the action for breach of statutory duty, the action for negligence tends to arise in cases concerning negligent failure to exercise statutory powers rather than statutory duties. In this sense the law of negligence tends to blur the distinction between the two, sometimes imposing a duty even where powers are being exercised. This blurring of the distinction between statutory powers and duties was adverted to in Chapter 2; there are of course situations in which statutory powers must be exercised if the purpose of a statute is to be fulfilled, and to that extent the existence of a statutory power imposes duties on the authority empowered. If the failure to exercise a power causes damage, the duty, arising as it does in negligence, is a duty of care, so that the question is one of fault; the standard imposed by the courts is that of reasonable care, not a standard of 'administrative-law reasonableness', nor a standard determined directly by reference to a statutory obligation which may be absolute in its terms. Properly, negligence should only be in issue in the case of a statutory duty where the duty is negligently performed; where the essence of the complaint is that the defendant has failed to perform the duty, the cause of action should be breach of statutory duty.

A duty of care may be a duty in public law, in the sense that special rules are applicable to the ascertainment or performance of the duty which are not applicable to private persons. Thus when a public authority is found liable in negligence, if the duty is one which is public in nature, the authority is in effect being compelled to act in a particular manner in relation to a public function. Since the possibility of an action in negligence to enforce a public duty, although not great, has been somewhat increased by recent decisions, the law of negligence has now attained some importance as a means of enforcing public duties.

A survey of the case law in this area indicates that a public duty of care, if we may coin such a phrase to describe the kind of duty under consideration here, can be imposed in a wide variety of situations which cover many of the duties considered in Chapter 2. The most common situations are those in which the authority has been found negligent in pursuance of a power or a duty to provide supervision of persons in its care or custody,[1] or in pursuance of a power or duty to inspect buildings.[2] Liability may, however, be imposed in other situations, for example where a highway authority is negligent in providing inadequate traffic signs[3] or barriers.[4] There are also a number of cases involving negligent misstatement.[5]

The crucial question of course is, to what extent does the public nature of a duty entail the application of special public-law principles which have the effect of modifying the classical doctrines of negligence in their application to public authorities?

The law has never taken the view that public duties are immune from consideration in an action for negligence, except that the Crown itself was immune from suit at common law. In nearly all jurisdictions

[1] *Home Office* v. *Dorset Yacht Co. Ltd.* [1970] AC 1004 (HL); see also *Holgate* v. *Lancashire Mental Hospitals Board* [1937] 4 All ER 19; *Thorne* v. *Western Australia* [1964] WAR 147; *Somerset County Council* v. *Kingscott* [1975] 1 All ER 326; *L.* v. *Commonwealth* (1976) 10 ALR 269; *Writtle (Vicar)* v. *Essex County Council* (1979) 77 LGR 656; *Teows* v. *MacKenzie* (1980) 109 DLR (3d) 473.

[2] *Voli* v. *Inglewood Shire Council* (1963) 110 CLR 74 (HCA); *Dutton* v. *Bognor Regis Urban District Council* [1972] 1 QB 373; *Anns* v. *Merton London Borough Council* [1978] AC 728 (HL); *Mount Albert Borough Council* v. *Johnson* [1979] 2 NZLR 234; *Kamloops City* v. *Neilsen* (1984) 10 DLR (4th) 642 (SCC); *Peabody Donation Fund* v. *Sir Lindsay Parkinson & Co. Ltd.* [1985] AC 210 (HL); *Sutherland Shire Council* v. *Heyman* (1985) 60 ALR 1 (HCA).

[3] *O'Rourke* v. *Schacht* [1976] 1 SCR 53 (SCC); *Bird* v. *Pearce* [1979] RTR 369; cf. also *Allison* v. *Corby District Council* [1980] RTR 111; *Barratt* v. *District of North Vancouver* (1980) 114 DLR (3d) 577 (SCC).

[4] *Malet* v. *Bjornson* (1981) 114 DLR (3d) 612.

[5] Space precludes discussion of the duty to take care in the making of a statement, but of course this is a duty the existence and content of which do not depend on whether the duty is a public one. The duty does, however, arise sometimes in the context of the exercise of statutory powers or duties e.g. the issuing of certificates: see *Ministry of Housing* v. *Sharp* [1970] 2 QB 223, and *L. Shaddock & Associates Pty. Ltd.* v. *Parramatta City Council* (1981) 36 ALR 385 (land charges certificate); *Rutherford* v. *Attorney-General* [1976] 2 NZLR 314 (vehicle certificate). For a full discussion of the cases see Aronson and Whitmore, *Public Torts and Contracts* (1982), pp. 108–14; see also *South Australia* v. *Johnson* (1982) 42 ALR 161 (HCA); *San Sebastian Pty. Ltd.* v. *Minister Administering the Environmental Planning and Assessment Act 1979* (1986) 68 ALR 161 (HCA).

this immunity has been removed by statute.[6] These statutes do not answer the question posed above but assume that public authorities can be dealt with in the same way as private persons, so that no public-law/private-law distinction is necessary. However, public authorities do all manner of things which a private person never does, and in relation to all these things they must make decisions which foreseeably may cause damage or loss to others. The untrammelled use of the classical Atkinian principles in relation to public authorities would obviously make nonsense of such decisions. A prison authority may be liable for failing to prevent the escape of prisoners who may foreseeably cause damage, but it is not negligence to make a conscious decision to keep an open prison in the interests of inculcating responsibility among prisoners, even if it is foreseeable that damage may be caused by adopting such a policy.[7] If public authorities are not to be subjected to liability merely for lawfully and properly exercising their discretion in a given manner, some special rule is required to give them immunity from the operation of the usual Atkinian principles.

(ii) *Development of public-law negligence liability*

The problem of special rules for public authorities was first encountered in the nineteenth century with the development of railways, ports, and other public works which were capable of conferring great benefits and of causing great losses in terms of property damage.

In *Mersey Docks & Harbour Board Trustees* v. *Gibbs*[8] the House of Lords found the Board liable for failing to remove a mud bank from the entrance to one of their docks, the plaintiff's cargo having been damaged when his ship collided with it. Similarly in *Geddis* v. *Bann*

[6] For details of these statutes, see Aronson and Whitmore, above n. 5, ch. 1; Street, *Governmental Liability* (1975), ch. 1; Hogg, *Liability of the Crown* (1971), pp. 2–11, 62–4. For discussion of liability of public authorities generally in negligence, see Aronson and Whitmore; Craig, *Administrative Law* (1983), pp. 534–44, and id., 'Negligence in the Exercise of a Statutory Power' (1978), 94 *LQR* 428; Harlow, *Compensation and Government Torts* (1982); Phegan, 'Public Authority Liability in Negligence' (1976), 22 *McGill LJ* 605; Seddon, 'The Negligence Liability of Statutory Bodies' [1978], 9 FLR 326; Oliver, '*Anns* v. *London Borough of Merton* Reconsidered' (1980), *CLP* 269; Bowman and Bailey, 'Negligence in the Realms of Public Law—a Positive Obligation to Rescue?' [1984], PL 277; Cohen and Smith, 'Entitlement and the Body Politic: Rethinking Negligence in Public Law' (1986), 64 *Can. BR* 1; Bailey and Bowman, 'The Policy/Operational Dichotomy—A Cuckoo in the Nest' [1986], *CLJ* 430.

[7] *Home Office* v. *Dorset Yacht Co. Ltd.*, above n. 1.

[8] (1866) LR 1 HL 93.

Reservoir Proprietors[9] the defendants were held liable for damage caused by failure to remove silt from a river, which caused water to overflow onto the plaintiff's land. In both cases the defendants were exercising statutory powers, and the damage was such as would not have been suffered if the defendants had refrained from exercising their powers. Lord Blackburn in *Geddis* saw no difficulty arising from the public[10] nature of the duty being imposed:

it is now thoroughly well established that no action will lie for doing that which the legislature has authorized, if it be done without negligence, although it does occasion damage to anyone; but an action does lie for doing that which the legislature has authorized, if it be done negligently. And I think that if by a reasonable exercise of the powers, either given by statute to the promoters, or which they have at common law, the damage could be prevented it is, within this rule, 'negligence' not to make such reasonable exercise of their powers.[11]

Thus although wide powers were conferred on utilities of this kind, they were not so wide as to empower indiscriminate infliction of loss on individuals. Losses would not be compensated only if they were inevitable.[12]

The problem with the useful but rather simplistic rule formulated by Lord Blackburn is that it does not deal with situations where the action is essentially an indirect attack on a conscious exercise of a discretion not to exercise powers. This is illustrated by *Sheppard* v. *Glossop Corporation*[13] in which the defendants were held not liable when the plaintiff was injured after missing his way while walking along a street in which, in pursuance of a decision to restrict lighting in the interest of economy, the lamplighter had extinguished the gas lamp earlier in the evening. Lord Blackburn's principle was held to be inapplicable;[14] it is indeed of no relevance where the damage is caused

[9] (1878) 3 App. Cas. 403, see *East Freemantle* v. *Annois* [1902] AC 213 (PC); *Great Central Railway Co.* v. *Hewlett* [1916] 2 AC 511 (HL). Many of the numerous cases applying the *Geddis* rule are set out by Mason J. in *Sutherland Shire Council* v. *Heyman* (1985) 60 ALR 1, 27–8. See also Craig, above n. 6, pp. 527–31, 534, and Aronson and Whitmore, above n. 5, pp. 157–62, who deal also with nuisance cases.

[10] Technically of course the case involved a private company, as did many public-duty cases in the 19th cent., but these days the functions of these companies (canal, railway, waterworks companies, etc.) are carried on by public bodies, and of course they acted under statutory powers as do the modern public utilities.

[11] Above n. 9, p. 455–6.

[12] This has been most apparent in nuisance cases: see Craig, above n. 9.

[13] [1921] 3 KB 132; cf. *Carpenter* v. *Finsbury Borough Council* [1920] 2 KB 195 (statutory *duty* to light entails liability).

[14] The dictum was later explained as covering situations in which the statutory power necessarily involves interfence with private property: *East Suffolk Rivers Catchment Board* v. *Kent* [1941] AC 74, 99 per Lord Romer. Lord Wilberforce in *Anns*, above

not by the empowered activity but by a failure to invoke the statute to remove a source of danger arising independently of that activity.

Sheppard's case brings us back to the nonfeasance/misfeasance distinction. While there are difficulties in applying the nonfeasance doctrine to cases involving a statutory duty, where there is after all an obligation of some kind to act, there can be no doubt that an authority whose only fault is that it decided to do nothing when it had only a power to act cannot be held liable even where its omission would foreseeably cause damage.[15] In this respect the argument that public authorities are like rescuers is convincing; the legislature has allowed the authority to decide whether it wishes to be a good Samaritan or a bad Samaritan, and it cannot be attacked for exercising its discretion in such a way as to be the latter. However, it is not clear whether the corollary of the 'rescuer' argument is also true, namely that because the authority is not obliged to act at all it is not obliged to take reasonable care if it does decide to act, so long as it does not cause any fresh damage.

The classic case applying the negligence equivalent of the nonfeasance doctrine is *East Suffolk Rivers Catchment Board* v. *Kent*.[16] In that case the defendants had a statutory power to carry out drainage works, and in exercise of this power they repaired a sea wall which had broken, flooding the plaintiff's land; however, the work was done so incompetently that they took 178 days to repair it, when with reasonable care, that is, using better equipment, more men and a better plan, it could have been repaired in 14 days. As a result the plaintiff's land was flooded for 164 days longer than it should have been. The House of Lords, Lord Atkin dissenting, held that the defendants were not liable, though the Court of Appeal, also with one dissentient, had held them liable. Although two members of the majority in the House of Lords based their decisions on the absence of causation, their judgments are formulated in such a way as to be essentially consistent with the reasons of the other two majority judges, who held, in the words of Lord Romer, that

> [w]here a statutory authority is entrusted with a mere power it cannot be

n. 2, p. 756, seems to restrict it further to private acts, which seems to miss the point made above n. 10 (see also *Fellowes* v. *Rother District Council* [1983] 1 All ER 513, 519 per Goff J.).

[15] This proposition, based on *Sheppard*, is still applicable (see below). By way of contrast and to show the extent of the reasoning in *Sheppard*, it is interesting to note that *Sheppard* was distinguished in *Farrell* v. *Northern Ireland Electricity Service* [1977] N.I. 39, in which the plaintiff also suffered an accident due to the absence of lighting, but this resulted from a mere failure to implement the authority's lighting policy.

[16] Above n. 14.

made liable for any damage sustained by a member of the public by reason of a failure to exercise that power. If in the exercise of their discretion they embark upon an execution of the power, the only duty they owe to any member of the public is not thereby to add to the damages that he would have suffered had they done nothing.[17]

This case is discussed further below.

A further opportunity to clarify the law came in *Home Office* v. *Dorset Yacht Co. Ltd.* in 1970.[18] It was this case which laid down the broad approach to Lord Atkin's neighbour principle[19] as a test of the existence of a duty of care, namely that Lord Atkin's principle 'ought to apply unless there is some justification or valid explanation for its exclusion'.[20] Some borstal trainees being supervised by three officers on an island escaped from the island at night after the officers had retired to bed without taking any precautions for ensuring that the boys did not escape, and damaged the plaintiffs' yacht. The case went up to the House of Lords on the preliminary question whether the Home Office owed a duty of care to the plaintiffs. They contended that they owed no duty because they were entitled to operate a system of light supervision in order to inculcate responsibility in borstal trainees, and that the imposition of a duty of care would prevent them from doing so. While the House of Lords acknowledged that it could not be negligence simply to operate such a system, it also held that if the act or omission occurred in pursuance of instructions which were *ultra vires* or if, as appeared to be the case, the officers simply failed to carry out their instructions, then liability would be decided on the usual principles of negligence.

The cases of *Dutton* v. *Bognor Regis Urban District Council*[21] and *Anns* v. *Merton London Borough Council*[22] both concerned allegations of

[17] Ibid. 102.

[18] Above n. 1.

[19] *Donoghue* v. *Stevenson* [1932] AC 562 (HL).

[20] [1970] AC 1027 per Lord Reid; see also *Anns*, above n. 2, pp. 751–2 per Lord Wilberforce. Recently the courts have cast doubt on this broad approach, and have emphasized the need for a further ingredient of proximity between the defendant and the plaintiff. See *Heyman*, above n. 9, p. 13 per Gibbs CJ, pp. 43–4 per Brennan J., and pp. 63–4 per Deane J. The English judges seem to have taken up the theme eagerly; see *Peabody*, above n. 2, p. 240 per Lord Keith; *Curran* v. *Northern Ireland Co-ownership Housing Association Ltd.* [1987] 2 WLR 1043, 1047 per Lord Bridge (HL); *Yuen Kun-yeu* v. *Attorney-General of Hong Kong* [1987] 2 All ER 705, 710 (PC); *Christchurch Drainage Board* v. *Brown*, *The Times*, 26 Oct. 1987; *Rowling* v. *Takaro Properties Ltd.* [1988] 1 All ER 163 (PC); *Jones* v. *Department of Employment* [1988] 2 WLR 493; *Hill* v. *West Yorkshire Chief Constable*, *The Times*, 29 Apr. 1988 (HL).

[21] Above n. 2.

[22] Ibid.

negligence against a local-authority building inspector in carrying out inspection of the foundations of a building which turned out to be insecure, damage being caused to the building, except that in *Anns* it was uncertain whether the inspector had carried out any inspection at all. In *Dutton* the Court of Appeal held that although the Council were only exercising a power, not performing a duty to inspect, their exercise of power to inspect buildings in their area was an assumption of control over building activities, so that they were under a duty to inspect with reasonable care. *Anns*, which before the House of Lords was an attempt by the Council to get the decision in *Dutton* overruled, was decided on a different basis, namely that the failure to inspect or failure to inspect carefully was a failure to take care at the operational level rather than the result of a policy decision by the Council, so that the authority was not immune from attack in an action for negligence.

These cases raise three interconnected issues which require further inquiry: (1) What is the difference between policy and operation? (2) What is the significance of the fact that the authority is exercising a statutory power rather than performing a statutory duty (the 'nonfeasance' problem)? And (3) what is the significance of the fact that the authority or the official is acting *ultra vires*?

(iii) *Policy and operation*

Lord Wilberforce for the majority in *Anns*, drawing the distinction between policy and operation, said this:

> Most, indeed probably all, statutes relating to public authorities or public bodies, contain in them a large area of policy. The courts call this 'discretion' meaning that the decision is one for the authority or body to make, and not for the courts. Many statutes also prescribe or at least presuppose the practical execution of policy decisions: a convenient description of this is to say that in addition to the area of policy or discretion, there is an operational area. Although this distinction between the policy area and the operational area is convenient, and illuminating, it is probably a distinction of degree; many 'operational' powers or duties have in them some element of 'discretion'. It can safely be said that the more 'operational' a power or duty may be, the easier it is to superimpose upon it a common law duty of care.[23]

[23] [1978] AC 754; for post-*Anns* examples of protected policy decisions, see *West* v. *Buckinghamshire County Council* (1984) 83 LQR 449 (decision not to place double white lines on a road); *Rigby* v. *Northamptonshire Chief Constable* [1985] 1 WLR 1242 (decision to use one kind of equipment rather than another safer kind); *Skuse* v. *Commonwealth* (1985) 62 ALR 108. In this last case a barrister was shot in a courthouse after the police had become aware of threats of violence by an angry litigant to two lawyers who appeared there, it being known that he might take revenge on anyone

This suggests that the distinction is a guide rather than a test, and that there may in certain instances be a common-law duty at the policy level and no duty at the operational level, a consequence which seems rather surprising in view of the introduction of a policy/operation distinction. Lord Wilberforce went on to say that the duty of the inspector in *Anns* might be discretionary with regard to the time and manner of inspection, and the techniques to be used, and that the plaintiff must show that the action taken was outside the limits of such discretion. Thus it would seem that on this view the crucial distinction is between a discretionary act and a non-discretionary or ministerial act, and that the policy/operation distinction is a description of the results generally obtained rather than an acid test of the duty of care. If so the position thus established seems in principle to be right, even if not very clearly formulated, and is indeed consistent with the purpose of maintaining the immunity which public authorities must enjoy in respect of the exercise of discretion. The term 'policy' suggests something very general in nature, distinct from implementation or 'operation' of policy in particular instances; it is not, however, only decisions which are policy decisions in this sense which are rendered immune by the *Anns* decision, but also legitimate exercises of discretion as to how to implement the policy, in particular decisions concerning the allocation of resources, such as that in *Sheppard*. To this extent the policy/operation distinction is confusing, and perhaps the terminology 'discretionary function' is nearer the mark.[24]

There is of course some difficulty with the precise content of the term 'discretion'. Aronson and Whitmore[25] consider that Lord Wilberforce has confounded two senses of discretion, namely policy decisions and professional judgments. It certainly seems unduly restrictive to say that the inspector's discretion with regard to the time, manner, and techniques of inspection render him immune. One needs to inquire further.

connected with the law, and that the risk was imminent. There was no policeman on duty, but the police station was only 100m away. On the facts this could be argued to be operational negligence. The court reasoned, however, that such basic functions of government did not sit easily with the standard of negligence.

[24] This is the term used in the Federal Tort Claims Act 1946 (US); see Aronson and Whitmore, above n. 5, pp. 36–60, for detailed discussion of the American cases. The distinction drawn in Canadian cases prior to Anns was that between 'administrative' and 'quasi-judicial' or 'legislative' functions: see e.g. *Wellbridge Holdings Ltd.* v. *Greater Winnipeg* (1970) 22 DLR (3d) 470 (SCC). In *Bowen* v. *City of Edmonton* (1977) 80 DLR (3d) 501, 519 it was said that *Wellbridge* and *Anns* are to the same effect.

[25] Above n. 5, p. 69. Their point appears to have been accepted by Mason J. in *Sutherland Shire Council* v. *Heyman*, above n. 9, p. 35 ('action or inaction that is merely

If, for example, the inspector used outmoded techniques he might either (1) be liable for his professional incompetence, or (2) not be liable because he was merely balancing the 'rival claims of efficiency and thrift'.[26] One can only decide which is the correct characterization of the inspector's conduct by inquiring what was the inspection policy which the authority by its resolutions or bylaws had decided to adopt; if the inspector's decision was dictated by this policy then he is immune, but if not then he may well be liable. If the inspector in *Anns* simply decided that the foundations were adequate, his decision would not be an immune policy decision, even if he could be said to be exercising discretion; similarly if the lamplighter in *Sheppard* had suddenly taken it into his head to extinguish the lamp.[27] One might, however, have difficulty in dealing with the following cases: (1) a doctor deciding whether to give leave to a mental patient,[28] and (2) a social worker deciding whether to put the record of a suspected arsonist before a remand home.[29] Presumably the answer is to ask whether the official is making a decision which is capable of being the subject of an application for judicial review;[30] if the official is acting *ultra vires* in the sense that he has no jurisdiction to make a decision of that kind, or the decision could not be the subject of review, then the ordinary Atkinian principles apply.[31] This would seem to be a logical consequence of *Anns*, though it did not hold as much.

Negligence in the formulation of a policy seems not to be actionable,[32] although if the negligence amounts to a failure to consider

the product of administrative direction, expert or professional opinion, technical standards or general standards of reasonableness').

[26] *Kent*'s case, above n. 14, p. 103 per Lord Romer.

[27] Cf. *Writtle (Vicar)* v. *Essex County Council*, above n. 1, where a social worker purporting to exercise such judgment was not acting within any discretionary power given by the law. The negligence of a commanding officer in formulating a plan to protect a bank against a terrorist attack would presumably be open to consideration in the courts on the usual principles because there is nothing which could be reviewed: see *Farrell* v. *Secretary of State for Defence* [1980] 1 WLR 172 (HL). There may of course be situations where negligent exercise of discretion will itself make the exercise of discretion invalid on the *Geddis* principle that statutory powers were not granted to be exercised negligently, or else on the basis that a relevant consideration, viz., safety, was not taken into account.

[28] See the *Holgate* and *Teows* cases, above n. 1.

[29] The *Writtle* case, above n. 1 and see *Takaro Properties Ltd.* v. *Rowling* [1978] 2 NZLR 314.

[30] See above Ch. 5.

[31] Both Lord Reid in *Dorset Yacht* (p. 1031) and Lord Wilberforce in *Anns* (p. 755) seem to presuppose the existence of an *ultra vires* act.

[32] *San Sebastian* above n. 5; and this is so even when the policy is specifically with regard to safety: *Sasin* v. *Commonwealth* (1984) 52 ALR 299.

relevant considerations[33] or is in relation to purely safety or technical matters there might be good reason to impose liability. In this sense it may be that the policy/operation approach is unduly restrictive.

Where the negligence is purely collateral or incidental to the implementation of policy, as in a case where an official loses a document or fails to inform a person of the granting of a planning permission, this is clearly negligence at the operational level.[34]

A Canadian decision suggests that a mere failure to implement a policy within a reasonable time can be operational negligence. In *Malet* v. *Bjornson*[35] a highway authority was aware of the inadequacy of a type of road barrier it had been using, and decided to replace it with a superior variety. In a particular district the engineer had discretion when to replace the barrier, and made a decision to replace; however, the decision was not implemented for several months, and in the meantime the very kind of accident feared occurred. The authority was held liable. Presumably, however, a deliberate decision to suspend the implementation of a policy would in itself be a policy decision in the *Anns* sense. One might, however, question the right of the court to decide what is a reasonable period within which to implement a policy decision, particularly where implementation depends on available resources.[36]

Another point of relevance here relates to the burden of proof. Lord Wilberforce said in *Anns* that it is for the plaintiff to show that 'action taken was not within the limits of a discretion *bona fide* exercised'.[37] However, this is inconsistent with the modern approach to Lord Atkin's neighbour principle, referred to above, which would seem to imply that the 'policy defence' is an exception to the rule and should therefore be proved by the party alleging that it applies. Furthermore, it is surely wrong to require the plaintiff to prove matters falling exclusively within the knowledge and competence of the authority,

[33] However, the Privy Council's *obiter* remarks in *Rowling* v. *Takaro Properties Ltd.* [1988] 1 All ER 163, 171–4 appear to contradict this.

[34] *Revesz* v. *Commonwealth* (1951) 51 SR (NSW) 63 seems to indicate otherwise, but would presumably be decided differently after *Anns*; cf. *Takaro*, above n. 33.

[35] Above n. 4.

[36] Since the authority is here exercising a statutory power and not performing duties, there is no obligation for the authority to act at all, so that it would seem to be reasonable to leave to the authority the decision when to implement its policy, in the light of its other commitments. In the case of a statutory duty delay can of course be challenged by mandamus as well as by an action for damages; see above Ch. 2.2. Arguably, however, there should be liability if the factor of reliance is present: see below.

[37] [1978] AC 755; cf. *Potter* v. *Mole Valley District Council*, *The Times*, 22 Oct. 1982.

especially when the precise form in which the policy is expressed (resolution, circular, bylaw, regulation, or simply accepted practice) may not be easily discoverable by the plaintiff. If the burden is placed on the plaintiff it may be tempting for the authority to invent spurious policies *ex post facto*.[38]

It is interesting to note that a number of cases decided during 1987 and 1988 have ignored the policy/operation distinction where one would have expected it to be invoked. In *Curran* v. *Northern Ireland Co-ownership Housing Association Ltd.*[39] the House of Lords refused to find a duty of care owed by an authority empowered to make improvement grants when an extension which was the subject of a grant was defectively constructed. In *Hill* v. *West Yorkshire Chief Constable*[40] the same court treated similarly an alleged duty owed by the police, while investigating the 'Yorkshire Ripper' murders, to a woman who was the last of his victims. The Privy Council in *Yuen Kun-yeu* v. *Attorney-General of Hong Kong*[41] found that no duty could be owed by the Commissioner of Deposit-taking Companies to depositors with a company, which the Commissioner had cogent reason to suspect was conducting its business to the detriment of depositors, when it went into liquidation and the depositors lost their money. In *Christchurch Drainage Board* v. *Brown*[42] the same court was able, without reference to 'operational negligence', to find a duty owed by the Board to an applicant for planning permission; the Board failed to pass on to the planning authority, as it habitually did, information relevant to the application concerning flooding, and as a result the applicant built a house too near a river, which was then flooded. This decision could just as easily, on its facts, have been made on the basis of reliance by the applicant on the board. Finally, in *Rowling* v. *Takaro Properties Ltd.* the same court opined that the policy/operation distinction 'does not provide a touchstone of liability, but is rather expressive of the need to exclude altogether those cases in which the decision under attack is of such a kind that a question whether it has been made negligently is unsuitable for judicial resolution . . .'.[43] These decisions seem to indicate that the policy/

[38] On principle it would not seem that a pre-existing policy is required, otherwise the authority will be precluded from adopting a legitimate policy response to the situation, which may raise some new issue or demand treatment outside the terms of the existing policy; see *Takaro Properties Ltd.* v. *Rowling*, above n. 29. For discussion of the burden of proof problem, see Seddon, above n. 6.

[39] Above n. 20.

[40] Ibid.

[41] Ibid.

[42] Ibid.

[43] [1988] 1 All ER 163, 172.

operation distinction is not in favour in the highest courts. None the less, it is clear that some means has to be found of defining the proper extent of public immunity from negligence actions.

Thus it seems that the policy/operation distinction, while useful as a rule for delimiting the extent of liability in respect of the exercise of statutory powers, contains considerable difficulties which still require careful consideration by the courts.

(iv) *Statutory powers and the nonfeasance problem*

In *Kent*'s case the House of Lords made it clear that to allow an action in this kind of case would mean that an authority attempting to strike a just balance between efficiency and thrift would be exposed to the risk of an action for damages.[44] Lord Atkin, however, found that the Board owed the plaintiff a duty of care on the basis that the relations between the two were 'much closer than the general relations of members of the public to a public authority',[45] because they were operating on the plaintiff's land with the intention of preventing further flooding thereon, and were doing work which the plaintiff could have done himself.

The arguments of Lord Atkin seem to miss the point. The Board could in the proper exercise of its discretion have stopped work at any time or transferred their attention to some other area. The existence of a duty owed to Kent would prevent them from exercising their discretion in this or indeed any other way which happened not to ameliorate the flooding of Kent's land.

The majority view on the other hand draws a sharp distinction between the exercise of statutory powers and the performance of statutory duties for the purpose of ascertaining the existence of a duty of care. The decision is therefore of great importance in the general understanding of the ambit of public duties.

The *Anns* case has made some inroads into the *Kent* decision, although in the last analysis the position is still unclear. It is in fact unfortunate that the 'nonfeasance' aspect of *Anns* has received so little attention in recent cases.

The Council had a power under the Public Health Act 1936 (England and Wales) to make bylaws for regulating the construction of buildings, although they were under a general duty to carry the Act into execution. Bylaws had been made by their predecessors some nine years before the plans of the building in question were passed.

44 [1941] AC 74, 86, 103. Presumably the word 'effectiveness' conveys the intention here better than 'efficiency'.

45 Ibid. 91.

The Council also had power under the Act to require a building which contravened the bylaws to be pulled down or altered. The bylaws required notice to be given of commencement of operations and the covering up of foundations. Lord Wilberforce answered the nonfeasance argument in the following way:

> there may be room, once one is outside the area of legitimate discretion or policy, for a duty of care at common law. It is irrelevant to the existence of this duty of care whether what is created by the statute is a duty or a power: the duty of care may exist in either case . . . to say that councils are under no duty to inspect, is not a sufficient statement of the position. They are under a duty to give proper consideration to the question whether they should inspect or not. Their immunity from attack, in the event of failure to inspect, though great, is not absolute.[46]

This approach has recently been followed in a Canadian case, *Kamloops City* v. *Neilsen*.[47] The City had exercised its powers of building control under the Municipal Act 1960 (British Columbia), requiring an occupancy permit to be obtained, which would only be granted if the building complied with the bylaws and relevant statutes; the owner had to give notice at various stages of construction. In addition the building inspector was under a duty to enforce the provisions. H built a house for his father (A) who was a city alderman. The plans were approved subject to requirements concerning the footings. The house was inspected and found not to comply with the plans. Two stop orders were imposed but H completed the house in defiance of the City, who took no further action against H. H then sold the house to A, and A later sold to the plaintiff, an innocent purchaser, who discovered that the footings were inadequate and incurred loss in remedying the defects.

A majority in the Supreme Court of Canada, following *Anns*, held that the City were under a duty to the plaintiff because they had not even considered prosecuting or seeking an injunction against H, and had appeared to drop the matter because an alderman was involved; to do nothing about the situation was not a policy decision which was open to them. The minority on the other hand held that *Anns* was inapplicable and denied that failure to enforce the law could give rise to an action for damages. Unlike the majority they were unable to say that the City had abused their discretion not to take proceedings, apparently placing the burden of proof on the plaintiff.

[46] [1978] AC 758, 755. Interestingly enough his Lordship treated the *Kent* case (p. 757) as one in which there was a decision well within the operational area. It is difficult, however, to see what *decision* was taken which could protect the authority. Cf. *Barratt* v. *District of North Vancouver*, above n. 3.

[47] (1984) 10 DLR (4th) 641.

The problem with this 'duty to consider' approach is similar to that encountered in cases of the *Thornton* variety in breach of statutory duty.[48] We do not know whether, if the duty to consider had been performed, the plaintiff's damage would have been prevented by the authority taking action. Thus it would seem that a duty of care should only arise if the authority, whether it decided not to act or merely omitted to act, had in effect no option in the exercise of its discretion but to act in a manner which would have prevented the damage occurring. If these conditions are not satisfied, then the court is both substituting its own opinion for that of the authority and assuming the element of causation. In other words the court has to embark on a judicial-review exercise before it can determine the negligence question. It would seem that in *Neilsen* there was inadequate evidence for the court to do this, but if the burden of proof lies with the authority, which, it is suggested above, it should, then the decision can be supported on the basis that the City had failed to show any policy decision to take no action, let alone a valid one, and that in the circumstances their only reason for not acting, had they considered the matter, was a bad one. For this reason the City were bound to take some action, and given the flagrant unlawfulness of H's behaviour, the house would certainly not have been built with inadequate footings if they had done so.

However, the case is an extreme one, and Lord Wilberforce's answer to the duty question, with respect, cannot be regarded as one of general application, because it only applies to those cases in which the authority can be said to have exhausted its discretion, or else failed to exercise any discretion where it has in effect no option but to act.[49] It seems almost as though Lord Wilberforce has discovered a duty of care by sleight of hand, and much the same is true of Lord Atkin in *Kent*.

It is suggested that a much more convincing answer to the nonfeasance argument is rather as follows.

While no one is under a duty to be a good Samaritan, there is a duty on public authorities who choose to exercise their powers to have regard to the fact that the general public may rely on their powers being exercised competently, as in the case where a highway authority fails to exercise its power to indicate the existence of a priority road.[50] In cases of this kind the reliance stems from the fact that the authority has, to the knowledge of the public, decided to

[48] See above Ch. 7.2.
[49] See above Ch. 1.3(iii).
[50] *Bird* v. *Pearce*, above n. 3.

exercise control over the activity or dangerous circumstance in question; members of the public are not therefore given a genuine opportunity to take their own precautionary measures; they may have no way of knowing that the authority has decided on limited inspection or limited enforcement or limited road markings, or some other form of control which amounts to inadequate amelioration of the danger, either generally or in particular instances. If the authority in its discretion decides to reduce its activity to such limited control it would be under a duty of care to inform the public who may otherwise rely on their exercising control, and under a duty, with regard to such control as it has undertaken, to exercise it competently. It is the factor of control combined with reliance which is therefore crucial, not the reviewability of a decision with regard to the extent to which an authority chooses to exercise its powers, nor the proximity of the plaintiff to the defendant.

To this extent the 'control' test propounded by Lord Denning MR in *Dutton* should, it is suggested, be regarded as relevant to ascertaining the existence of a duty of care.[51] Thus an authority which assumes control over a given activity or danger[52] owes a duty of care to those

[51] Above n. 2, pp. 392 (Lord Denning MR), and 403 (Sachs LJ).

[52] One would not wish to restrict unduly the categories of matters which could be brought within the control/reliance syndrome. It would clearly cover building control and control of prisoners or mental patients (a particularly good example of control: can one seriously imagine that the negligently caused escape of a psychotic killer would not result in liability merely because the authority concerned had detained him by exercising statutory powers rather than duties?); see the cases cited above n. 1. Presumably control over safety precautions is also subject to a duty of care; in *Sasin* v. *Commonwealth*, above n. 32, the alleged negligence, in relation to approval of a seat belt used in aircraft, was held not to be subject to a duty of care because it was within the policy area (cf. *Kwong* v. *R.*, below n. 53), but it was suggested that the authority might be liable if the plaintiff had relied on the approval. The principle would also cover works designed to deal with dangerous natural forces, as in *Kent*'s case and *Fellowes* v. *Rother District Council*, above n. 14, provided the plaintiff actually relies on the exercise of power. In *Birch* v. *Central West County District Council* (1969) 119 CLR 652 (HCA) the authority was held liable in respect of fluctuations in electricity supply causing damage which it knew would be caused thereby; a case of this kind could be based convincingly on the control/reliance test. In *Fellowes* it was held that there is no rule that there is no liability for exercise of statutory powers except where fresh damage is caused, but the case is perhaps better seen as a simple application of the *Geddis* rule; a coast protection authority repairing a groyne lowered its height causing the plaintiff's land to be washed away by the sea. Lord Wilberforce (*Anns*, p. 758) objects that the control test puts the duty too high, without explaining why. Aronson and Whitmore, above n. 5, p. 68, argue similarly that the control test could apply where the Treasurer devalues the currency, causing financial loss, because he has control over the rate of exchange; uncontrolled currency, however, is not dangerous, and it would be difficult to argue that the public rely on the Treasurer to control it; in any case the Treasurer would in that

who may be caused damage by the thing controlled because the public place reliance on such control. However, to the extent to which the authority decides not exercise its power to control, its decision is immune even though damage is foreseeable; thus the control test must, to this extent, be regarded as subject to the policy defence. On this basis Anns can recover because the Council had decided to exercise full control and had done so for nine years but failed to take reasonable care, but Kent cannot recover because no reliance could be placed on the authority having assumed control over drainage works;[53] indeed Lord Porter, who agreed with Lord Romer, specifically left open the possibility of liability if there was such reliance.[54]

If one takes this approach one needs to look carefully at the purpose of the statutory powers. Are they conferred with the intention that they should be exercised but in a manner which the authority is left to decide, or are they merely conferred on the basis that the authority will do whatever it can with limited resources to alleviate the danger? In the former instance there must come a point at which control and reliance are present to the extent that members of the public will forego relying on their own remedies; in the latter it is unlikely that reliance will ever occur because members of the public are unlikely to have a remedy more effective than the authority's.

In *Sutherland Shire Council* v. *Heyman*,[55] a majority in the High Court of Australia adopted just such a test of reliance, but in such a way as to deny liability in relation to facts which would seem to fall squarely within the *Anns* principle. The facts were slightly different from *Anns*. The application for a building permit was made under the Local Government Act 1919 (NSW), a code which had the following features. The Council were empowered to control or regulate the

instance be protected by the policy defence, to which the control test would be subject in the manner indicated in the text. The control/reliance factor has been evident in recent decisions of the Court of Appeal in New South Wales: see *Wollongong City Council* v. *Fregnan* [1982] 1 NSWLR (failure to warn of danger of slippage of land in exercise of building control), and *Sutherland Shire Council* v. *Heyman* [1982] 2 NSWLR 618 (revd. (1985) 60 ALR 1 (HCA)).

[53] Cf. *Administration of Papua New Guinea* v. *Leahy* (1961) 105 CLR 6 (HCA); *Kwong* v. *R.* (1978) 96 DLR (3d) 214 (affd. (1979) 105 DLR (3d) 576 (SCC).

[54] [1941] AC 107; du Parcq LJ, whose judgment in the Court of Appeal was heavily relied on by the majority in the House of Lords, went so far as to concede, at [1940] 1 KB 338, that 'the law would perhaps . . . seem more satisfactory in some hard cases, if a body which chose to exercise its powers were regarded as being in exactly the same position as one upon which an Act of Parliament imposed a duty'; he also conceded the reliance argument (ibid. 339). The 'hard cases' are clearly those where there is control and reliance.

[55] Above n. 52. For discussion see Todd, 'The Negligence Liability of Public Authorities: Divergence in the Common Law' (1986) 102 *LQR* 370.

erection of buildings but buildings were required to be erected to the satisfaction of the Council in accordance with the Act and the approved plans. There was provision under the Act for obtaining a certificate that a building had been so erected. The Council could (and did in this case) prohibit any occupation of a building which had not been passed by them. The application related to a house to be erected on a steeply sloping site; it was granted subject to the builder giving notice at various stages of construction, including prior to laying the foundations, which were required to be to a specific depth. The Council also required the submission of a check survey when the footings of the house were commenced. The builders gave notice only on completion of the frame of the house, which was inspected and found to be adequate; the footings could have been inspected at this time and would or should have been found to be inadequate, but it was not certain whether they had in fact been inspected. No check survey was submitted, and the house was occupied without being passed by the Council. Some five years later the plaintiffs purchased the house, which was found to have defects resulting from inadequate footings. The plaintiffs incurred losses resulting from the repairs which had to be carried out.

The New South Wales Court of Appeal,[56] applying *Anns*, found unanimously that the Council were liable. The High Court unanimously upheld the Council's appeal, holding that they were under no duty to the plaintiffs, Gibbs CJ and Wilson J. on the ground that the Council had not breached their duty to consider whether to exercise their powers, Mason and Brennan JJ. on the ground that the plaintiffs had not relied on the authority having inspected, and Deane J. on the ground that there was not sufficient proximity between the plaintiffs and the Council because the plaintiffs had not relied on their having inspected. The last three judges named refused to follow *Anns*.

The decision is a curious one because the majority, stressing the importance of reliance, applied the concept in a specific and narrow sense, and as a means of limiting the extent of the *Anns*-type duty rather than as a means of justifying it. It is interesting to note that Lord Wilberforce in *Anns* regarded the control test as 'putting the duty too high';[57] the reliance test in *Heyman* seems, on the contrary, to impose an arbitrary limit. The New South Wales code seems equally if not more stringent than the English code, both with regard to the builder and the Council. Although the Council were required to

[56] Ibid.

[57] [1978] AC 758.

satisfy themselves that the building complied with the Act and their own requirements under the Act, all they did by way of enforcing the law in this instance was to inspect the building once very superficially, and this in spite of the special danger arising from the nature of the site, the builders' flagrant, indeed almost total, disregard of their obligations under the Act and the permit, and the fact that it was effectively only by inspection that the Council could hope to enforce those obligations. Yet according to the majority the only crucial consideration was that the Council had done nothing to induce belief in the structure on the part of the plaintiffs, for example by issuing a certificate of compliance or responding to an inquiry.

What kind of reliance then is relevant, on this view, to the existence of a duty?

Mason J. expressed the duty as depending on 'foreseeability of the plaintiff's reasonable reliance',[58] but made it clear also that the reliance he had in mind was specific reliance by the plaintiff. Brennan and Deane JJ attached great importance to the fact that no certificate of compliance had been obtained. In other words the basis of the duty is not the fact that the public rely on building control to ensure so far as possible that in the interest of the community generally buildings are required to comply with certain standards, but the fact that a particular member of the public has made specific inquiry about the soundness of a particular building, a species of liability which is catered for anyway by the law relating to negligent misstatements.

The results of such a narrow approach seem quite clear, but none the less undesirable. If, for example, a highway authority fails to replace a 'give way' sign as a result of which I collide with a vehicle proceeding along a priority road,[59] I can (presumably) recover damages because I specifically relied on the priority road being indicated, but the driver of the vehicle with which I collided cannot recover because he did not. Similarly if the authority negligently inspects my vehicle and I therefore refrain from testing the brakes myself I can recover damages in the ensuing accident but not my passenger who may well rely on me but not the vehicle inspector.[60]

[58] (1985) 60 ALR 30–1.

[59] Cf. *Bird* v. *Pearce* [1979] RTR 369. It is perhaps possible that physical damage case would be treated differently from economic-loss cases; in *Muirhead* v. *Industrial Tank Specialities Ltd.* [1985] 3 All ER 705 a 'real reliance' test was applied to a case of economic loss; but if reliance were developed as a general test in economic-loss cases, including those against public authorities, it is not at all clear how cases of physical damages brought against public authorities would be dealt with. For economic loss and public authorities, see *San Sebastian*, above n. 5.

[60] Cf. *Rutherford* v. *Attorney-General*, above n. 5.

It is suggested that the effect of *Heyman* is really a travesty of the *Anns* duty. A better use of the control/reliance factor would be to determine the duty as follows. Where damage arises from failure to exercise a statutory power, the public's reliance on the authority having exercised the degree of control which it purports to have exercised will establish a duty of care, subject to the authority's policy immunity. In other words the authority is free to adopt whatever control policy it validly chooses to adopt, but a high level of control will entail a duty to take reasonable care to ensure that that level of control is maintained; if the level is reduced no liability results from such reduction unless the authority fails to take reasonable steps to ensure that the public does not rely on the higher level of control being exercised. This, it is suggested, is the best way of distinguishing between 'launching an instrument of harm' and 'refusing to become an instrument of good'.[61]

A good example of a decision along these lines is the decision of the House of Lords in a Scottish case, *Anchor Line* (*Henderson Brothers*) *Ltd.* v. *Dundee Harbour Trustees*[62] in 1922, a case which, curiously, was not cited in any of the cases mentioned above. The trustees had statutory powers to licence and appoint pilots and to make bylaws for the purpose of providing a pilotage service for the harbour. They exercised these powers and published the bylaws and pilotage rates. The pilots were obliged to be available at a buoy at the river estuary, but habitually refused to do so, waiting instead further up the estuary, because the trustees would not provide a steam cutter at the buoy, but only a sailing cutter, which the pilots considered inadequate. A vessel approaching the harbour found that no pilot was available at the buoy and proceeded, colliding with an unmarked sunken wreck. The trustees were liable for failing to mark the wreck, but were also found liable for failing to take reasonable steps (for example disciplinary measures) to ensure that the pilots complied with the bylaws. They had represented to the public that they maintained an adequate pilotage system, but had failed to maintain it. 'Something much more energetic was required to awaken the pilots to a sense of their duty' and to reverse the 'pretty steady eclipse of the code'.[63]

A similar line was taken in *Bird* v. *Pearce*,[64] a post-*Anns* decision of

[61] *Bird* v. *Pearce*, above n. 59, p. 374, per Eveleigh LJ, quoting Cardozo CJ.

[62] (1922) 38 TLR 299. The writer is indebted to Mr George Wei for bringing this case to his attention. Admiralty and Queen's Bench barristers are presumably very distinct breeds of men.

[63] Ibid. 308, 312 per Lords Finlay and Dunedin respectively.

[64] Above n. 59, pp. 373, 378–9. Interestingly enough, *Sheppard*'s case, above n. 13, was distinguished (p. 379) on the ground that Sheppard had not relied on the lamp being lit. The result would therefore be different today if Sheppard could show

the Court of Appeal, in which the highway authority in resurfacing a road obliterated and failed to replace a white line indicating a priority road. In *Yuen Kun-yeu*[65] the Privy Council appeared to approve the reliance argument in principle, but rejected its application to the facts: it was open to the depositors to seek their own counsel as to the soundness of a deposit-taking company by use of public documents and the advice of experts; the Commissioner did not have day-to-day control of the company, but only a quasi-judicial power to deregister; reliance on him was therefore neither reasonable nor foreseeable.

To summarize, the approach suggested here is as follows. There is no duty of care in respect of a decision, validly made, not to exercise powers at all. However, if a decision is made to exercise powers in circumstances which satisfy the control/reliance test, then there is a duty to take reasonable steps to ensure that the public which relies on the exercise of control suffers no damage as a result of failure to exercise control in a particular instance. It is suggested that this position (1) does not involve trespassing on the authority's discretion; (2) provides a fair and reasonable approach to public liability suited to the conditions of modern society with its high degree of regulation; (3) is consistent with the proper relationship between statutory powers and duties indicated in Chapter 1; and (4) does not involve open-ended liability or allow unwarranted recovery from public funds for private misfortune.

(v) *Illegality and negligence*

In the light of the issues discussed above, the next question is, to what extent is there a correlation between negligence and illegality?

Both *Dorset Yacht* and *Anns* seem to suggest that the official must be acting *ultra vires* before the usual negligence rules apply.[66] Does this mean that there is liability wherever there is an *ultra vires* act or omission, and that there can be no liability where there is no such illegality?

According to the Privy Council in *Dunlop* v. *Woolahra Municipal Council*,[67] illegality does not entail liability in negligence, a proposition which seems irrefutable. An unlawful act is entirely consistent with

reliance, but the authority would presumably be protected if it took reasonable steps to inform the public of its change of policy.

[65] [1987] 2 All ER 705 (PC).).

[66] See above n. 31.

[67] [1982] AC 158; see also Aronson and Whitmore, above n. 5, pp. 84–6, 99–103; Craig, above n. 6, pp. 541–4. It is also the burden of the Privy Council's decision in the *Takaro* case, above n. 33, that there can be no easy equation of illegality and negligence.

the exercise of reasonable care; in *Dunlop* the Council had consulted its solicitor before passing resolutions which turned out to be *ultra vires*. Presumably, however, an authority which does not take reasonable care to ensure that its decision on which the plaintiff relies is within its power will be liable.[68] This is presumably particularly true where the illegality involves a failure to take into account a relevant consideration, viz., the possibility of damage being caused, although illegality of this kind cannot of course automatically give rise to a finding of negligence.

At the operational level there are circumstances where the negligence itself will take the official outside his discretion. This was the case in *Dorset Yacht* and *Anns*, and is of course implicit in the *Geddis* principle.[69]

(vi) *Public duties of care: some conclusions*

The policy/operation distinction appears to deal with the bulk of the reasons why public authorities should not be subjected to the Atkinian duty of care, and the control/reliance test, it is suggested, deals with the distinction between statutory powers and statutory duties in this context. It is also, however, true that there are some public functions which can never attract a duty of care because of the identity of the defendant as, for example, a judge,[70] or because of the nature of the powers exercised, such as the prerogative power of granting a pardon.[71] These specific immunities were established before the policy/operation distinction, though there seems to be no good reason why the distinction could not be applied satisfactorily in these areas too.

There is also the problem of the standard of care. Is it reasonable to impose the same standards on a public body as on a private body performing the same or analogous activities?

It is suggested that it is reasonable. The policy/operation distinction takes care of the argument that the courts cannot interfere with the proper exercise of discretion by the authority, and therefore the court cannot be accused of substituting its decision for that of the authority by finding that a particular act or omission is negligent, even though the court has to weigh the utility of the authority's activity against the cost of its taking precautions. It would be intolerable if the courts,

[68] *Port Underwood Forests Ltd.* v. *Malborough County Council* [1982] 1 NZLR 343; *Takaro*, above n. 33.

[69] This may help to explain cases such as the *Writtle* and *Holgate* cases referred to above nn. 28–9.

[70] For discussion of immunity of persons acting judicially, see Aronson and Whitmore, above n. 5, pp. 138–47.

[71] *Hanratty* v. *Butler (Lord)*, *The Times*, 13 May 1971.

having found that a duty is owed in a particular case, were none the less to excuse the authority on the basis of a policy decision as to the allocation of resources. The essence of the duty problem is precisely whether there are good reasons for excluding the ordinary principles. It would of course be possible to deal with the problem by reference to the appropriate standard of care,[72] but this would merely produce the same result by a different means. It might be objected that the authority should be able to balance efficiency with thrift in relation to its purely private as well as its public duties of care. However, constitutional considerations of equality before the law seem to preclude such an argument, and no great difficulty seems to have been encountered in subjecting public authorities to the same standards as private institutions, which also, one might say, have to balance efficiency with thrift. In some situations the very nature of the function may entail accession to public policy arguments: for example a fire engine answering an emergency call[73] or a police car chasing a criminal need not observe the 'ordinary standards'.[74]

In conclusion it is interesting to note that nearly all cases of public liability in negligence involve the exercise of powers to provide or to enforce the law, and have the common feature that the authority is being asked in these cases to prevent or mitigate a danger arising other than by the exercise of the powers themselves. The overall effect of the law discussed in this section is that there are many situations where the existence of a power in effect compels the authority, at least in so far as it does not decide on limited exercise of the power, to perform duties which, by reason of its own decision, it has imposed on itself.

[72] Lord Morris' speech in *Dorset Yacht*, above n. 1, pp. 1035 et seq. is to this effect.

[73] *Watt* v. *Hertfordshire County Council* [1954] 2 All ER 368. There are of course innumerable instances where the ordinary principles of negligence apply to public authorities. After *Anns* and *Dorset Yacht* it seems unlikely that the courts will negative or restrict the duty save for very compelling reasons: see e.g. *Groves* v. *Commonwealth* (1982) 40 ALR 193 (HCA).

[74] *Marshall* v. *Osmond* [1982] 2 All ER 610.

8

Summation

At almost every point in the law relating to judicial review of public duties one experiences a sense of disappointment at the inconclusiveness of the doctrines propounded, and in some instances the sheer timidity of the decisions. A revision of the questions discussed in this book will reveal that doctrinally the law of public duties has been encrusted with many principles which almost seem designed to avoid rather than create a law of public duties.

It will be useful at this point to highlight some of the problems indicated in Chapters 1 to 7.

1. The first problem is that of the implication of statutory duties, particularly in instances where only a power is granted by the statute. There is some difficulty in knowing when a power must be exercised. In spite of the principles developed in cases such as *Padfield*, there still seems to be a quite unjustifiable distinction between permissive and obligatory wording. The common law lacks a concept of *competence liée*, which could be supplied by judges being prepared to hold that a discretion has been exhausted rather than hiding behind the inadequate 'mandamus to decide in accordance with law' formula, and by clarifying the relevance of a right to the exercise of a power. This problem is particularly evident in relation to the duty to determine and in the area of negligent failure to exercise statutory powers.

2. A restrictive view has sometimes been taken of the kinds of duty which the courts will enforce through the administrative-law process. A broad view of 'public law' based on the applicability of public-law reasoning rather than the legal source of the duty would avoid pitfalls for the litigant. This problem is evident mainly in the context of contractual duties and the duties of extra-statutory bodies.

3. The law has not developed principles to lay down the content of the important duty with regard to applications requiring determination. This is an area where general principles are both necessary and easily formulated.

4. Duties to provide have been restrictively interpreted, both where the duty is simply to provide and where the duty depends on a decision as to entitlement. The tort of breach of statutory duty could

play an important part here, but a recasting of the tort is required. Principles to give content to such duties need to be developed: in particular the relevance of equality, inconvenience, and duress needs some clarification.

5. The duty to enforce the law is one which the courts are singularly unwilling to enforce. This unwillingness creates an unjustifiable area of immunity amounting virtually to a prerogative of suspension of the laws. A more radical approach is needed.

6. The law of mandamus is still encrusted with the remains of quite unnecessary restrictions such as the demand and refusal requirement, the unavailability of mandamus against the Crown, and the alternative-remedy rule in *Pasmore*'s case. These restrictions have been reduced in their practical effect but need to be done away with completely.

7. The differences between the various remedies for enforcing public duties, particularly with regard to standing, and the lack of opportunity to combine remedies in some instances, can be solved adequately through the modern statutory review procedures. These reforms should allow duties and not just decisions to be reviewed, and should allow flexibility in choice and cumulation of remedies, in particular with regard to adding a claim for damages where appropriate.

8. Standing to enforce public duties has been a particularly troublesome area. This problem too is soluble in the context of statutory review, but unless the courts allow a citizen's action to develop, there is a danger that many duties will be enforceable only in theory, because no one is able to satisfy the standing requirements.

9. The tort of breach of statutory duty has been developed in such a way as to render public statutory duties virtually immune from actionability. The old public duty/special damage rule would open up this tort as a suitable principle of public-duty liability.

10. The *Anns* principles in relation to liability for failure to exercise a statutory power offer a potentially useful basis for liability, but contain many confusions which remain to be clarified.

Although some optimism has been expressed in this book with regard to the potential within the law as it is for developing an effective law of public duties, there can be no doubt that the courts have consistently used various devices for avoiding review of public duties. These have been tabulated in some detail in Chapters 1 to 7 and summarized above, but it is important to try to understand why this attitude has been taken.

Of great importance is the common-law tradition which was backward (though not as backward as one might think if one looked only at nineteenth-century cases) in terms of development of principles of public law, until the renaissance of administrative law from the early 1960s onwards.

More importantly, one senses in public-duty cases a deliberate attempt to prevent the development of such doctrines as were already in existence; in this connection it seems clear that in the face of administrative developments in the nineteenth and early twentieth centuries the courts did not feel either equipped experientially or justified constitutionally in 'interfering' with the process of development. It may well be that these familiar considerations have been even more marked in the area of public duties than in other areas of administrative law, because this area more than any other involves actual coercion: there is no way round a writ of mandamus except into prison. In addition public-duty cases generally involve a citizen asking for something to be done for him rather than for his right to be protected from interference; in such cases there is no obvious substratum of constitutional rights to be enforced. Public-duty cases therefore involve a speculative as well as a coercive element. There is always the uneasy feeling that by coercing authority in favour of one citizen one is possibly depriving the rest of society of a benefit, or at least that these matters are best left with the administration with its superior knowledge and political accountability.

These factors are substantial and not imaginary. However, they have been decisive for far too long. The law has now developed to a stage where the nature, and proper extent of control, of administrative discretion are much clearer and better understood than previously. The courts are capable both of assessing the proper limits of administrative activity, and of deciding in which areas they are not qualified to act. As a result one can see in the field of public duties that there have been many developments which have served to dispel much of the nineteenth-century fog and gloom. In the area of remedies in particular it would seem that much of the desirable reform has already taken place; in standing there is cause for hope that it will take place; in negligence there is a workable body of doctrine which will no doubt be refined.

In the other areas mentioned the law is still very backward doctrinally. One cannot say that the law on implication of statutory public duties makes a great deal more sense than it did fifty or a hundred years ago; the matter of performance of public duties seems always to be a matter of judicial review of the exercise of discretion rather than judicial review of a public duty, and peters out with rather lame references to reasonableness; the law of breach of statutory duty is in a state of utter confusion.

As a result the law of public duties still presents a displeasing aspect not just to the lawyer but to the litigant who is faced, not just with the usual hazards of litigation, but with a succession of formidable obstacles of uncertain height. The administration benefits to the extent of a large area of immunity which, if it represented a general picture of administrative law, would cause a public outcry. The renaissance of administrative law and the winds of the new judicial activism have left the law of public duties only partially and uncertainly reformed.

It is true that in these areas it is sometimes difficult to lay down any hard and fast rules. None the less it is possible for the courts to develop doctrine by indicating the ways in which various statutory contexts or individual rights will be regarded or certain facts analysed. A jurisprudence of public duties is badly needed to deal with the kinds of case which seem to arise these days, to which the old concepts of administrative law may be inapplicable. The courts often seem unaware of the need to articulate such a jurisprudence, and even unaware of relevant cases which require discussion. Even the statutory reforms of administrative law seem to ignore the particular remedial requirements of public-duty cases. Yet very few of the reforms which have been suggested in this book require legislative intervention and none of them requires any radical departure from currently held theories of administrative law. The remedies lie in the judges' own hands; the precedents and the doctrinal equipment are available, and need only to be developed.

Currently administrative law appears to be going through a conservative phase in which the proper limits of judicial review are being considered and perhaps even being shaped. This is a necessary phase, but not one which should have any important effect on the matters discussed in this book. The law of public duties does not need a conservative phase, because it has never had a radical phase; the period of judicial activism in administrative law has left it largely untouched and undeveloped. The last remedy it needs is for the judges to say that the performance and enforcement of public duties is an administrative, not a judicial, matter; that would leave us no wiser or better than we were a century ago. Perhaps the only area where a tenable theory of judicial restraint in relation to review of administrative acts might contradict the position advocated in this book would be that relating to discretionary refusal of a remedy where an administrative remedy exists (the *Pasmore* problem). There is perhaps something to be said generally for the courts requiring an administrative remedy to be sought before, or even instead of, embarking on judicial review. This general approach is not, however, one which should be applied to public duties and default powers; the problems involved in the use of default powers have been discussed in Chapters 3 and 4 and the courts have an important role to play both in ensuring that public duties are observed even where superior authorities refuse to act, and in ensuring that when they do act their default powers are exercised reasonably and with a proper regard for the constitutional position of the inferior authority.

The law of public duties is in a state of flux. A host of cases has been reported even since this book was commenced. Many of these raise new issues concerning the duties of public authorities and the rights of individuals. Given the stimulus of crucial cases and the general recognition of the importance of developing administrative law, the courts may be forced willy-nilly to attempt to solve many of the outstanding problems. It is, however, unlikely that, in an area of the law which has proved controversial ever since Coke delivered his famous judgment in *Bagg*'s case nearly four hundred years ago, the really fundamental issues will ever go away. The author will be content if he has succeeded in drawing attention to these and indicating what approaches are appropriate at this very interesting stage in the history of administrative law.

Appendix: Supreme Court Act 1981 (England and Wales), section 31

(1) An application to the High Court for one or more of the following forms of relief, namely—
  (a) an order of mandamus, prohibition or certiorari,
  (b) a declaration or injunction under subsection (2); or
  (c) an injunction under section 30 restraining a person not entitled to do so from acting in an office to which that section applies,

shall be made in accordance with rules of court by a procedure to be known as an application for judicial review.

(2) A declaration may be made or an injunction granted under this subsection in any case where an application for judicial review, seeking that relief, has been made and the High Court considers that, having regard to—
  (a) the nature of the matters in respect of which relief may be granted by orders of mandamus, prohibition or certiorari;
  (b) the nature of the persons and bodies against whom relief may be granted by such orders; and
  (c) all the circumstances of the case

it would be just and convenient for the declaration to be made or the injunction to be granted, as the case may be.

(3) No application for judicial review shall be made unless the leave of the High Court has been obtained in accordance with rules of court; and the court shall not grant leave to make such an application unless it considers that the applicant has a sufficient interest in the matter to which the application relates.

(4) On an application for judicial review the High Court may award damages to the applicant if—
  (a) he has joined with his application a claim for damages arising from any matter to which the application relates, and
  (b) the court is satisfied that, if the claim had been made in an action begun by the applicant at the time of making the application, he would have been awarded damages.

(5) If, on an application for judicial review seeking an order of certiorari, the High Court quashes the decision to which the application relates, the High Court may remit the matter to the court, tribunal or authority concerned, with a direction to reconsider it and reach a decision in accordance with the findings of the High Court.

(6) Where the High Court considers that there has been undue delay in making an application for judicial review, the court may refuse to grant—
  (a) leave for the making of the application; or
  (b) any relief sought on the application,

if it considers that the granting of the relief sought would be likely to cause substantial hardship to, or substantially prejudice the rights of, any person or would be detrimental to good administration.

(7) Subsection (6) is without prejudice to any enactment or rule of court which has the effect of limiting the time within which an application for judicial review may be made.

Table of Cases

Table of Statutes

1. *England and Wales*

4. *Other Jurisdictions*

Bibliography

Aldous, G., and Alder, J., *Applications for Judicial Review: Law and Practice* (1985), London, Butterworths.
Allen, C. K., *Legal Duties* (1931), Oxford, Clarendon Press.
Aronson, M., and Franklin, N., *Review of Administrative Action* (1987), Sydney, Law Book Company.
—— and Whitmore, H., *Public Torts and Contracts* (1982), Sydney, Law Book Company.
Austin, J., *The Province of Jurisprudence Determined*, with introduction by H. L. A. Hart (1954), London, Weidenfeld and Nicholson, Australian Law Reform Committee Discussion Paper, No. 4 (1977), Sydney, Australia Law Journal.
Bailey, S. H., and Bowman, M. J., 'The Policy/Operation Dichotomy—a Cuckoo in the Nest' [1986] *Cambridge Law Journal*, 430.
Bailey, S. H., Cross, C. A., and Garner, J. F., *Cases and Materials in Administrative Law* (1977), London, Sweet and Maxwell.
Bayne, P., *Freedom of Information* (1984), Sydney, Law Book Company.
Beatson, J., ' "Public" and "Private" in English Administrative Law' (1987), 103 *Law Quarterly Review*, 34.
Blom-Cooper, L., 'The New Face of Judicial Review: Administrative Changes in Order 53' [1982], *Public Law*, 250.
Borrie, G. J., and Lowe, N., *Law of Contempt*, 2nd edn. (1983), London, Butterworths.
Bowman, M. J., and Bailey, S. H., 'Negligence in the Realms of Public Law—a Positive Obligation to Rescue?' [1984], *Public Law*, 277.
Branson, N., *Poplarism 1919–1925: George Lansbury and the Councillors Revolt* (1979), London, Lawrence and Wishart.
Brown, L. N., and Garner, J. F., *French Administrative Law*, 3rd edn. (1983), London, Butterworths.
Buckley, R. A., 'Liability in Tort for Breach of Statutory Duty' (1984), 100 *Law Quarterly Review*, 204.
Campbell, E. M., 'Private Claims on Public Funds' (1969), *University of Tasmania Law Review*, 138.
—— 'Judicial Review and Appeals as Alternative Remedies' (1982), 9 *Monash University Law Review*, 14.
Cane, P. F., 'The Function of Standing Rules in Administrative Law' [1980], *Public Law*, 303.
—— 'Ultra Vires Breach of Statutory Duty' [1981], *Public Law*, 11.
—— 'Standing, Legality and the Limits of Public Law' [1981], *Public Law*, 322.
—— 'Public Law and Private Law Again: *Davy* v. *Spelthorne B. C.*' [1984], *Public Law*, 16.
—— 'Public Law and Private Law: Some Thoughts Prompted by *Page Motors Ltd.* v. *Epsom and Ewell Borough Council*' [1984], *Public Law*, 202.
—— 'Public Law and Private Law: A Study of the Analysis and Use of a

Legal Concept', ch. 3 of Eekelaar, J. and Bell, J., eds., *Oxford Essays in Jurisprudence*, 3rd series (1987), Oxford, Clarendon Press.
Cohen, D., and Smith, J. C., 'Entitlement and the Body Politic: Rethinking Negligence in Public Law (1986), 64 *Canadian Bar Review*, 1.
Craig, P. P., 'Negligence in the Exercise of a Statutory Power' (1978), 94 *Law Quarterly Review*, 428.
—— *Administrative Law* (1983), London, Sweet & Maxwell.
Cranston, R., *The Legal Foundations of the Welfare State* (1985), London, Weidenfeld and Nicholson.
Cripps, Y., 'Jurisdiction, Remedies and Judicial Review' (1984), 42 *Cambridge Law Journal*, 214.
Davies, K., *Local Government Law* (1983), London, Butterworths.
Davis, K. C., *Administrative Law Treatise*, 3rd edn. (1983), San Diego, University of San Diego.
—— *Police Discretion* (1975), St Paul, Minnesota, West Publishing Co.
De Smith, S. A., *Constitutional and Administrative Law*, 5th edn. by H. Street and R. Brazier (1985), Harmondsworth, Penguin Books.
—— *Judicial Review of Administrative Action* (1959), 4th edn. by J. M. Evans (1980), Stevens & Sons, London.
Denning, Lord, *The Discipline of Law* (1979), London, Butterworths.
Dias, M., *Jurisprudence*, 4th edn. (1976), London, Butterworths.
Dickens, B. M., 'Discretion in Local Authority Prosecutions' [1970], *Criminal Law Review*, 618.
Driver, C., 'The Judge as Political Pawnbroker: Superintending Structural Change in Public Institutions' (1979), 65 *Virginia Law Review*, 43.
Ellesmere, Lord, attrib., *Observations on the Lord Coke's Reports* (1710?), ed. G. Paul.
Enright, C., *Judicial Review of Administrative Action* (1985), Sydney, Branxton Press.
Finn, P. D., 'A Road not Taken: the Boyce Plaintiff and Lord Cairns' Act' (1983), 57 *Australian Law Journal*, 493, 591.
Fleming, J. G., *The Law of Torts*, 6th edn. (1983), Sydney, Law Book Company.
Flick, G. R., *Federal Administrative Law* (1983), Sydney, Law Book Company.
Forsyth, C. F., 'Beyond *O'Reilly* v. *Mackman*: the Foundations of Procedural Exclusivity' [1985], *Cambridge Law Journal*, 415.
Fricke, G. L., 'The Juridical Nature of the Action upon the Statute' (1960), 76 *Law Quarterly Review*, 240.
Garner, J. F., and Jones, B. L., *Administrative Law* (6th edn., 1985), London, Butterworths.
Gordon, R. J. F., *Judicial Review: Law and Procedure* (1985), London, Sweet & Maxwell.
Gould, B., 'Anisminic and Jurisdictional Review' [1970], *Public Law*, 358.
Griffith, J. A. G., *Central Departments and Local Authorities* (1966), London, George Allen and Unwin.
Griffiths, J., 'Legislative Reform of Judicial Review of Commonwealth Administrative Action' (1978), 9 *Federal Law Review*, 42.

—— 'Mickey Mouse and Standing in Administrative Law' [1982], *Cambridge Law Journal*, 6.
Grubb, A., 'Two Steps Towards a Unified Administrative Law Procedure [1983], *Public Law*, 190.
Harlow, C., 'Administrative Reaction to Judicial Review' [1976], *Public Law*, 116.
—— 'Public and Private Law: Definition Without Distinction' (1980), 43 *Modern Law Review*, 241.
—— *Compensation and Government Torts* (1982), London, Sweet and Maxwell.
—— 'Public Interest Law in England: the State of the Art', ch. 4 of Dhavan, R., and Cooper, J., eds., *Public Interest Law* (1986), Oxford, Basil Blackwell.
Harlow, C., and Rawlings, R., *Law and Administration* (1984), London, Weidenfeld and Nicholson.
Harris, J. W., *Legal Philosophies* (1980), London, Butterworths.
Harrison Moore, W., 'Misfeasance and Nonfeasance in the Liability of Public Authorities' (1914), 30 *Law Quarterly Review*, 276, 415.
Hawkins, W., *Hawkins' Pleas of the Crown*, 8th edn. by Curwood, J. (1824), London, S. Sweet.
Hearn, W. E., *The Theory of Legal Rights and Duties* (1883), Melbourne, George Robertson.
Henderson, E. G., *Foundations of English Administrative Law* (1963), Fairfield, North Jersey, Augustus Keller.
Hogg, P. W., 'Judicial Review of Action by the Crown Representative' (1969), 43 *Australian Law Journal*, 215.
—— *Liability of the Crown* (1971), Sydney, Law Book Company.
Hohfeld, W., *Fundamental Legal Conceptions as Applied to Judicial Reasoning* (1919), New Haven, Yale University Press.
Holdsworth, Sir W. S., *A History of English Law*, 4th edn. (1936), Cambridge, Cambridge University Press.
Hood, C., *The Tools of Government* (1983), London, Macmillan.
Hotop, S. D., *Principles of Australian Administrative Law*, 6th edn. (1985), Sydney, Law Book Company.
Howard, C., *Australian Criminal Law*, 4th edn. (1982), Melbourne, Law Book Company.
Howell, G., 'An Historical Account of the Rise and Fall of Mandamus' (1985), 15 *Victoria University of Wellington Law Review*, 127.
Jain, M. P., *Administrative Law of Malaysia and Singapore* (1980), Singapore, Malayan Law Journal.
Jenks, E., 'The Prerogative Writs in English Law' (1923), 32 *Yale Law Journal*, 523.
Kyrou, E., 'Locus Standi of Private Individuals Seeking Declaration or Injunction at Common Law' (1982), 13 *Melbourne University Law Review*, 453.
Law Commission Report, No. 20 (1969), London, HMSO.
Law Commission Report, No. 73 (1976), London, HMSO.
Law Commission Report, No. 76 (1976), London, HMSO.

Law Commission Working Paper, No. 40 (1970), London, HMSO.

Lee, H. P., *Emergency Powers* (1984), Sydney, Law Book Company.

Lloyd of Hampstead, Lord, and Freeman, M. D. A., *Lloyd's Introduction to Jurisprudence* (5th edn., 1985), London, Stevens & Sons.

Loughlin, M., *Local Government in the Modern Constitution* (1985), London, Sweet & Maxwell.

—— *Local Government in the Modern State* (1986), London, Sweet & Maxwell.

Lucas, W. W., 'The Immunity of the Crown from Mandamus' (1909), *Law Quarterly Review*, 290.

Lustgarten, L., *The Governance of Police* (1986), London, Sweet & Maxwell.

Moody, S. R., and Tombs, J., *Prosecution in the Public Interest* (1982), Edinburgh, Scottish Academic Press.

New Zealand Public and Administrative Law Reform Committee, *Standing in Administrative Law* (1978), Wellington, Government Printers.

Oliver, D., '*Anns* v. *London Borough of Merton* Reconsidered' (1980), *Current Legal Problems*, 269.

Pearce, D. J., *Statutory Interpretation in Australia*, 2nd edn. (1981), Sydney, Butterworths.

Peiris, G. L., 'The Doctrine of Locus Standi in Commonwealth Administrative Law' [1983], *Public Law*, 52.

Phegan, P., 'Breach of Statutory Duty as a Remedy Against Public Authorities' [1974–6], *University of Queensland Law Journal*, 158.

—— 'Public Authority Liability in Negligence' (1976), 22 *McGill Law Journal*, 605.

Polyviou, P., *The Equal Protection of the Laws* (1980), London, Duckworths.

Rawlings, H. F., 'Jurisdictional Review after Pearlman' [1979], *Public Law*, 404.

Report of the Commonwealth Administrative Review Committee (Parl. Paper No. 144, 1971), Canberra, Australian Government Printers.

Report of the Royal Commission on Criminal Procedure, Cmnd. 8092 (UK), London, HMSO.

Robinson, G. E., *Public Authorities and Legal Liability* (1925), London, London University Press.

Rogers, W. V. H., *Winfield and Jolowicz on Tort*, 12th edn. (1984), London, Sweet & Maxwell.

Rubinstein, A., *Jurisdiction and Illegality* (1965), Oxford, Clarendon Press.

Samuel, A., 'Public and Private Law: a Private Lawyers's Response' (1983), 46 *Modern Law Review*, 558.

Sawer, G., *Nonfeasance Revisited*' (1955), 18 *Modern Law Review*, 541.

—— 'Nonfeasance under Fire' (1966), 2 *New Zealand Universities Law Review*, 115.

—— Seddon, N., 'The Negligence Liability of Statutory Bodies' [1978], 9 *Federal Law Review*, 326.

Sharpe, R. J., *Injunctions and Specific Performance* (1983), Toronto, Canada Law Book Company.

Spry, I. C. F., *Equitable Remedies* (1971), Sydney, Law Book Company.

Stanton, K. M., *Breach of Statutory Duty in Tort* (1986), London, Sweet & Maxwell.

Stein, L. A., ed., *Locus Standi* (1979), Sydney, Law Book Company.

Street, H., *Governmental Liability* (1953, repr. 1975), Cambridge, Cambridge University Press.

Sykes, E. I., Lanham, D. J., and Tracey, R. R. S., *General Principles of Administrative Law*, 2nd edn. (1984), Sydney, Butterworths.

Tapping, T., *Mandamus* (1848), London, W. Bennings & Co.

Taylor, G. D. S., 'Individual Standing and the Public Interest: Australasian Developments' (1983), 2 *Civil Justice Quarterly*, 353.

Thayer, E. R., 'Public Wrong and Private Action' (1914), 27 *Harvard Law Review*, 317.

Thio, S. M., *Locus Standi and Judicial Review* (1971), Singapore, Singapore University Press.

Todd, S., 'The Negligence Liability of Public Authorities: Divergence in the Common Law' (1986), 102 *Law Quarterly Review*, 370.

Trindade, F. A., and Cane, P. F., *The Law of Torts in Australia* (1985), Sydney, Oxford University Press.

Vining, E., *Legal Identity* (1978), London, Yale University Press.

Wade, Sir H. W. R., *Administrative Law*, 5th edn. (1983), Oxford, Clarendon Press.

—— 'Procedure and Prerogative in Public Law' (1985), 101 *Law Quarterly Review*, 180.

Walker, C., 'Review of the Prerogative: the Remaining Issues' [1987], *Public Law*, 62.

Whitmore, H., and Aronson, M., *Review of Administrative Action* (1978), Sydney, Law Book Company.

Wilcox, A. F., *The Decision to Prosecute* (1972), London, Butterworths.

Williams, G., *Crown Proceedings* (1948), London, Stevens & Sons.

—— 'Discretion in Prosecuting' [1956], *Criminal Law Review*, 222.

—— 'The Effect of Penal Legislation in the Law of Tort' (1960), 23 *Modern Law Review*, 233.

—— 'Letting off the Guilty and Prosecuting the Innocent' (1985), *Criminal Law Review*, 115.

Woolf, Sir H., 'Public Law—Private Law: Why the Divide?' [1986], *Public Law*, 220.

Yardley, D. C. M., 'Prohibition and Mandamus and the Problem of Locus Standi' (1957), 73 *Law Quarterly Review*, 534.

Young, P. W., *Declaratory Orders*, 2nd edn. (1984), Sydney, Butterworths.

Zamir, I., *The Declaratory Judgment* (1962), London, Butterworths.

Index